CATIONIC
SURFACTANTS

SURFACTANT SCIENCE SERIES

CONSULTING EDITORS

MARTIN J. SCHICK
Consultant
New York, New York

FREDERICK M. FOWKES
(1915–1990)

1. Nonionic Surfactants, *edited by Martin J. Schick* (see also Volumes 19 and 23)
2. Solvent Properties of Surfactant Solutions, *edited by Kozo Shinoda* (out of print)
3. Surfactant Biodegradation, *by R. D. Swisher* (see Volume 18)
4. Cationic Surfactants, *edited by Eric Jungermann* (see also Volumes 34 and 37)
5. Detergency: Theory and Test Methods (in three parts), *edited by W. G. Cutler and R. C. Davis* (see also Volume 20)
6. Emulsions and Emulsion Technology (in three parts), *edited by Kenneth J. Lissant*
7. Anionic Surfactants (in two parts), *edited by Warner M. Linfield* (out of print)
8. Anionic Surfactants: Chemical Analysis, *edited by John Cross* (out of print)
9. Stabilization of Colloidal Dispersions by Polymer Adsorption, *by Tatsuo Sato and Richard Ruch*
10. Anionic Surfactants: Biochemistry, Toxicology, Dermatology, *edited by Christian Gloxhuber* (see Volume 43)
11. Anionic Surfactants: Physical Chemistry of Surfactant Action, *edited by E. H. Lucassen-Reynders* (out of print)
12. Amphoteric Surfactants, *edited by B. R. Bluestein and Clifford L. Hilton* (out of print)
13. Demulsification: Industrial Applications, *by Kenneth J. Lissant*
14. Surfactants in Textile Processing, *by Arved Datyner*
15. Electrical Phenomena at Interfaces: Fundamentals, Measurements, and Applications, *edited by Ayao Kitahara and Akira Watanabe*
16. Surfactants in Cosmetics, *edited by Martin M. Rieger*
17. Interfacial Phenomena: Equilibrium and Dynamic Effects, *by Clarence A. Miller and P. Neogi*
18. Surfactant Biodegradation, Second Edition, Revised and Expanded, *by R. D. Swisher*

ADDITIONAL VOLUMES IN PREPARATION

CATIONIC SURFACTANTS

Analytical and Biological Evaluation

edited by

John Cross
University of Southern Queensland
Toowoomba, Queensland, Australia

Edward J. Singer
Mobil Oil Corporation
Princeton, New Jersey

CRC Press
Taylor & Francis Group
Boca Raton London New York

CRC Press is an imprint of the
Taylor & Francis Group, an **informa** business

CRC Press
Taylor & Francis Group
6000 Broken Sound Parkway NW, Suite 300
Boca Raton, FL 33487-2742

First issued in paperback 2019

© 1994 by Taylor & Francis Group, LLC
CRC Press is an imprint of Taylor & Francis Group, an Informa business

ISBN-13: 978-0-8247-9177-3 (hbk)
ISBN-13: 978-0-367-40208-2 (pbk)

Library of Congress Cataloging-in-Publication Data

Cationic surfactants: analytical and biological evaluation / edited by John Cross,
     Edward J. Singer.
          p.   cm. -- (Surfactant science series; v. 53)
          Includes bibliographical references and index.
          ISBN 0-8247-9177-0 (acid-free)
          1. Surface active agents.   I. Cross, John.   II. Singer, Edward J.
     III. Series.
     TP994.C382   1994
     541.3'3--dc20   93-48585
                                                                           CIP

Visit the Taylor & Francis Web site at
http://www.taylorandfrancis.com

and the CRC Press Web site at
http://www.crcpress.com

Preface

Of the cationic surfactants utilized in today's marketplace, quaternary ammonium compounds (quats) represent by far the largest proportion. American Oil Chemists' Society newswriter Anna Gillis has described quats as "the ultimate workhorse of the surfactant industry. They are, for example, used to decolorize sugar, kill bacteria in water beds, and loosen stubborn oil in oil wells. They are found in a huge range of commercial preparations from underarm deodorants to fiber glass resins. Mostly, however, they are used to keep the family laundry soft." Although the actual tonnage utilized each year falls well below that of anionic and nonionic surfactants, the market is well established and growing steadily. Some 200 million pounds (90,000 metric tons) of quats are produced each year in the United States alone.

One of the earliest treatises devoted solely to this class of compounds was Volume 4 of Marcel Dekker's outstandingly successful Surfactant Science Series, *Cationic Surfactants*, edited by Eric Jungermann, a substantial work published in 1970 containing over 600 pages. Over 20 years later, there is a much wider range of speciality uses to which they may be put. Advances in techniques by which they may be examined have been considerable, and, of course, the literature concerning their various properties has mushroomed. Any attempt to update Volume 4 in a single volume would have produced a book of totally unwieldy dimensions. Consequently, the topics have been subdivided into *Cationic Surfactants*: *Organic Chemistry* (Vol. 34), *Cationic Surfactants*: *Physical Chemistry* (Vol. 37), and *Cationic Surfactants*: *Analytical and Biological Evaluation* (this volume).

The analytical segment of this book is aimed at practicing chemists and the new breed of environmental scientists who possess a general background in chemistry, including the basic concepts of classical and modern chemical analysis. Similarly, in addressing both short- and long-term interactions of cationic surfactants with biological systems, the biological section relies on the basic working knowledge of scientists and engineers, and focuses on their need to understand the systems with which they work, be they skin, health, or wastewater. Detailed familiarity with cationic surfactants is not assumed, an introductory chapter being included for that purpose.

It is not intended that this book should function as a laboratory manual, although appropriate experimental procedures are included. It is, rather, a critical guide to the literature and methodology currently available, a tour through the array of weapons that the practicing scientist can bring to bear on any particular problem. Seldom will there be any one "correct" approach to that problem. In the case of analytical work, so much depends on the nature of the sample and on the detail required by the end user of the result.

The authors comprise an international collection of experienced academics and practitioners. They have not been required to operate within any preconceived framework or to confine their treatment to a specified set of surfactants: rather, they have been encouraged to present their material in the manner that best suits their own areas of expertise.

Our sincere thanks go to the authors, who have provided their contributions so readily, and to their employers for supporting their efforts; to the Series Editor, Dr. Martin Schick, who has worked and liaised tirelessly on our behalf; to Brenda Cross for her patience and many late nights chained to an uncooperative word processor; to the Marcel Dekker organization for their smooth guidance along the production chain; and finally, to the many hundreds of researchers, past and current, whose published works have made this volume possible.

John Cross
Edward J. Singer

Contents

Contributors

Robert S. Boethling Office of Pollution Prevention & Toxics, U.S. Environmental Protection Agency, Washington, D.C.

M. Bos Chemical Technology, University of Twente, Enschede, The Netherlands

John Cross Faculty of Sciences, University of Southern Queensland, Toowoomba, Queensland, Australia

Hans P. Drobeck Consultant, Rensselaer, New York

Dale L. Fredell Klenzade Research & Development, Ecolab, Inc., St. Paul, Minnesota

Henry T. Kalinoski Analytical Chemistry Section, Unilever Research U.S., Edgewater, New Jersey

Bruce Paul McPherson Analytical Sciences, Colgate-Palmolive Company, Piscataway, New Jersey

Gwilym James Moody School of Chemistry and Applied Chemistry, University of Wales College of Cardiff, Cardiff, Wales

Foad Mozayeni Applied Analytical Support, Akzo Chemicals, Inc., Dobbs Ferry, New York

Henrik T. Rasmussen   Analytical Sciences, Colgate-Palmolive Company, Piscataway, New Jersey

Edward J. Singer   Environmental Health and Safety Department, Mobil Oil Corporation, Princeton, New Jersey

J. D. R. Thomas   School of Chemistry and Applied Chemistry, University of Wales College of Cardiff, Cardiff, Wales

James Waters   Ecotoxicology Section, Unilever Research Port Sunlight Laboratory, Wirral, Merseyside, England

CATIONIC SURFACTANTS

I
Introduction

Introduction to Cationic Surfactants

JOHN CROSS *School of Sciences, University of Southern Queensland, Toowoomba, Queensland, Australia*

II.   Some Direct Applications
    A.   Adsorption at solid surfaces
    B.   Double-layer effects
    C.   Mineral flotation
    D.   Organic dyes
    E.   Corrosion inhibitors
    F.   Road priming
    G.   Bituminous spray priming
    H.   Phase transfer catalysis
    I.   ...
    J.   Corrosive agents

III.   Origins of Characterization
    A.   Mixtures
    B.   Isomerism
    C.   ...

IV.   ...

V.   ...

References

1

Introduction to Cationic Surfactants

JOHN CROSS Faculty of Sciences, University of Southern Queensland, Toowoomba, Queensland, Australia

I. INTRODUCTION

The term *surfactant*, or *surface-active agent* to use the full title, may be
defined conveniently for the purpose of this book as any substance which,
when mixed with a solvent, will congregate at interfaces rather than in the
bulk of the solution. This behavior is in contrast to the majority of other
solutes, which to maximize the solute–solvent interaction, concentrate in
the bulk of the solution rather than at the interfaces.

In the course of an involvement with such compounds that spans nearly
40 years, this author has read countless volumes and papers about them
and can find no more apt introduction than that provided by Connie Hendrickson in her preface to Book III of *The Sprouse Collection of Infrared
Spectra* [1]:

> In working with surfactants, no course or training program takes the
> place of the old "on-the-job" method. A surfactant has an oil-like, or
> lipid-soluble portion connected to a water-soluble portion. It doesn't
> know its head from its tail! A surfactant designed for an aqueous en-
> vironment reacts altogether differently in an oil environment. This alone
> makes predictability difficult—not to mention the effects of surfactant
> interactions! The first rule I learned when dealing with surfactants is to
> throw away all the rules I already knew:
> Like dissolves like.   OUT!
> Heat it up and it dissolves easier.   OUT!
> Add more and it thickens linearly.   OUT!

Surfactants owe such unique and often confusing properties to the presence of two distinctly different sections in each molecule: one or more
hydrophobic alkyl chains coupled to an ionic or highly polar hydrophilic
group. The molecules are sufficiently large (molecular weights are typically
300 to 400) for these two sections to act independent of each other. For
example, the hydrophobic end may adsorb onto a grease spot while the
hydrophile remains in the aqueous phase, thus creating a bridge between
the two otherwise immiscible phases.

When the hydrophile consists of a positively charged group, the molecule
is known as a cationic surfactant. Almost invariably, this is centered around
one or more nitrogen atoms. Although corresponding analogs are to be
found that contain sulfur, phosphorus, or arsenic, they are considerably
more expensive than their nitrogenous counterparts and are used relatively
rarely.

Quaternary ammonium surfactants, *quats* for short, retain their cationic
character at any pH (unless molecular breakdown occurs). Some typical

structures are:

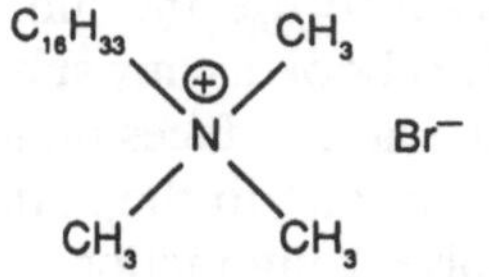

(1) Cetyltrimethylammonium bromide

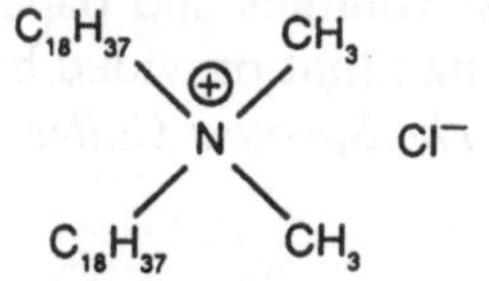

(2) Distearyldimethylammonium chloride

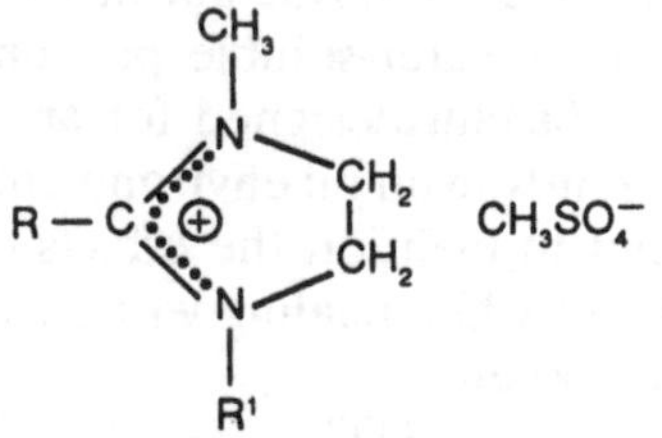

(3) Dialkylmethylimidazolium methylsulfate

Also included in this treatise are fatty amine compounds and amine oxides. A typical fatty primary amine has a pK_b value of approximately 3.4. Although the simple equilibrium between the free base (RNH_2) and its conjugate acid (RNH_3^+) is complicated by adsorption of both species from solution and by micelle formation, as a rough guide the concentration of the cationic form is equal to that of the free base at a pH value of 10.6 and greater than it at lower values. Fatty amines are therefore justifiably included as cationic surfactants in their own right in addition to being intermediates in the synthesis of quats.

Amine oxides are produced by the action of hydrogen peroxide on amines. Although less basic than their parent amines, they also acquire cationic properties in acidic solutions; for example:

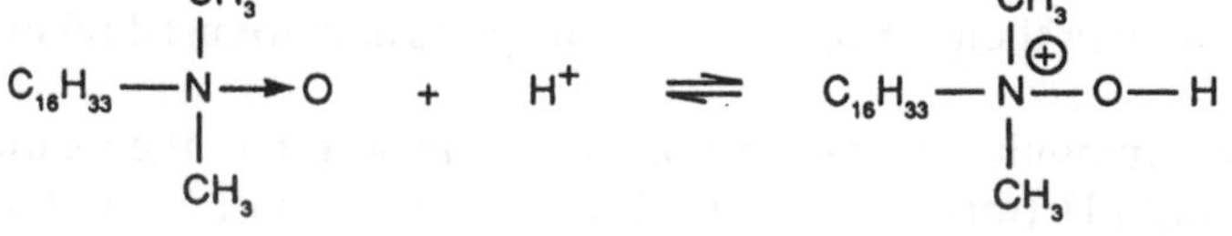

They show good foaming properties, are nonirritating to the skin, and find wide application in cosmetic, cleaning, and antistatic preparations.

Cationic surfactants first became important more than 50 years ago when their unique bacteriocidal properties were reported [2,3]. Although their share of the overall surfactant market is only about 8% (worldwide consumption of cationic surfactants was estimated to exceed 250,000 metric tons in 1986 [4] and U.S. quat consumption has been forecast to reach some 130,000 metric tons by 1995 [5]), they have a firm place in a wide range of applications. To quote Anna Gillis [6]: "In the industry they are considered the ultimate workhorses. They decolorize sugar, kill bacteria growing in waterbeds, and help pull the last driblets of oil from drilling wells. They serve as ingredients in products ranging from underarm deodorants to fiberglass for sailboats. Mostly, they're used to keep the laundry soft. These jacks-of-all-trades are quaternary ammonium compounds, or quats."

In quantitative terms, this utilization translates into:

Fabric softeners	66%
Asphalt additives	2%
Biocides	8%
Textiles	1%
Coated clays	16%
Oil-field chemicals	6%

Including nonquaternary compounds in the equation expands the scope considerably, particularly to include agents for mineral flotation and corrosion inhibition.

II. SOME MAJOR APPLICATIONS

Extensive reviews of a wide range of cationic surfactants and their numerous applications have been prepared by Jungermann [7] and more recently by Richmond [8] and by Rubingh and Holland [9]. Biocidal properties are covered extensively in later chapters of this book. Suffice it to mention here that quats may be found in a multitude of products, including cosmetics, pharmaceuticals, topical antiseptics, sanitizers, mildew preventives, and algicides. Synergistic combinations are common.

In the next few pages we provide a brief overview of some of the principal uses of cationic surfactants based on their unique physicochemical properties: these are dominated by the consequences of adsorption from solution onto a variety of solid surfaces.

A. Adsorption at Solid Surfaces

Adsorption at solid surfaces provides the key to a wide spectrum of applications for both quaternary and nonquaternary cationic surfactants. The majority of minerals and a high proportion of organic substances present surfaces that have high energy and are hydrophilic and polar in nature. For example, minerals with a high silica content possess surface —OH groups that engage readily in ion exchange with cationic surfactants leaving the solid with a hydrophobic coating:

$$—Si—O—H + R_4N^+ = —Si—O—R_4N + H^+$$

Similarly, a variety of organic substrates, such as wool, cotton, skin, and hair, present negatively charged sites, arising mainly from carboxylate groups, at which adsorption of the cation can occur.

An interesting analytical application of this phenomenon is found in a qualitative test described by Hummel [10]. A clean (soda) glass surface, immersed in a solution of a cationic surfactant and then rinsed, will display a greasy appearance and uneven wetting, whereas other types of surfactants will result in a typical "clean," well-wetted appearance.

The adsorption process is encouraged by pH values greater than the isoelectric point, or the point of zero charge (PZC), of the substrates: these are not unduly high (e.g., quartz, 1.5; ilmenite clay, 5.6; corundum, 9.4; wool, 3.5; and polyamide, 5.2) [4,11]. By way of contrast, adsorption of anionic surfactants is favored at pH values below the isoelectric points. Incidentally, it is this phenomenon that helps make assessment of environmental samples so difficult, since the cationic surfactants bind strongly to clay and other particulate substrates and hence are removed from solution. Detailed reviews of the adsorption process have been made by Ginn [12] and more recently by Ingram and Ottewill [13] and Fuerstenau and Herrara-Urbina [14].

B. Fabric Softening

The use of cationic surfactants as fabric softeners exceeds that of all other applications combined. Laughlin [15] defines fabric softeners as "products which impart to clothing and fabrics a feel or handle which is soft and pleasant during wear and use." Supplementary benefits include making the fabric easier to dry (since the fiber surfaces are now hydrophobic), reducing the tendency to build up a triboelectric charge, making ironing easier and leaving the fabric with a pleasant odor. Arising from the same adsorption phenomenon, the enhanced antistatic effect they impart to synthetic textiles is a great aid during the spinning and weaving processes. For similar reasons, hair-conditioning preparations also include cationic surfactants.

The fabric-softening market is dominated by three structural types: dialkyldimethylammonium salts (4), imidazolinium salts (5), and (to a lesser extent) diamido quaternary ammonium salts (6):

R—N(+)(CH₃)(CH₃)—R   Br⁻   (4)

(5) imidazolinium structure with R—C, CH₃, CH₂, CH₂, CH₃SO₄⁻, CH₂CH₂NHCOR

$RCONHCH_2CH_2N^{(+)}(CH_3)(CH_2CH_2O)_nH \cdot CH_2CH_2NHCOR$ CH₃SO₄⁻   (6)

They are all adsorbed almost quantitatively onto fabrics from dilute solution but are also easily desorbed by the action of the detergent in a subsequent wash.

There has been some dispute over the structure of the quaternized imidazolinium compounds, but the version shown in (5) is now accepted as correct [16–18]. It is not a true quaternary in the sense that the positive charge is believed to be delocalized over the N—C—N combination in the ring rather than concentrated on one particular nitrogen atom. The applicability of these substances to the detergency market is affected markedly by the nature of the substituents.

1. Unsaturation in the Alkyl Chain(s)

Although unsaturation has comparatively little effect on the general behavior of a fatty material, it has a substantial effect on the physical properties by disrupting the regularity of the fatty chain. This lowers the melting point of the parent fat or oil and of the surfactant, making it easier to disperse and at the same time improves the rewettability of the treated garment. For example, for an amidoimidazolinium methosulfate made up as a 75% mixture in isopropanol [19]:

R = oleyl (typically, 75% monounsaturation) gives a clear amber liquid at room temperatures.

R = tallow (typically, 40% monounsaturation) gives an opaque liquid requiring heating to higher than 40°C to achieve clarification.

R = hydrogenated tallow (zero unsaturation) gives a nonpourable paste.

2. Alkyl Chain Length

As an overiding factor to the trend noted above, however, performance testing of the softening efficiency of tetraalkyl quats shows that monoC$_{18}$ < diC$_{12}$ < monoC$_{22}$ [4]. The C$_{22}$ analog does not form a stable aqueous dispersion and the diC$_{18}$ (tallow or hydrogenated tallow) compounds are

Introduction to Cationic Surfactants 9

the most commonly used: the coconut oil analogs (mainly C_{12}) are relatively ineffective.

3. Hydrophilic Substituents

The incorporation of hydroxyl, ethoxy, and other such polar groupings within the molecules tend to enhance the dispersibility of the product and improve the finish that it provides, but reduce the softening ability [20].

Three distinctive modes of application of softeners in the home laundry have been addressed. The traditional method of employment has been in the rinse cycle, when the copious amounts of anionic surfactant with which it would react quantitatively have been removed. Softeners may also be applied within the heated tumble drier. The ideal softener is one that could be incorporated with the main wash cycle without mutual nullification with the washing surfactant. To produce the latter has been a long-term major challenge to the laundry products industry. In this regard, during the period 1965–1975, 11 patents were filed relating to cationic surfactant–containing detergents: during the next 10-year period the number had risen to 81 [21]. Mixtures of cationic surfactants with nonionic surfactants have been evaluated and progress has been made with mixtures of specific cationic and anionic surfactants. For further discussion on these, the reader is referred to a treatise by Rubingh [21].

Although quats find substantial application due to their biocidal properties, the large amounts of these substances that pass down the laundry sinks into the sewage systems of the world do not appear to cause any specific problems in the general purification and nitrification capacity of biological treatment plants [22]. This may in part be due to the fact that the hydrophile:lypophile balance (HLB) of quats used for softening is generally significantly lower than that required for optimal damage to bacteria (i.e., the alkyl chains are significantly longer). In addition, they are substantially deactivated by reaction with anionic surfactants and by adsorption and ion-exchange processes occurring on particulate matter. There is no evidence of their accumulation in higher aquatic life-forms such as fish [23]. These aspects are discussed in detail in a later chapter.

C. Mineral Flotation

Very few of the ores mined today are ready for immediate processing to win the targeted metal or compound. They are more likely to comprise several different minerals, only one or two of which may be of commercial interest. The detailed separation technology required for any one particular ore body is unique but will almost certainly involve crushing and grinding as a first step. This will then be followed by one or more separational treatments which could be based on differences in physical properties such as specific gravity and magnetic susceptibility of the various components but will frequently use the process of flotation.

The essential step in flotation is the adsorption of a surfactant from solution onto the polar surfaces of the mineral particles, making them hydrophobic. The exterior fatty chains are not readily accommodated in the water phase but will penetrate the surface of air bubbles that are forced into the suspension. If the net density of the mineral particle–air bubbles combination is less than that of the solution, the former will float to the surface and are thus separated from minerals that do not adsorb the surfactant.

Cationic surfactants have an important role in this technique, particularly for nonsulfide ores. Fatty primary amines are the most popular choice. As mentioned above, typical K_b values for such compounds are around 3.4. Hence at pH 10.6 the concentration of the protonated species would equal that of the unprotonated amine and at 8.6 would be 100 times greater. Such pH values are considerably higher than the zero-charge pH for most minerals. Hematite (Fe_2O_3), for example, has a PZC value around 7 and can be floated successfully with octodecylamine anywhere in the pH range 3 to 11 [11]. A full discussion of flotation has been given by Fuerstenau and Herrara-Urbina [14].

D. Organophilic Clays

An organophilic clay is the result of the displacement of inorganic cations, such as sodium and magnesium, from the surfaces of clay particles by large organic cations. The latter are, in practice, almost exclusively quaternary ammonium ions with at least one long chain and most commonly dimethyldi(hydrogenated-tallow)ammonium (DMDHTA) or dimethyl-(hydrogenated-tallow)benzylammonium (DMHTBA). The most suitable clays are the bentonite (smectic) types with a high cation-exchange capacity: they are converted from hydrophilic clays that swell in water into hydrophobic (organophilic) clays that swell in organic solvents to produce gels. The annual demand for organophilic clays, which typically contain 40% by weight of organic cation, is around 40,000 tons. High-strength gels usually require the modified clay to be dispersed in organic liquids that contain a specified percentage of highly polar materials such as a low-molecular-weight alcohol or ketone. The polarity range of suitable organic dispersants and the ease of dispersion are affected appreciably by the nature of the other substituents around the nitrogen atom (e.g., benzyl or 2-hydroxyethyl groups provide a more versatile surfactant than do hydrogen or methyl groups).

Typical applications include:

1. *Lubricants*. Lubricating oil gelled with 5 to 12% organophilic clay provides a grease that is particularly useful at high temperatures.
2. *Paints*. Up to 1% organophilic clay helps to control the viscosity and reduce the tendency of pigments to settle out in the can. During times when oil-well drilling is at a low ebb, the paint industry becomes a major consumer.
3. *Printing inks*. Organophilic clays reduce the tendency of printing ink to fly into a mist of tiny droplets (which would represent a respiratory hazard), thus enabling a higher printing rate to be used.
4. *Oil-well drilling mud*. Drilling muds are used to cool and lubricate the bit and string (the unprotected portion of the shaft above the bit). They are mostly aqueous systems, thickened by bentonite or water-dispersible polymers. In many cases, however, for drilling into hot, deep zones or through water-sensitive formations, a mud based on water-in-oil emulsions is preferable. Originally, this was usually an emulsion of diesel fuel and brine thickened by a DMDHTA-based organophilic clay, but the recent trend to more environmentally friendly oils with a minimal aromatic content has necessitated the use of cations with a higher degree of hydrophobic substitution, typically methylbenzyldi(hydrogenated-tallow)ammonium and methyltri-(hydrogenated-tallow)ammonium ions.

Comprehensive reviews of organophilic clays have been prepared by Tatum [24] and Mardis [25].

E. Corrosion Inhibitors

Surfactants find use as corrosion inhibitors in a variety of situations, such as acid pickling baths, crude oil production, temporary rust prevention, cutting oils, and steam systems. The mechanism underlying the protection given undoubtedly involves adsorption on the metal surface, but this is not the only factor involved, since other nonsurfactant additives show a synergistic effect.

In the preparation of steel for permanent rust prevention treatments such as galvanizing, it is essential that all traces of oxide be removed from the surface. The agent required for the task must attack the rust but not the bright metal surface beneath. Hydrochloric acid is a popular choice, but at the 60 to 90°C temperatures employed, it also attacks the metal, resulting in the release of hydrogen bubbles: these, in turn, give rise to an undesirable acidic spray. Cationic surfactants are very useful additives to such a bath. First, they aid acid penetration into the fissures of the oxide scale, thereby accelerating the derusting process. Second, they are adsorbed onto the freshly exposed base metal surface, presenting a barrier to further attack by the acid and by atmospheric oxygen when removed from the bath. Primary amines with a C_{12}–C_{18} chain are sometimes used, but quaternary ammonium salts, particularly trialkylbenzylammonium, alkylquinolinium, and alkylisoquinolinium, are most effective, provided that the temperature does not exceed 70°C.

In oil drilling and recovery operations, corrosion at depth is frequently attributed to the combination of sulfate ions and sulfate-reducing bacteria. Temperatures at depth may be high, in excess of 100°C. To provide a range of inhibitors for various conditions, advantage can be taken of variations in the relative solubilities of different types of surfactants in both oil and water. The major products used are long-chain fatty amines. As their acetate salts, they are water soluble, but in conjunction with a hydrophobic anion, such as that from tall oil fatty acid, they are oil soluble. A more detailed summary has been published by Porter [26].

F. Road Paving

The use of cationic surfactants to improve the bond between asphalt/ bitumen and crushed rock to provide a hard-wearing road surface is a special application of their mineral-bonding power.

The surface of most crushed rocks, such as granite, is quite hydrophilic and will frequently carry an adsorbed film of water. There is very little attraction between this and the nonpolar bitumen matrix into which the aggregate is to be encased; hence the adhesion between the two components is less than desired.

The polar sections of cationic surfactants, typically alkylamines and polyamine salts, are adsorbed strongly onto the mineral surface, even to the extent of displacing the water adsorbed. The protruding hydrophobe chain is free to diffuse into the bitumen, and thus a vastly improved bond between the latter and the crushed rock is created.

Generally, the bitumen is emulsified with water and up to 2% of surfactant. This mixture bonds well to the aggregate, even in the rain. It can also be used to waterproof concrete and to act as a base for preparations used to protect the bodywork on the undersides of automobiles [27]. Commonly used surfactants for such purposes are $R(tallow)NHCH_2CH_2$-CH_2NH_2 as its hydrochloride salt and its quaternized counterpart [4].

G. Electrostatic Spray Painting

Not all applications depend on adsorption onto solid surfaces: the remaining examples are in that category. Conventional spray painting is not an ideal process: a considerable amount of paint is lost as a fine mist, and some awkward areas do not receive an adequate coverage. The efficiency can be improved by use of an electrostatic spraying technique. In essence, the atomized paint particles are given a high electrostatic charge as they leave the paint gun and are attracted toward the object to be painted, which has been grounded.

Paints to be used in this manner necessarily require high electrical conductivity. Nonpolar solvents such as xylene which are typically used in many paints are far too resistive and require mixing with considerable quantities of a much more polar solvent before they become sufficiently conductive. The presence of a quaternary ammonium salt with at least one long alkyl chain will achieve the same effect but at a concentration of only about 2%. A particularly useful example is the stearoyl diaminopropane–based surfactant $C_{17}H_{35}CONH(CH_2)_3N^+(CH_3)_2C_2H_5 \cdot C_2H_5SO_4^-$, which is readily soluble in xylene (whereas the lauroyl derivative is not). The use of halide salts is not recommended since they are less soluble than those whose anions contain some degree of organic content and also induce corrosion beneath the paint film. Combinations of quats and polar solvents are particularly effective in producing a conductive paint [28].

H. Phase-Transfer Catalysts

There are many organic syntheses carried out in which one reactant is dissolved in an aqueous solution and the other in a water-immiscible organic phase. Reaction can only occur directly at the interface between the two solvents or after one of the solutes has passed through the interface and entered the other phase. Examples are the formation of polycarbonates by reaction between phosgene (in dichloromethane) and sodium salts of dihydric phenols (in water), and the synthesis of nitriles by the attack of the cyanide ion (in water) upon an alkyl halide.

Agents that accelerate such reactions are known as phase-transfer catalysts. Quaternary ammonium salts rate highly in this regard, being claimed to show better selectivity, greater rate increases, and to cost less than other types of phase-transfer catalyst. Common examples are salts of tetrabutylammonium, benzyltriethylammonium, hexadecyltrimethylammonium, and trialkyl(C_8–C_{18})methylammonium [29,30].

I. Solvent Extraction of Metals

In the same vein as phase-transfer catalysis comes the important hydrometallurgical process of solvent extraction. Any metal capable of forming an anionic complex can potentially be separated from those that are not by formation of a relatively nonpolar fatty ammonium salt that can be extracted into an organic solvent such as kerosene.

For example, the trialkylammonium ion, R_3NH^+, can be used to separate cobalt from nickel because in a solution rich in chloride ions, cobalt will form the complex ion $CoCl_4^{2-}$, leaving the uncomplexed nickel behind. Recovery of the cobalt is achieved by stripping the organic phase with aqueous sodium carbonate.

The technique is widely used in the commercial production of copper, cobalt, nickel, tungsten, molybdenum, vanadium, and uranium: some 85% of the world's production of the latter is processed by such means.

The formation of some metal anion complexes necessitates the metal being in a high oxidation state, which may make it a strong oxidant (e.g., vanadate), or the solution itself may contain other oxidizing agents. The stability of the fatty amine compound to oxidative attack from such species is obviously important. Tridodecylamine is a proven performer: amines of lower molecular weight or with branched chains are less successful.

Quaternary ammonium salts are also useful extraction agents, the most popular being a methylated trialkylamine. Tricaprylmethylammonium chloride, for example, is versatile both as an extractant and a phase-transfer catalyst. The advantage of a quaternary compound lies in the fact that it carries a permanent positive charge, so that its potential as an extractant is thus not pH dependent; this enables alkaline solutions to be extracted. For the same reason, however, subsequent stripping of the metal-rich organic phase becomes correspondingly more difficult.

One of the requirements of an extractant is that it should not promote formation of an emulsion between the organic and aqueous phases. High-purity trialkylamines allow rapid settling, but the presence of small amounts of nitriles and amides (artifacts of the conversion of fatty acids into amines) add considerable stability to the emulsion.

An informative review of amines in solvent extraction has been given by House [31]. He provides a comprehensive list of metal-based anions that are amenable to extraction. This could provide the basis of alternatives to colorimetric methods for trace analysis of cationic surfactants via an atomic absorption spectroscopic determination of the stoichiometrically extracted amount of metal.

J. Bioactive Agents

No general treatise on the applications of quats would be complete without mention of their valuable role in the public health sector, where they may be found as agents to increase the solubility or bioavailability of drugs or more probably in the roles of disinfectants and antiseptic agents.

They exhibit a broad spectrum of microbiological activity over a wide range of pH and are used in many branches of industry, agriculture, hospitals, housekeeping, and so on. On a personal-care level, the reader could well have used them to control dandruff, to relieve a sore throat, or to prevent a cut from turning septic. They are effective agents against a variety of bacteria, fungi, and viruses, while presenting comparatively little threat to higher lifeforms at the concentration levels at which they are normally to be found.

Considering humankind's ruthless vendetta on the biological equilibrium of our planet and the extraordinary ability of microorganisms to adapt rapidly to new situations, both hostile and beneficial, it seems inevitable that new classes of quats will be developed and tested in this area for many years to come. The biological activity of quats is the subject of three subsequent chapters in this volume; thus further discussion here would be inappropriate.

III. ORIGINS OF CATIONIC SURFACTANTS

A. Alkyl Chains

All cationic surfactants contain at least one long alkyl chain, which forms the basis of the hydrophobic section of the molecule. In many cases it will be sufficient for the analyst to identify the hydrophilic portion only and report, for example, that "the surfactant is an alkyldimethylbenzylammonium compound" and to provide no further detail of the alkyl group. In other circumstances, a fuller description may be requested, necessitating a detailed chromatographic analysis. The complex nature of the alkyl chain composition is highlighted in the next few pages by considering the various sources of the hydrophobes and the results of their subsequent processing.

1. Animal Fats and Vegetable Oils

The traditional major source of hydrophobe for the surfactant industry is undoubtedly the fatty acid triglycerides, from both animal and vegetable sources. They yield linear, readily biodegradable alkyl chains. In nearly all cases the principal-component acids are higher members of the series, typically:

Saturated acids:
Lauric acid   (dodecanoic, $C_{11}H_{23}COOH$)
Myristic acid   (tetradecanoic, $C_{13}H_{27}COOH$)
Palmitic acid   (hexadecanoic, $C_{15}H_{31}COOH$)
Stearic acid   (octadecanoic, $C_{17}H_{35}COOH$)

Unsaturated acids:
  Oleic acid   (8-octadecenoic, $C_{17}H_{33}COOH$)
  Linoleic acid   (8,11-octadecadienoic, $C_{17}H_{31}COOH$)

Lower members of the series [e.g., caproic acid (C_8 saturated) and caprylic acid (C_{10} saturated)] appear in small amounts in oils such as palm kernel and coconut oils. Of the overall average of acids produced in the United States from agricultural sources, 42% are saturated and 58% unsaturated [32].

The distribution of chain lengths of the saturated and unsaturated acids was once believed to constitute a "fingerprint" by which the original triglyceride source could be identified, but subsequent experience has shown considerable variation due to different strains of the animal or plant, different seasons, and the environmental factors. Typical analyses of the major components of some common fats and oils are given in Table 1. The classic work of Hilditch and Williams [33] should be consulted for further details.

Oils and fats of the highest quality are usually directed toward edible-product industries. Lower-quality oils, in which any enzymic hydrolysis resulting in free-acid, mono- or diglyceride formation is of little consequence, are the usual source materials for the surfactant industry. Oils with an appreciable content of polyunsaturated acids, such as linoleic, are not suited without hydrogenation. In such acids the hydrogen atoms on the number 10 carbon atom

  —CH=CH—CH$_2$—CH=CH—
   8   9   10   11   12

are particularly susceptible to peroxide formation upon exposure to air, leading to degradation and solidification. It is upon such a process that the polyunsaturated drying oils used in traditional oil-based paints depend.

Acids derived from the saponification of the parent triglyceride followed by acidification are normally contaminated with residual glyceride esters and other colored and odorous materials. They are frequently purified by vacuum distillation of either the free acid or its methyl ester, a process in which a certain amount of fractionation will occur, resulting in modification of the original fatty chain distribution. However, distillation is used primarily for separating cuts from the homologous series and does not readily separate the unsaturated acids from their saturated counterparts, since their vapor pressure–temperature characteristics are very similar. Due to the presence of alkenyl groups which inhibit the ability of the fatty chain to pack down neatly into a crystalline form, the unsaturated compounds have lower melting points than those of their saturated counterparts and thus tend to modify the overall melting point of the derivatives and other phys-

TABLE 1 Typical Percent Fatty Acid Distribution Among Some Common Oils and Fats

Fatty acid			Source							
Name	Carbon no.	Unsaturated groups	Coconut	Palm kernel	Olive	Groundnut	Cottonseed	Sunflower	Mutton tallow	Beef tallow
Caprylic	8	—	8	3						
Caproic	10	—	7	5						
Lauric	12	—	46	47						
Myristic	14	—	15	15	0.4					
Palmitic	16	—	9	7	18	6	20	6	2	5
Stearic	18	—	2	2		3	2	2	25	28
Arachidic	20	—				6			27	21
Oleic	18	1	6	14	68	60	24	25	41	40
Linoleic	18	2	1	1	12	25	50	66	4	3

Source: Ref. 33.

ical properties, such as texture. Overall modification of the alkenyl chain content may be made by blending oils from two or more sources, by fractional recrystallization from a polar solvent, or by the technique of pressing. In the latter process, which is labor intensive and permits only a relatively slow throughput, the acid mixture is cooled to approximately 5°C, whereupon partial solidification occurs to yield a soft "cake" that is pressed to remove the liquid phase, which is rich in unsaturated materials. This operation is usually repeated twice on the remaining solid. Typically, "triple-pressed" tallow fatty acids have their stearic and palmitic acid contents raised from 24 and 27% to 45 and 50%, respectively.

2. Wood-Pulping Industry

Another major source of fatty acids are the tall oils, by-products of the wood-pulping industry. The digestion of pulped softwoods with sodium hydroxide yields the sodium salts of mixed fatty acids and rosin acids (basically, polycyclic terpene carboxylic acids). Acidification and washing leave the free acids in the form of a dark oil known as tall oil, some 18,000 tons of which were produced in the United States alone in 1978 [34]. The product at this stage is too crude for further use and requires additional treatment by acid washing and/or by fractional distillation to yield fatty acids containing approximately 40% or 5% rosin acids, respectively. Such mixtures are suitable for conversion to surfactants.

3. Petroleum-Based Feedstocks

Competitive to the sources noted above are the synthetic hydrophobes arising from by-products of the massive petroleum and natural gas industries, which can be processed to yield olefins, alcohols, paraffins, and aromatic hydrophobes [35]. Such sources are particularly important in China and many countries from the former Soviet block and Europe [32]. Hydrocarbon chains of suitable length can be generated by polymerization of small alkenes or by cracking of very large molecules.

(a) Polymerization of Alkenes. Long-chain α-olefins are frequently prepared by polymerization of ethylene via the Ziegler process using a metal catalyst. The result will be a mixture of chain lengths, each containing an even number of carbon atoms (e.g., C_{10}, C_{12}, C_{14}, etc.). Similarly, the products of polymerization of propene and isobutene are equally predictable and are highly branched.

An advantage of having a hydrophobe prepared via this route is that fractional distillation of the alkene mixture can produce a relatively pure hydrophobe, free from homologs. Fractionation of low-molecular-weight hydrocarbons is much more readily achieved than for molecules in which a polar group (e.g., —COOH) is included since the latter tend to dominate physical properties such as vapor pressure and boiling point. The di(isobutyl) chain in (1,1,4,4-tetramethylbutyl)phenoxyethoxyethyl-dimethylbenzylammonium chloride is completely free of mono- or tri(isobutyl) homologs, and this substance is frequently used as a primary standard for the quantitative analysis of quats.

(b) Cracking of Petroleum Waxes. The alternative to building a hydrocarbon chain from small, discrete units is to subject very high molecular weight petroleum waxes to a catalytic steam cracking process. This inevitably gives rise to a mixture of olefins containing both odd and even numbers of carbon atoms.

In the case of fabric softeners, the favored hydrophobe source is undoubtedly tallow fatty acids which have been hydrogenated and typically contain C_{14}, C_{16}, and C_{18} saturated acids in the ratio 5:30:65 [19]. Although this saturation is accompanied by difficulties in dispersing the derived surfactants into water, the enhanced action as a softener more than compensates for the additional problems caused. Hydrogenation, also known as *hardening*, is so common that many authors and marketing agencies feel it unnecessary to mention that it has been done. Consequently, as used in the literature, *tallow* could refer to either the natural or hydrogenated versions.

B. Fatty Amines

1. Synthesis of Fatty Amines

Fatty acids are the main source of alkyl chains in the overall surfactant industry. For incorporation into cationic surfactants they are most commonly converted to amides via their ammonium salts and then to nitriles [36]:

$$\underset{\substack{\text{ammonium}\\\text{salt}}}{RCOO\!-\!NH_4^+} \xrightarrow{\text{dehydration}} \underset{\text{amide}}{RCONH_2} \xrightarrow{\text{dehydration}} \underset{\text{nitrile}}{R\!-\!C\!\equiv\!N}$$

The nitrile may be (fractionally) distilled before the next step, in which it is catalytically hydrogenated, first to an imine and then to an amine:

$$\underset{\text{nitrile}}{R\!-\!C\!\equiv\!N} \xrightarrow{\text{hydrogenation}} \underset{\text{imine}}{RCH\!=\!NH} \xrightarrow{\text{hydrogenation}} \underset{\text{amine}}{RCH_2NH_2}$$

During this step the imine may condense with the amine to liberate ammonia and yield a larger imine with two alkyl chains:

$$RCH\!=\!NH + H_2NCH_2R \rightleftharpoons RCH\!=\!NCH_2R + NH_3$$

This larger imine then becomes hydrogenated to yield a secondary amine:

$$RCH\!=\!NCH_2R \xrightarrow{\text{hydrogenation}} RCH_2NHCH_2R$$

In practice, if a primary amine is the desired product, the hydrogenation is carried out in the presence of ammonia and a small amount of water to suppress the condensation reaction. Conversely, if a secondary amine is targeted, the ammonia liberated during the condensation is removed from the system continually. Inevitably, however, no single product is obtained exclusively, and a primary amine can be expected to contain small amounts of secondary or even tertiary amine [37].

2. Quaternization

The amine may be quaternized by reaction with a variety of reagents, such as dimethyl sulfate, chloromethane, bromomethane, benzyl chloride, or ethylene oxide/acid. For example, a secondary amine may be quaternized with 2 mol of chloromethane in the presence of a base to remove the hydrogen chloride produced:

$$R_1R_2NH + 2CH_3Cl \longrightarrow R_1R_2(CH_3)_2N^+Cl^- + HCl$$

The dialkyldimethylammonium chloride thus produced will be a far more complicated mixture than the parent alkyl chains alone. For example, Laughlin [15] has provided the following data on the composition of di(hydrogenated-tallow)dimethylammonium chloride.

A typical alkyl chain distribution in hardened tallow acid is:

	Weight fraction	Molar fraction
C_{18}	0.675	0.649
C_{16}	0.265	0.283
C_{14}	0.040	0.048
C_{17}	0.015	0.015
C_{15}	0.005	0.006

The major components of the dialkyldimethylammonium salt that arises from such a combination is given by:

R_1	R_2	Molar fraction
18	18	0.421
18	16	0.367
16	16	0.080
18	14	0.062
16	14	0.027
18	17	0.020
		0.977

To complicate matters further, dialkyl salts are not the only products. Mono- and trialkyl quats are also to be found, and a typical spread between these three types is given by:

	Wt %
$R_2(CH_3)_2N^+Cl^-$	85
$R(CH_3)_3N^+Cl^-$	12
$R_3CH_3N^+Cl^-$	3

Thus a trade product that is nominally labeled as "100 percent active" may actually contain significant amounts of a dozen or more different substances.

Adding confirmation to these observations, Okumura et al. [38] subjected di(hydrogenated-tallow)dimethylammonium chloride to repeated recrystallization from acetone before drying under vacuum and over phosphorus pentoxide. High-performance liquid chromatographic (HPLC) analysis showed only 63% di(octadecyl)dimethylammonium chloride; the remainder consisted of mono- and tri-C_{18} salts, di- and tri-C_{14} and C_{16} salts, and some unquaternized material.

C. Amidoamines and Imidazolines

Quaternary ammonium salts based on amidoamines provide an example of a family of quats in which the marketed product could contain a range of different molecular types. Consider first the formation of a quat from N,N—dimethylaminopropylamine:

$$H_2NCH_2CH_2CH_2N(CH_3)_2 \xrightarrow{RCOOH} RCONHCH_2CH_2CH_2N(CH_3)_2$$
$$\text{diamine} \qquad\qquad\qquad \text{amidoamine}$$
$$\xrightarrow{(CH_3)_2SO_4} RCONHCH_2CH_2CH_2N^+(CH_3)_3 \cdot CH_3SO_4^-$$
$$\text{quat}$$

Apart from the variations in the nature of R, discussed previously, there seems little opportunity for much further diversification since there is only one amino group with a replacable hydrogen atom.

When the polyamine contains two active hydrogen atoms, the scope for different reaction products widens. 1,6-Diaminohexane, for example, reacts with an equimolar amount of fatty acid to give rise to a mixture:

$H_2NCH_2CH_2CH_2CH_2CH_2CH_2NH_2$	Unchanged diamine	25%
$RCONHCH_2CH_2CH_2CH_2CH_2CH_2NH_2$	Monoamide	50%
$RCONHCH_2CH_2CH_2CH_2CH_2CH_2NHOCR$	Diamide	25%

This particular diamine is itself produced by the hydrogenation of adiponitrile and could contain by-products such as bishexamethylenetriamine or 1,2-diaminocyclohexane.

Should the diamine be 1,2-diaminoethane or based thereon, the (reversible) process of cyclization from monoamide to imidazoline complicates the reaction dynamics. 1,2-Diaminoethane itself has been shown to yield diamine, monoamide, and diamide in ratios in the order of 30:37:30 when reacted with an equimolar amount of fatty acid.

Another diamine, aminoethylethanolamine,

$$H_2NCH_2CH_2NHCH_2CH_2OH \qquad\qquad (\underline{7})$$

has been shown to react with fatty acids to produce four major products. Just as in the previous example, cyclization plays an important role:

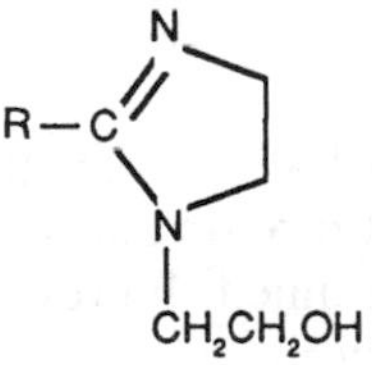

Imidazoline (48%) (8)

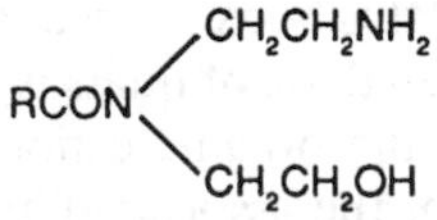

Tertiary amide (5%) (9)

RCONHCH₂CH₂NHCH₂CH₂OH

Secondary amide (20%) (10)

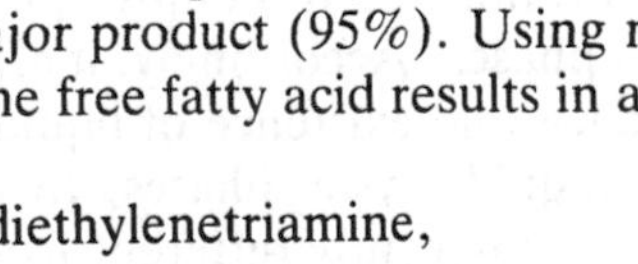

Diamide (20%) (11)

If water is not removed from the system, subsequent hydrolysis reactions result in the secondary amide as the major product (95%). Using methyl esters as the acylating agent instead of the free fatty acid results in a much more complex mixture of products.

As a final example in this category, diethylenetriamine,

$$H_2NCH_2CH_2NHCH_2CH_2NH_2 \qquad\qquad (\underline{12})$$

has three potential sites for amide formation and can give rise to eight possible structures from reaction with a fatty acid. Typical product distribution from a 1:1 mixture is:

Monoamide (primary)   38%
Monoamide (secondary)
1,3-Diamide   35%
1,2-Diamide
1,2,3-Triamide

Monoalkylimidazoline ⎱
Dialkylimidazoline ⎰ 2%
Unreacted diamine   20%

In practice, the secondary amino hydrogen is much less reactive than its primary counterparts and there are only three major components to the reaction products. The 1,3-diamide is particularly well suited for cyclization to an imidazoline and subsequent quaternization with dimethyl sulfate to yield a stable amidoimidazolinium salt of the type favored for fabric softening. For more details on the composition of these and other mixtures obtained during the synthesis of amidoamines and imidazolines, readers are referred to recent reviews by Friedli [39] and Earl [40] and references therein.

IV. EFFECT OF THE COUNTERION

Throughout this book the reader will find that almost 100% of the attention paid to the analysis of quats is directed toward identification and/or quantification of the organic cation. It should not be assumed, however, that the nature of the associated anion is of no importance.

The thermal stability of the salt, for example, varies with the associated counterion. In this regard, alkyltrimethylammonium chlorides are stable to heating up to about 130°C, whereupon they begin to decompose, mainly into alkyldimethylamines and chloromethane. The corresponding bromide and iodide salts are significantly less stable [41].

More important, the anion has considerable effect on the phase equilibria between the surfactants and water. The anion is an important site for hydration for the entire quaternary salt within the liquid and liquid crystal phases. Accordingly, it affects such factors as the limit of the Krafft boundary, the existence of liquid–liquid miscibility gaps, and the stability of the liquid crystal phases. In these matters, for example, acetate salts behave significantly different from halide salts [41,42].

In more dilute solutions, the anion has considerable impact on the shape and size of the micelle [43]. The smaller the radius of the hydrated ion, the greater effect it will have on contraction of the Helmholtz double layer and hence on the magnitude of the critical micelle concentrations (CMCs). In this regard, the CMC values of dodecylammonium and dodecylpyridinium salts increase through the series chloride, bromide, nitrate, iodide [44]. The CMC values of tetradecyltrimethylammonium bromide measured in the presence of various organic anions decreased in the order bromide, formate, acetate, butyrate, isovalerate, benzoate, caprylate, there being about a two-order-of-magnitude difference between the first and last of this series [45]. Other parameters, such as viscosity, are also affected by the nature of the anion. The solubility of quats in nonaqueous solvents and the ability of such solutions to solubilize water are additional factors that are very anion dependent [43].

The identification of the anion is not addressed specifically in this book. Normal qualitative and quantitative tests are applicable, assuming due regard for the inherent tendency of the surfactant cation to adsorb and suspend in general and to precipitate with large inorganic anions (see Chapter 5).

The most common anions associated with quats are chloride and methosulfate, arising from the use of chloromethane and dimethylsulfate (respectively) as quaternizing agents. It is not uncommon to use argentiometric titration of halide as a quality control measure for some halide quats.

V. EXAMPLES OF CATIONIC SURFACTANTS

Presented below are a few examples of the nominal structures of cationic surfactants taken from the hundreds presented in the Surfactant Science Series, Vols. 4 and 34, and in the McCutcheon Annual [7,8,46]. In all cases, R R', and R" represent long alkyl chains, except where stated otherwise.

Alkyltrimethylammonium bromide

Dialkyldimethylammonium bromide

Alkyldimethylbenzylammonium (benzalkonium) chloride

Dialkylmethylbenzylammonium chloride

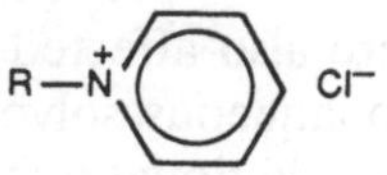

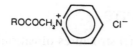 **Alkylpyridinium chloride**

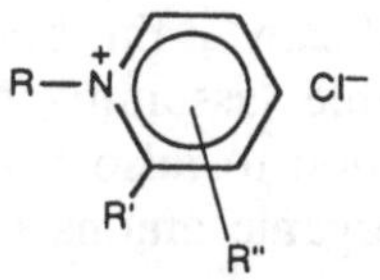

Substituted alkylpyridinium chloride
   $R = C_4–C_{18}; R' = C_1–C_{13};$
   $R'' = H, C_1, C_2$

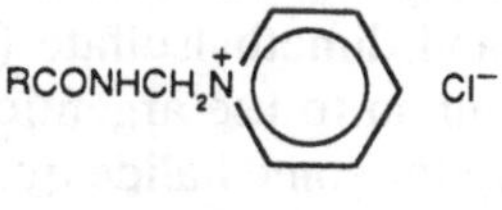

Alkylamidomethylpyridinium chloride

Carbalkoxypyridinium chloride

Alkylquinolinium chloride

Alkylisoquinolinium chloride

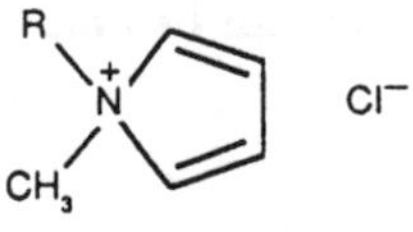

N,N-**alkylmethylpyrollidinium chloride**

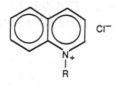 **Amidoamine from diethylenetriamine (DETA)**

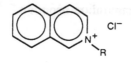 **Imidazoline from dehydration/cyclization of above**

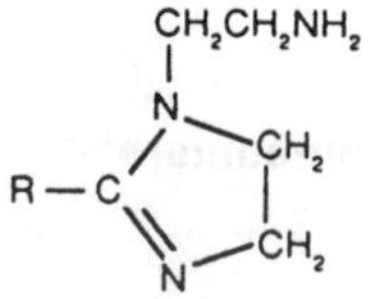

CH₂CH₂NHOCR structure — Amidoimidazolinium quat

$$\text{CH}_2\text{CH}_2\text{NHOCR}$$

Amidoimidazolinium quat from acylation and quaternization of above [e.g., 3-methyl-1-(tallowamidoethyl)-2-tallow imidazolinium methyl sulfate]

(Structure: R—C imidazolinium ring with N—R¹, $CH_3SO_4^-$)

$$\text{RCONHCH}_2\text{CH}_2\overset{\oplus}{\text{N}}\text{CH}_2\text{CH}_2\text{NHCOR}$$

with CH_2CH_2OH and CH_3 substituents, $CH_3SO_4^-$

Methylbis(tallowamidoethyl)-2-hydroxyethylammonium methyl sulfate

$$\text{RCONHCH}_2\text{CH}_2\overset{\oplus}{\text{N}}\text{CH}_2\text{CH}_2\text{NHCOR}$$

with $(CH_2CH_2O)_nH$ and CH_3 substituents, $CH_3SO_4^-$

Poly(oxyethylated) version of above

$$\text{RCONHCH}_2\text{CH}_2\text{CH}_2\overset{\oplus}{\text{N}}\text{CH}_2(\text{phenyl})$$

with CH_3, CH_3 substituents, X^-

(Tallowamidopropyl)dimethylbenzylammonium halide

$$\text{RCOOCH}_2\text{CHCH}_2\overset{\oplus}{\text{N}}\text{—CH}_3$$

with CH_3, CH_3 substituents, RCOO, Cl^-

Ditallow ester of 2,3-dihydroxytrimethylammonium chloride (example of the new diester quats that degrade readily)

REFERENCES

1. C. Hendrickson, in *The Sprouse Collection of Infrared Spectra*, Book III (D. L. Hansen, ed.), Elsevier, Amsterdam, 1988.
2. G. Domagk, *Dtsch. Med. Wascht. 61*: 828 (1935).
3. G. Domagk, U.S. patent 2,108,765 (1938).
4. A. D. James, P. H. Ogden, and J. M. Wates, in *Industrial Applications of Surfactants* (D. R. Karsa, ed.), Special Publication 59, Royal Society of Chemistry, London, 1987, p. 250.

5. C. A. Houseman, *J. Am. Oil Chem. Soc. 63*: 1395 (1986).
6. A. Gillis, *J. Am. Oil Chem. Soc. 64*: 887 (1987).
7. E. Jungermann, ed., *Cationic Surfactants*, Marcel Dekker, New York, 1970.
8. J. M. Richmond, ed., *Cationic Surfactants: Organic Chemistry*, Marcel Dekker, New York, 1990.
9. D. N. Rubingh and P. M. Holland, eds., *Cationic Surfactants: Physical Chemistry*, Marcel Dekker, New York, 1991.
10. D. Hummel, *Identification and Analysis of Surface-Active Agents*, Interscience, New York, 1962.
11. H. L. Shergold, in *Industrial Applications of Surfactants* (D. A. Karsa, ed.), Special Publication 59, Royal Society of Chemistry, London, 1987, p. 269.
12. M. E. Ginn, in *Cationic Surfactants* (E. Jungermann, ed.), Marcel Dekker, New York, 1970, pp. 341, 373.
13. B. T. Ingram and R. H. Ottewill, in *Cationic Surfactants: Physical Chemistry* (D. N. Rubingh and P. M. Holland, eds.), Marcel Dekker, New York, 1991, p. 87.
14. D. W. Fuerstenau and R. Herrara-Urbina, in *Cationic Surfactants: Physical Chemistry* (D. N. Rubingh and P. M. Holland, eds.), Marcel Dekker, New York, 1991, p. 407.
15. R. Laughlin, in *Cationic Surfactants: Physical Chemistry* (D. N. Rubingh and P. M. Holland, eds.), Marcel Dekker, New York, 1991, p. 449.
16. B. C. Trivedi, A. J. Digioia, and P. J. Menardi, *J. Am. Oil Chem. Soc. 58*: 754 (1981).
17. S. Takano and K. Tsuji, *J. Am. Oil Chem. Soc. 60*: 870 (1983).
18. W. M. Lindfield, *J. Am. Oil Chem. Soc. 61*: 437 (1984).
19. J. A. Ackerman, *J. Am. Oil Chem. Soc. 60*: 1166 (1983).
20. R. Puchta, *J. Am. Oil Chem. Soc. 61*: 367 (1984).
21. D. N. Rubingh, in *Cationic Surfactants: Physical Chemistry* (D. N. Rubingh and P. M. Holland, eds.), Marcel Dekker, New York, 1991, p. 469.
22. P. Gerike, W. K. Fisher, and W. Jasiak, *Water Res. 12*: 1117 (1978).
23. L. Huber, *J. Am. Oil Chem. Soc. 61*: 377 (1984).
24. J. P. Tatum, in *Industrial Applications of Surfactants* (D. R. Karsa, ed.), Special Publication 59, Royal Society of Chemistry, London, 1987, p. 288.
25. W. S. Mardis, *J. Am. Oil Chem. Soc. 61*: 382 (1984).
26. M. R. Porter, in *Recent Developments in the Technology of Surfactants* (M. R. Porter, ed.), Elsevier, London, 1990, p. 163.
27. W. M. Lindfield, in *Cationic Surfactants* (E. Jungermann, ed.), Marcel Dekker, New York, 1970, p. 53.
28. B. Davis, in *Industrial Applications of Surfactants* (D. R. Karsa, ed.) Special Publication 59, Royal Society of Chemistry, London, 1987, p. 307.
29. F. E. Friedli, T. L. Vetter, and M. J. Bursik, *J. Am. Oil Chem. Soc. 62*: 1058 (1985).
30. Anon., *Chem. Eng. News 65*: 27 (Nov. 2, 1987).
31. J. E. House, *J. Am. Oil Chem. Soc. 61*: 357 (1984).
32. H. Fineberg, *J. Am. Oil Chem. Soc. 56*: 805A (1970).

33. T. P. Hilditch and P. N. Williams, *The Chemical Constitution of Natural Fats*, Chapman & Hall, London, 1964.
34. R. Logan, *J. Am. Oil Chem. Soc.* *56*: 777A (1979).
35. K. R. Smith, in *Cationic Surfactants: Organic Chemistry* (J. M. Richmond, ed.), Marcel Dekker, New York, 1990, p. 145.
36. R. A. Reck, in *Cationic Surfactants: Organic Chemistry* (J. M. Richmond, ed.), Marcel Dekker, New York, 1990, p. 163.
37. R. A. Reck, *J. Am. Oil Chem. Soc.* *56*: 796A (1979).
38. O. Okumura, N. Yokoi, K. Yamada, and D. Saiko, *J. Am. Oil Chem. Soc.* *60*: 1699 (1983).
39. F. E. Friedli, in *Cationic Surfactants: Organic Chemistry* (J. M. Richmond, ed.), Marcel Dekker, New York, 1990, p. 51.
40. G. W. Earl, in *Cationic Surfactants: Organic Chemistry* (J. M. Richmond, ed.), Marcel Dekker, New York, 1990, p. 101.
41. R. G. Laughlin, in *Cationic Surfactants: Physical Chemistry* (D. N. Rubingh and P. M. Holland, eds.), Marcel Dekker, New York, 1991, p. 23.
42. B. W. Ninham, D. F. Evans, and G. J. Wei, *J. Phys. Chem.* *87*: 5020 (1983).
43. R. Zana, in *Cationic Surfactants: Physical Chemistry* (D. N. Rubingh and P. M. Holland, eds.), Marcel Dekker, New York, 1991, p. 60.
44. J. Koublek, *Tenside Deterg.* *19*: 74 (1982).
45. P. Kothwala, N. V. Sastry, and P. Bahadur, *Tenside Deterg.* *22*: 182 (1985).
46. *McCutcheon's Emulsifiers and Detergents*, McCutcheon Division, Manufacturing Confectioner Publishing Co., Glen Rock, N.J., annually.

II
Biological Evaluation

EDWARD J. SINGER   Environmental Health and Safety Department, Mobil Oil Corporation, Princeton, New Jersey

The high toxicity of cationic surfactants for microorganisms which is the basis for their effective germicidal and deodorizing activity, is neatly complemented by their low toxicity for man and other mammalian species. Such low mammalian toxicity is an important factor in their broad use as disinfecting or sanitizing agents.

While true that exposure to high doses of certain cationic surfactants can result in life-threatening effects on the nervous system, extraordinary conditions are necessary for such exposures, so that such cases are quite rare.

The primary concern for human safety has not been neurotoxicity, but rather the injury that might result from exposure of the skin and eye or other mucous membranes to elevated levels. Such elevated levels could potentially be available in over-the-counter antibacterial or cosmetic products, but long commercial history has verified the concentrations which are safe for the consumer population, and careful manufacturing controls have assured that these levels are not exceeded.

In view of their marked potency against microorganisms, it is somewhat paradoxical that in heavy commercial usage the cationics have produced no damage to sewer systems, which depend heavily on microbiological degradation processes. Such is the case not only because concentrations of cationics in sewage influent are low, but also because of "neutralizing" interaction with anionics present in the influent, inactivation by sorption onto sewage sludge, and acclimatization by the sludge microorganisms to cationics.

Environmentally, strong sorption of the positively charged cationics to the negatively charged soils or sediments can inhibit biodegradation. This same binding, however, tends to make the cationics ecologically unavailable, so that they are rarely reported in groundwater in spite of their solubility.

Conversely, in the laboratory cationic surfactants can be quite toxic to aquatic invertebrates including benthic (bottom-feeding) organisms, while in nature sedimentary binding of these chemicals tends to lower bio-availability and hence toxicity to these organisms.

Quite appropriately, much attention has been dedicated in this section to the methods used–no result being more valid than the method by which it is produced. Hence, there is extensive discussion of in vitro approaches in toxicology, of the techniques for estimating eco-availability of cationics in various systems, and of the validation of the effectiveness of sanitizing agents.

Because each author has tried to provide an integrated picture within his area of expertise, some overlap of information and concepts presented in the various chapters of this publication is inevitable. However, this is hardly undesirable–*Repetitio mater studiorum est.*

2

Biological Properties and Applications of Cationic Surfactants

DALE L. FREDELL Klenzade Research & Development, Ecolab, Inc., St. Paul, Minnesota

I. INTRODUCTION

The intrinsic properties of cationic surfactants as represented by quaternary ammonium compounds have resulted in a variety of uses and a high level of popularity for these chemicals among users. Quaternary ammonium compounds (QACs) possess surface-active properties, detergency, and antimicrobial properties, including activity against bacteria, fungi, and viruses. Quaternary ammonium compounds are odorless or are of low odor. Quaternary ammonium compounds have little or no deleterious adverse effect on a variety of synthetic materials, including plastics, rubber, ceramic, and stainless steel. Additionally, the human toxicity of QAC is relatively low. Thus QACs have been used in laundry applications as fabric softeners and antistatic agents. Quaternary ammonium compounds have been used as antiseptics for injuries to the skin and as preservatives in skin product formulations.

The largest area of application of quaternary ammonium compounds is for sanitization and disinfection. At low concentrations, QACs tend to be bacteriostatic, while at higher concentrations they are bactericidal. The generalized functionalities, advantages, and disadvantages are listed in Table 1.

One apparent deficiency of QACs is their antimicrobial activity against gram-negative bacteria. Quaternary ammonium compounds do possess antimicrobial efficacy against a wide range of organisms; however, gram-

TABLE 1 Properties of Quaternary Ammonium Compounds as Antimicrobials

Advantages	Disadvantages
Broad-spectrum kill	Reduced activity against some
Provide residual static activity	gram-negative bacteria
Low order of human toxicity	
Nonirritating to the skin	Residual film may be undesirable
Noncorrosive	
Long shelf life	Aerosols are irritating
Stable-use solutions	
Temperature stable	Foam precludes use as clean-in-place (CIP) sanitizer
Easily dispensed	
Easily measured	Moderate cost
Possess detergency	
Moderate soil tolerance	Reduced activity in hard water
Control off-odors	
Effective at varying pH	

negative bacteria, particularly those organisms that belong to the Enterobacteriaceae family, have a marked resistance to quaternary ammonium compounds. In formulation, the use of additives and builders such as chelators, buffers, and detergents help to minimize the reduced activity against gram-negative bacteria.

The above "deficiency" noted, QACs are still the preferred choice of antimicrobial by many sanitarians and health care professionals because of other attributes. The fact that quaternary ammonium compounds are surfactants allows the user to apply them in a variety of ways. Foam applications are often used in environmental sanitation, whereas one-step cleaning/disinfection is often used for floors, walls, and other environmental surfaces in health care facilities. When QACs are used as disinfectants, the use concentrations are typically 400 to 500 ppm and almost always below 1000 ppm, levels that are not acutely toxic [1] and do not cause surface corrosion or deterioration of materials of construction. Quaternary ammonium compounds are an important component of many sanitation programs.

Antiseptics, sanitizers, and disinfectants should be selected with care. Personnel may assume that because a chemical is listed as an antimicrobial, it will kill organisms under all conditions. This is not the case. Indeed, the compounds do not eradicate surface contamination, but rather, reduce surface contamination to a level unlikely to be hazardous [2]. Sanitation is more than the use of chemicals with antimicrobial properties. Sanitation is the continual maintenance of premises, equipment, and processes. Sanitation is a philosophy and a strategy driven by people and their ideas.

II. CHEMISTRY

The quaternary ammonium compounds are produced by the nucleophilic substitution reaction of tertiary amines by an alkyl halide, benzyl chloride, or similar material. The general structure of QAC can be expressed as

$$\left[\begin{array}{c} R_1 \diagdown \quad \diagup R_3 \\ N \\ R_2 \diagup \quad \diagdown R_4 \end{array} \right]^+ X^-$$

Each of the R_1–R_4 groups represents a covalently bonded alkyl group, and the X represents the anion, which is a halide, usually a chloride atom. The nitrogen atom, with the attached radicals, creates a positively charged cationic group, the functional portion of the molecule. The R groups may vary in structure and in chain length. The R groups may be saturated or unsaturated, cyclic or acyclic, branched or nonbranched, and aromatic or substituted aromatics.

There are two important milestones in the development of quaternary ammonium compounds as germicides [1]. The first came in 1915 when Jacobs published eight papers on the structure, preparation, and antimicrobial activity of QACs [3–10]. These writings are generally accepted as the first to correlate the quaternary ammonium compound chemical structure with antimicrobial activity. It was not until 1935 that the second milestone occurred, when Domagk reported on the antimicrobial activity of the long-chain quaternary ammonium salts [11]. His research revealed that a large aliphatic residue attached to the quaternary nitrogen greatly improved activity. This discovery led to further research and an increase in the number of publications and patents on QACs, and that activity extends to the present time. As stated previously, the QAC cationic structure is the functional moiety. QAC activity is related to the structure and size of the radicals that are bonded to the nitrogen atom.

Table 2 relates how the length of the carbon chain affects bactericidal activity. In a study comparing the activity of a homologous series of alkylbenzyldimethylammonium chlorides, it was determined that compounds having alkyl side chains with 12 to 16 carbon atoms exhibited the greatest

TABLE 2 Effect of Length of Carbon Chain on Bactericidal Activity

Chain length	Bactericidal test (minimum concentration that kills in 10 min but not in 5 min in ppm)		
	Staphylococcus aureus ATCC 6538	*Salmonella typhosa* ATCC 6539	*Pseudomonas aeruginosa* ATCC 15442
8	3000	4500	6000
9	800	1400	2500
10	450	300	1200
11	160	130	400
12	45	40	120
13	25	20	50
14	15	12	40
15	25	20	70
16	30	25	200
17	170	15	360
18	450	60	1000
19	330	90	1300

Source: Ref. 1.

bactericidal activity [12]. Maximum bactericidal activity was observed for the 14-carbon compound. The reader will note from Table 2 that as chain length increases from 8 carbon atoms to 14 carbon atoms, the level of quaternary ammonium compound (in ppm) necessary for kill decreases. At that point, as the carbon chain continues to increase, the level of quaternary ammonium compound needed for kill begins to increase again. This is due, in part, to the decreased solubility of the compound. Thus most quaternary ammonium compound formulations contain higher percentages of C_{12} to C_{16} components, usually richest in C_{14}. Earlier commercially available QACs were limited to various mixtures of fatty acid chain lengths. With improvements and efficient fractionization of fatty acids, it is now possible to control the chain distribution more precisely. Recent research [13] confirms the importance of the alkyl chain length to the activity of quaternary ammonium compounds. The researchers conclude that from a practical viewpoint, C_{12}-benzalkonium chloride is the most effective component of the benzalkonium chloride homologs. The C_8 and C_{10} homologs of benzalkonium chloride had very weak bactericidal activity, while the C_{16} and C_{18} homologs showed variable bactericidal activity toward the test bacteria [13]. A separate study evaluated two sets of homologs of polycationic quaternary ammonium compounds [14], which contained different spacer structures to develop the oligomers. They found that activity depends strongly on molecular weight, and that both bacteriostatic and bactericidal activity increase in the order monomer $\simeq$ dimer $<$ trimer $<$ tetramer $<$ polymer.

Petrocci [1] has classified quaternary ammonium compounds into eight different categories, depending on the representative R groups. Quaternary ammonium compound germicides can range from the relatively simple monoalkyltrimethylammonium chloride to the more complex polymeric quaternary ammonium salts. By varying the alkyl chains and the other substituent groups, various properties can be obtained. For example, high-molecular-weight polymeric quaternary ammonium salts are unique in that unlike traditional quaternary ammonium germicides, they are low foaming and thus are suitable antimicrobial additives for industrial water treatment. The Petrocci classifications of quaternary ammonium compounds and examples of chemical structures are presented in Fig. 1.

III. TEST METHODS FOR BIOLOGICAL ACTIVITY

Since the late nineteenth/early twentieth century, several methods of evaluating the disinfectant capacity of various chemicals have been reported [15–19]. Joseph Lister used phenol as a means to reduce postsurgical infections, and phenol became the standard by which other chemicals were

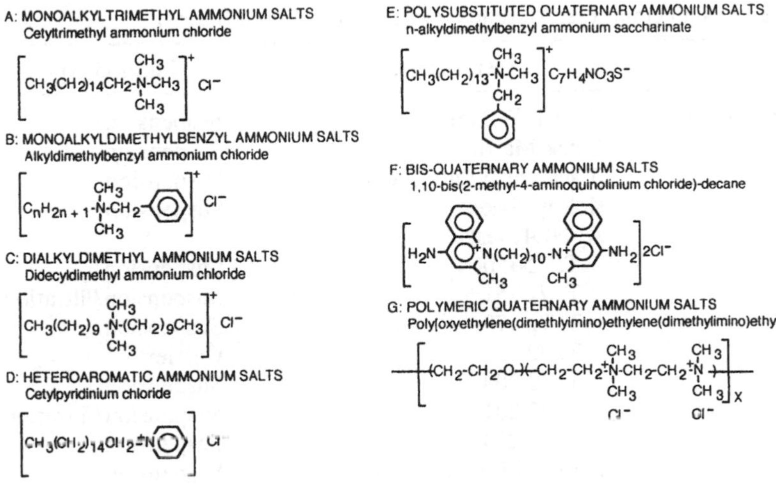

A: MONOALKYLTRIMETHYL AMMONIUM SALTS
Cetyltrimethyl ammonium chloride

$$\left[CH_3(CH_2)_{14}CH_2\text{-}\overset{\overset{\displaystyle CH_3}{|}}{\underset{\underset{\displaystyle CH_3}{|}}{N}}\text{-}CH_3 \right]^+ Cl^-$$

B: MONOALKYLDIMETHYLBENZYL AMMONIUM SALTS
Alkyldimethylbenzyl ammonium chloride

$$\left[C_nH_{2n+1}\text{-}\overset{\overset{\displaystyle CH_3}{|}}{\underset{\underset{\displaystyle CH_3}{|}}{N}}\text{-}CH_2\text{-}\bigcirc \right]^+ Cl^-$$

C: DIALKYLDIMETHYL AMMONIUM SALTS
Didecyldimethyl ammonium chloride

$$\left[CH_3(CH_2)_9\text{-}\overset{\overset{\displaystyle CH_3}{|}}{\underset{\underset{\displaystyle CH_3}{|}}{N}}\text{-}(CH_2)_9CH_3 \right]^+ Cl^-$$

D: HETEROAROMATIC AMMONIUM SALTS
Cetylpyridinium chloride

$$\left[CH_3(CH_2)_{14}CH_2\text{-}N^+\bigcirc \right] Cl^-$$

E: POLYSUBSTITUTED QUATERNARY AMMONIUM SALTS
n-alkyldimethylbenzyl ammonium saccharinate

$$\left[CH_3(CH_2)_{13}\text{-}\overset{\overset{\displaystyle CH_3}{|}}{\underset{\underset{\displaystyle CH_2}{|}}{N}}\text{-}CH_3 \right]^+ C_7H_4NO_3S^-$$
(with benzene ring attached below CH_2)

F: BIS-QUATERNARY AMMONIUM SALTS
1,10-bis(2-methyl-4-aminoquinolinium chloride)-decane

$$\left[H_2N\text{-}\bigcirc\bigcirc\text{-}\overset{+}{N}(CH_2)_{10}\text{-}\overset{+}{N}\bigcirc\bigcirc\text{-}NH_2 \right] 2Cl^-$$
(with CH_3 on each N)

G: POLYMERIC QUATERNARY AMMONIUM SALTS
Poly[oxyethylene(dimethylimino)ethylene(dimethylimino)ethylene dichloride]

$$\left[\text{-}(CH_2\text{-}CH_2\text{-}O)\text{-}CH_2\text{-}CH_2\text{-}\overset{\overset{\displaystyle CH_3}{|}}{\underset{\underset{\displaystyle CH_3}{|}}{\overset{+}{N}}}\text{-}CH_2\text{-}CH_2\text{-}\overset{\overset{\displaystyle CH_3}{|}}{\underset{\underset{\displaystyle CH_3}{|}}{\overset{+}{N}}}\text{-} \right]_x$$
Cl^- Cl^-

FIG. 1 Quaternary ammonium compound classification.

measured. As early as 1913, the U.S. Public Health Association published reports on the standardization of disinfectant testing [20]. Several countries, including Australia, France, Germany, and the United Kingdom, adopted standardized test methods as official procedures for disinfectant evaluation. As yet, there is no single method that is recognized internationally as the procedure of choice. A list of various test methods and the country in which they are used is shown in Table 3. Table 4 lists disinfectant test methods used in the United States.

In the United States, disinfectants are defined as antimicrobials that eliminate specific vegetative microorganisms but not necessarily bacterial spores. Disinfectants are used primarily in health care applications. The primary test used to register disinfectants in the United States is the Association of Official Analytical Chemists' (AOAC) Use Dilution Method [21]. In this test, specified organisms are dried on stainless steel penicylinders which are used to simulate a contaminated environmental surface. The penicylinders, with dried organisms, are exposed to the disinfectant solution for 10 min and are then transferred to a subculture medium containing a germicide neutralizer. Results are determined on a turbidity basis: growth or nongrowth in the subculture media. As many as 1 million organisms can be dried on the penicylinder, and theoretically, the survival of one organism from the disinfectant challenge will result in a turbid, positive tube. To become an accepted disinfectant, 59 of 60 penicylinders

TABLE 3 Disinfectant Test Methods (Outside the United States)

Country	Test method	Method type
United Kingdom	Rideal–Walker	Suspension
	Chick Martin	Suspension
	Black and White Fluids	Suspension
	Kelsey–Sykes	Suspension
France	AFNOR tests	
	T. 72-150	Suspension
	T. 72-151	Suspension/filtration
	T. 72-170	Suspension
	T. 72-190	Carrier
	T. 72-200	Suspension
	T. 72-201	Suspension/filtration
	T. 72-230	Suspension
Germany	D.G.H.M. quantitative	Suspension
	D.G.H.M. quantitative	Suspension
	D.G.H.M. carrier	Carrier
Netherlands	5-5-5	Suspension
Australia	Therapeutic Goods and Cosmetic Act (TGA)	Suspension
New Zealand	New Zealand Standard 8302	Suspension

TABLE 4 Disinfectant Test Methods in the United States

Test methods	Method type
AOAC[a]	
Use dilution	Carrier
Fungicidal	Suspension
Tuberculocidal	Carrier
Sporicidal	Carrier
Germicidal spray products	Carrier
ASTM[b]	
Virucidal E 1052-85	Suspension
Virucidal E 1053-85	Carrier

[a]Association of Official Analytical Chemists.
[b]American Society for Testing and Materials.

must show no growth. Thus, at least in theory, the survival of two organisms on two separate penicylinders could result in a test failure, even though tens of millions of organisms have been inactivated.

The AOAC Use Dilution Method has been criticized in recent years for lack of interlaboratory reproducibility and lack of standardization of test variables. Activities to review the method, identify variables, and improve test validity have been undertaken; however, collaborative studies confirmed that the method has intrinsic variability. At this writing various groups are working to develop a meaningful and reproducible test. These working groups include the U.S. Environmental Protection Agency (USEPA), which oversees disinfectant registration in the United States. One new method, the Hard Surface Carrier Test [22], has become the procedure of choice even though it has many of the inherent variabilities of the method it may replace.

In the United States, sanitizers are evaluated by the AOAC Germicidal and Detergent Sanitizer Test. Sanitizers are defined as antimicrobials that reduce microbial contaminants in the inanimate environment to levels considered safe by public health standards. The method has been developed from the Chambers method [23] and has been adopted by the USEPA for registration of sanitizing rinses for food-processing and environmental surfaces. Practically, 1 mL of test culture is inoculated into 99 mL of diluted sanitizer. Successful sanitizers will achieve a 99.999% (or 5-log order) reduction in the number of organisms within 30 s.

The American Society for Testing and Materials (ASTM) has developed a procedure for evaluating sanitizers to be used on precleaned, nonporous, non-food-contact surfaces [24]. A 99.9% (or 3-log order) reduction in the number of organisms within 5 min is required for acceptance. The method uses inoculum dried on glass slides.

The aforementioned test procedures are designed with bacteria as the test organism. The ASTM has also developed virucidal test procedures [25,26] to evaluate disinfectant activity against this group of bioagents. Further, there are methods to evaluate fungicidal activity, preservation capacity, and antimicrobial laundry additive preparations. The reader is advised to consult Beloian et al. [27] for a more detailed list and comparison of test methods for evaluating antimicrobials.

IV. ANTIMICROBIAL PROPERTIES

Quaternary ammonium compounds have a broad spectrum of antimicrobial activity with reported activity, against both gram-positive and gram-negative bacteria, yeast, mold, viruses, and protozoans. Because of the variety of quaternary ammonium compounds and test condition variations, such as microbial density, presence of organic matter, temperature, exposure times, and so on, it is difficult to make general comments regarding efficacy. Additionally, QACs are typically formulated with additives such as builders and surfactants which modify their intrinsic properties. Because of these additives, the antimicrobial profile is probably enhanced and often is different from that of the pure quaternary ammonium compound itself. Nevertheless, Table 5 presents a comparison of minimum inhibitory concentrations for 10 quaternary ammonium compounds, from which it can be seen that QACs generally have greater activity against gram-positive bacteria than against gram-negative bacteria.

A. Antibacterial Properties

Early on, it was recognized that gram-positive bacteria differ from gram-negative bacteria in susceptibility to the action of QACs [28]. In this study, QACs exhibited marked bactericidal activity on the gram-positive bacteria and less pronounced activity on the gram-negative bacteria. The three gram-positive bacteria—*Staphylococcus aureus* and isolates of the genera of *Lactobacillus* and *Streptococcus*—were killed within 10 min at a 1:6000 dilution (167 ppm) by most of the cationics tested. This work not only reveled differences in activity between gram-positive bacteria and gram-negative bacteria, but also among genus and species of gram-negative bacteria. For example, of the gram-negative bacteria, *Proteus vulgaris* was less susceptible to cationics than was *Escherichia coli* [28].

Campylobacter jejuni, a gram-negative organism, has become an important human pathogen. Wang et al. [29] studied the effects of various disinfectants on the viability of *C. jejuni* and found that a 1:50,000 dilution (20 ppm) of alkylbenzyldimethylammonium chloride killed the organism within 1 min.

A 1:8000 dilution (125 ppm) of QACs for a 1-min exposure period was found to be inhibitory on the growth of *Legionella pneumophila*, the organism associated with Legionnaires' disease [30]. Other researchers found alkyldimethyldichlorobenzylammonium chloride to be bacteriostatic at less than 1 ppm, and bactericidal at 62 ppm against this organism [31].

The quaternary ammonium compound n-alkyldimethyldichlorobenzylammonium chloride achieved greater than a 4-log decrease in cell numbers of *Listeria monocytogenes* when tested at a concentration of 50 ppm [32]. This preparation was found to be equally effective in vitro and on both smooth and pitted stainless steel. Surfaces that are pitted or have other defects can harbor microorganisms and typically are more difficult to sanitize or disinfect.

TABLE 5 Minimum Inhibitory Concentration (ppm)[a] of Quaternary Ammonium Compounds for Several Bacteria and Fungi

	Organism				
Compound[b]	Escherichia coli	Pseudomonas aeruginosa	Bacillus subtilis	Staphylococcus aureus	Aspergillus niger
One R group is fatty acid					
Benzalkonium chloride	200	300	3	4	60
Dodecyltrimethylammonium chloride	500	500	5	5	500
Tetradecyltrimethylammonium chloride	150	100	1.5	5	50
Hexadecyltrimethylammonium chloride	5000	5000	5	5	5000
Dodecylbenzyldimethylammonium chloride	750	750	2	2	75
Two R groups are fatty acids					
Dioctyldimethylammonium chloride	40	75	20	20	225
Didecyldimethylammonium chloride	225	750	>0.7	7	75
Ditetradecyldimethylammonium chloride	2250	>2250	225	750	2250
Three R groups are fatty acids					
Tri(octyldecyl)methylammonium chloride	500	500	3	5	150
Tridodecylmethylammonium chloride	>5000	>5000	500	1500	1500

[a]Values in ppm are for bacteriostatic or fungistatic activity, not cidal.
[b]Compounds studied were highly purified. The general structure is $R_4N^+X^-$; the name of the major alkyl group(s) present is given in the table.
Source: Ref. 45.

The antimicrobial properties of low levels of QACs were evaluated against *Listeria innocua* and *Listeria monocytogenes* in solution and dried on surfaces. When the organism was in suspension, 40 ppm of dimethylbenzylammonium chloride was found to be effective but was not sufficient to kill the organisms dried on hard surfaces [33].

B. Antifungal Properties

Quaternary ammonium compounds also have fungicidal activity. A mixture of *n*-alkyl (60% C_{14}, 30% C_{16}, 5% C_{12}, 5% C_8)-dimethylbenzylammonium chloride and *n*-alkyl (68% C_{12}, 32% C_{14})-dimethylethylbenzylammonium chloride was found to be effective at 488 ppm against the fungus *Trichophyton mentagrophytes* [34]. A mixture of benzalkonium chloride and cetyldimethylethylammonium chloride was effective against *Candida albicans*, *Trichophyton rubrum*, *Epidermophyton floccosum*, and *Fusarium* sp. but was not effective against other fungi, including *Mucor* sp., *Aspergillus niger*, and *Microsporum canis* [35].

In a separate study, the fungistatic and algistatic activity of several quaternary ammonium compounds was determined [36]. While this study determined static, not cidal activity, the evaluation of 164 compounds provides the interested reader with valuable comparative data. Selected data from this study are shown in Table 6. Again due to the diversity of quaternary ammonium compounds and the milieu of formulations, it is difficult to make accurate generalizations about the efficacy of the QACs themselves.

C. Antiviral Properties

Quaternary ammonium compounds are virucidal for lipophilic viruses such as the myxovirus, paramyxovirus, adenovirus, herpesvirus, and poxvirus groups [37–39]. Of particular significance is the fact that in virucidal evaluations sponsored by our laboratory, the human immunodeficiency virus type 1 was shown to be susceptible to disinfection by quaternary ammonium compounds at typical use concentrations (D. L. Fredell, 1987, unpublished results).

Laboratory evaluations of various disinfectants have been conducted against animal viruses [40]. Among the disinfectants tested, two contained quaternary ammonium compounds as the active antimicrobial agent. The two QACs are both alkyldimethylbenzylammonium chlorides with different alkyl chains, one containing 67% C_{12}, 25% C_{14}, 7% C_{16}, 1% C_{18} and the other 60% C_{14}, 30% C_{16}, 5% C_{12}, 5% C_{18}. As might be expected, their virucidal activity was similar, in that each inactivated pseudorabies virus

TABLE 6 Inhibiting Concentrations (ppm) of Fatty Nitrogen Compounds for Some Bacteria and Fungi

	Benzethonium chloride	Benzalkonium chloride	Dodecyl-trimethyl-ammonium chloride	Dododecyl-benzyldimethyl-ammonium chloride	Cocobenzyl-dimethyl-ammonium chloride	Didecyldi-methyl-ammonium chloride
Bacteria						
Escherichia coli	1000	200	500	750	225	225
Pseudomonas fluorescens	300	300	500	750	225	750
Bacillus subtilis	3	3	5	2	2	0.7
Staphylococcus aureus	3	4	5	2	2	7
Fungi						
Aspergillus niger	300	60	500	75	20	75
Chaetomium globosum	30	10	50	7	7	7
Myrothecium verrucaria	300	40	500	150	20	20
Trichoderma viridae	200	80	500	75	20	20

Source: Ref. 1.

and transmissible gastroenteritis virus (TGEV), while both failed to inactivate porcine parvovirus. Pseudorabies virus and TGEV are lipophilic viruses that have a lipid envelope and are sensitive to most disinfectants. Hydrophilic viruses such as porcine parvovirus are nonenveloped and are resistant to some disinfectants [40].

The efficacy of QACs against hepatitis A virus dried on surfaces has been compared to that of other commonly used disinfectants [41]. Of five QAC-based formulations tested, only one was effective against hepatitis A virus. Inasmuch as the effective formulation had a pH of 0.42 (hydrochloric acid was present at a level of 23%) the efficacy of this quaternary ammonium compound formulation apparently was due to the high acidity and not to the action of the quat. The types of QAC tested and the results are presented in Table 7.

The efficacy of n-alkyldimethylbenzylammonium chloride (50% C_{14}, 40% 10% C_{12}, C_{16}) against bacteriophage also has been studied [42]. The investigators applied the QAC as an aerosol to surfaces inoculated with streptococcus phages 144F and 18-16. Whereas 500 ppm QAC provided some inactivation, little or no inactivation was observed at the 1000- and 2000-ppm levels. In an attempt to improve activity, the chelating agent ethylenediaminetetraacetic acid (EDTA) was incorporated at 200 ppm into the 2000-ppm solution of QAC and the combination applied as an aerosol. However, the inactivation of bacteriophage was no greater than that of QAC without EDTA. In contrast, other researchers found QAC to inactivate bacteriophage of lactic acid bacteria, although not as quickly as hypochlorite [43].

TABLE 7 Virucidal Activity of Selected Quaternary Ammonium Compound (QAC) Disinfectants Against Hepatitis A Virus

QAC	pH	Log_{10} virus inactivated
Alkyl (50% C_{14}, 40% C_{12}, 10% C_{16} = dimethylbenzylammonium chloride	0.4	>4.0
Alkyl (60% C_{14}, 30% C_{16}, 5% C_{12}, 5% C_{18}) = dimethylbenzylammonium chloride and alkyl (68% C_{12}, 32% C_{14}) dimethyethylbenzylammonium chloride	6.3	<1.0
Alkyl (40% C_{12}, 50% C_{14}, 10% C_{16}) = dimethylbenzylammonium chloride	7.6	<1.0
Alkyl (50% C_{14}, 40% C_{12}, 10% C_{16}) = n-n-dimethyl n-benzylammonium chloride	9.2	<1.0

Source: Adapted from Ref. 41.

D. Other Biocidal Properties

The lethality of quaternary ammonium compounds on the parasite *Plasmodium falciparum*, in vitro, has also been demonstrated [44].

Because quaternary ammonium compounds have demonstrated algistatic activity [36] and have little deleterious effect on materials of construction, they are often used to treat water used in cooling towers. Quaternary ammonium compounds are also registered with the USEPA as swimming pool algicides [1], although this is not a major application for these compounds. The algistatic concentrations of selected QACs are presented in Table 8.

E. Inhibitory Properties

Cords [45] reviewed the antimicrobial activity of the quaternary ammonium compounds, including the minimum inhibitory concentration (MIC) against several bacteria and fungi. A summary of the MIC values is presented in Table 5.

Due to the great amount of interest in quaternary ammonium compounds during the 1930s and 1940s, some startling claims were made about the chemicals' properties. One of the more unfortunate reports was that quaternary ammonium compounds were both sporicidal and tuberculocidal. There is little doubt that such observations were a result of inadequate neutralization of the QACs, as the distinction between inhibition and kill was not always made. As a result, the active antimicrobial QAC was allowed to be transferred into the subculture medium, resulting in biostatic action. Quaternary ammonium compounds are now known to *inhibit* sporulation, but they are only static agents for mycobacteria and do not inactivate these biological forms [46]. Tuberculocidal formulations exist that contain QACs in combination with other active ingredients, such as alcohol; however, the tuberculocidal activity is due to the presence of the additional active ingredient.

F. Antimicrobial Resistance

The development of resistance to antibacterial agents is often associated with therapeutic pharmaceuticals, but antimicrobial compounds used for preservation and disinfection have also been implicated. Among those compounds are iodines, phenolics, anionics, and quaternary ammonium compounds.

Resistance to the antimicrobial properties of QAC has been widely reported, beginning in the early 1950s. Among the early reports of demonstrated resistance was that of *Serratia marcescens* to benzalkonium chloride [47]. *S. marcescens* resistance was increased by repeated culture in

TABLE 8 Algistatic Concentrations (ppm) of Selected Quaternary Ammonium Compounds

QAC	Algae			
	Anabaena cylindrica	*Chlorella vulgaris*	*Oscillatoria tenuis*	*Stigeoclonium* sp.
Benzethonium chloride	1	3	1	1
Benzalkonium chloride	1	1	0.6	0.7
Dodecyltrimethylammonium chloride	5	50	0.5	5
Dododecylbenzyldimethylammonium chloride	0.2	0.2	0.2	0.2
Didecyldimethylammonium chloride	0.2	2	0.7	0.7

Source: Ref. 1.

media containing benzalkonium chloride. Nishikawa et al. [48] screened organisms isolated from the soil and found several with resistance to benzalkonium chloride. They identified one organism, *Enterobacter cloacae,* which grew in the presence of 10% QAC. They did not find evidence that the bacteria metabolized the benzalkonium chloride, and like earlier investigators, found that resistance was lost when the organisms were cultured in the absence of benzalkonium chloride.

Reports of QAC-resistant bacteria have not been limited to the research laboratory. A pseudobacteremia was reported in a hospital setting, the cause being the use of diluted benzalkonium chloride as an antiseptic prior to venipuncture [49]. Several investigators have described bacteremias due to resistant bacteria [50–53].

Nosocomial transmission of *Serratia marcescens* was reported in a veterinary facility and associated with the use of benzalkonium chloride as an antiseptic [54]. Other investigators have reported bacteremia associated with contaminated benzalkonium chloride antiseptic [55–57].

Not all of the reports of QAC resistance are associated with gram-negative bacteria. Gram-positive staphylococci have been isolated from meat- and poultry-processing plants and found to possess some resistance to QAC [58], their resistance increasing when they were exposed to sublethal concentrations.

Species of the yeast genus *Saccharomyces* are associated with spoilage of soft drinks. Neumayr et al. [59] found that a quaternary ammonium compound had excellent activity against vegetative cells of the genus, but destruction of the yeast ascospores was delayed, most notably at lower concentrations of QAC.

There are several explanations for organisms surviving exposure to chemicals designed to be lethal to the organism. Russell and Gould [60] reviewed two possible mechanisms for such resistance: first, intrinsic resistance, and second, acquired resistance. Intrinsic resistance is related to the structural and chemical composition of the outer layers of the cells, which may provide an effective barrier to the entry of antibacterial agents. Acquired resistance results from genetic changes in the bacterial cell and arises either by mutation or by the acquisition of genetic material from another cell. Russell and Gould suggest that enterobacteria resistance is associated with the outer cell layers, which limit penetration of the chemicals. Other investigators have agreed with that hypothesis and found that the phospholipid content and fatty and neutral lipid content of the bacterial cell wall of certain gram-negative bacteria increased the resistance to benzalkonium chloride [61]. Similar mechanisms of QAC resistance involving modification of the lipid composition of the outer membrane have been proposed for *Pseudomonas aeruginosa* [62].

Other authors [54,55] report that the loss of antimicrobial activity of benzalkonium chloride, when in contact with cotton fibers or sponges, has contributed to nosocomial outbreaks. Prolonged exposure to weak solutions of benzalkonium chloride can also allow organisms to develop resistance [63].

Attachment to surfaces also appears to be an important factor in microorganisms surviving exposure to QAC [58]. Dhaliwal et al. [64] evaluated the efficacy of two different quaternary ammonium-based disinfectants at three concentrations against biofilms of *E. coli*, *Listeria monocytogenes*, *Staphylococcus aureus*, and *Salmonella enteritidis* on six substrates. As might be expected, there were differences in efficacy between the two QACs, as well as differences in susceptibility among the different organisms, and the type of surface also had an effect on results. In another study, an acidic quaternary ammonium compound was found to be more effective than a neutral-pH quaternary ammonium compound on a *L. monocytogenes* biofilm [65].

V. MODE OF ACTION

The germicidal activity of quaternary ammonium compounds is directly related to the chemical properties of cationic surfactants. Those properties include a reduction in surface tension and formation of ionic aggregates, which results in changes in conductivity and solubility [1]. Since quaternary ammonium compounds are positively charged, the cations are attracted to negatively charged materials such as bacterial proteins. Proteins comprise a large portion of the bacterial cell, both structural components of the cell and bacterial enzymes [66]. Disorganization and denaturation of essential proteins [67] and release of nitrogen- and phosphorus-containing cellular constituents [68] were among the first explanations for the mode of antibacterial activity of the quaternary ammonium compounds.

Electrophoresis has been used to study the interaction of various chemicals with bacteria. Dyar and Ordal [69] investigated the effect of cetylpyridinium chloride on 10 different gram-positive and gram-negative bacteria. They found that as the concentration of the QAC was gradually increased, the mobility of the cells toward the cathode decreased, and eventually, the charge was reversed and these cells moved toward the anode. McQuillen [70] confirmed the findings of Dyar and Ordal using the gram-negative bacteria *E. coli*. McQuillen, however, found the gram-positive bacteria *Staphylococcus aureus* and *Streptococcus faecalis* to react differently. McQuillen explained that the difference was due to release of molecules from the cell due to changes in cell permeability resulting from interaction with the QAC. Hotchkiss [71] showed that membrane damage was occurring by demonstrating that nitrogen- and phosphorus-containing compounds leaked from staphylococci when treated with QACs. Salton [72] demonstrated that the bactericidal activity of cetrimide was related to leakage of purine or pyrimidine components of the cell.

Baker et al. [73] studied the effects of various synthetic detergents on aerobic and anaerobic respiration of glucose by a variety of bacteria. They concluded that the cationic detergents were effective inhibitors of respiration in both gram-positive and gram-negative bacteria. Alkyldimethyl-benzylammonium chloride was found to inhibit the cytochrome–cytochrome oxidase system in certain yeast, demonstrating an effect on enzyme systems [74]. Working with six different cationics and yeast cells, Armstrong [75] found that the initial effect was on the cell membrane, followed by inactivation of cellular enzymes.

Hugo [66] presents a good overview of the mechanism of antimicrobial action, discussing five possible modes of cidal action for QACs:

1. Direct effects on proteins: denaturation and disruption
2. Effects on metabolic reactions
3. Effects on cell permeability and membrane damage
4. Stimulatory effect on the glycolysis reactions
5. Effect on the enzyme system, maintaining a dynamic cytoplasmic membrane

Hugo concludes that damage to the cytoplasmic membrane, with resulting changes in cell permeability, is the primary mode of antimicrobial activity for quaternary ammonium compounds.

Lawrence [76] also reviewed studies evaluating the effects of cationic surfactants on microorganisms. Other articles [77,78] mention the effects of quaternary ammonium compounds on microbial enzymes, proteins, nucleic acids, and toxin production.

VI. FACTORS AFFECTING ACTIVITY

In general, the rate of antimicrobial effect increases with increasing temperature, although the effect is more dramatic with some compounds than with others. As temperature is increased arithmetically, the rate of bacterial kill increases geometrically. The effect of temperature increase on the rate of bactericidal activity, at stated conditions of a fixed concentration of QACs and standardized microbial population, is expressed quantitatively as a temperature coefficient [79].

Thus

$$\Theta^{10} = \frac{\text{time required to kill at } x^\circ\text{C}}{\text{time required to kill at } (x + 10)^\circ\text{C}}$$

or

$$\Theta^{T_2-T_1} = \frac{t_1}{t_2}$$

where T_1 and T_2 are two temperatures differing by 10°C, and t_1 and t_2 are the corresponding lethal times [79].

The disinfecting capabilities of n-alkyl (40% C_{12}, 50% C_{14}, 10% C_{16})-dimethylbenzylammonium chloride were reported to increase four to sevenfold with increasing temperature (4 to 50°C) [80]. Johns [81] also stated that the antimicrobial properties of quaternary ammonium compounds are enhanced with increasing temperature.

The bactericidal properties of n-alkylbenzyldimethylammonium chloride have been tested at a concentration of 400 ppm QAC at pH 3.5 over the temperature range 20 to 55°C [82]. The bactericidal properties increased from no detectable kill at 1 min at 20°C to a five-decimal-place reduction in 1 min at 55°C.

Figure 2 shows the effect of exposure time and temperature on QAC disinfection. These data show that effective disinfection can be achieved at approximately one-third the QAC concentration by increasing the exposure time from 10 min to 30 min. Hence, by optimizing exposure period and exposure temperature, less germicide will be required.

The presence of organic soil on the surface to be treated will affect the antimicrobial properties of the QAC. First, the soil will act as a barrier or

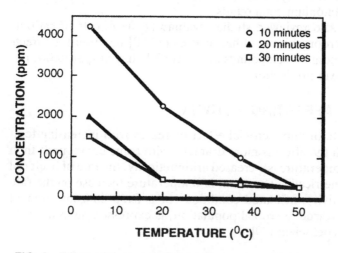

FIG. 2 Effect of time and temperature on quaternary ammonium compound disinfection. AOAC Use Dilution Method (21), *Pseudomonas aeruginosa* ATCC 15442. (Adapted from Ref. 80.)

protectant for the target organism, not allowing the QAC to come in intimate contact with the cell. Second, the soil may interact with the QAC, bonding it and rendering it incapable of exerting lethal effects. Literature reports, however, have conflicting data regarding the effects of organic soil on the antimicrobial activity of quaternary ammonium compounds.

As cationic surfactants, quaternary ammonium compounds do possess detergency. This detergency is sufficient for lighter soils but not for a heavy level of organic soiling. Some of the conflict in reported results is probably due to the wide variety of compounds available and the variety of compounds and conditions used to generate these data. Various disinfectants and their tolerance to an organic soil of sterile yeast suspension were tested [83]. At equal concentrations of 0.25%, quaternary ammonium compounds and phenolic compounds were found to have similar soil tolerance levels. Other investigators [84,85] demonstrated that QAC antimicrobial activity was not greatly affected by the presence of 10 to 16% serum present as an artificial soil load incorporated into the bacterial inoculum. The capability of QACs to tolerate soil makes them the sanitizer of choice in applications where cleaning performance is required.

Quaternary ammonium compounds generally are considered to be most effective at pH values in the alkaline range. For the most part, this is probably true. However, the relation of pH to antimicrobial effectiveness is more complex than is generally believed. For example, it has been reported that antimicrobial activity was constant over the pH range 3 to 10 [86]. Soike et al. [87] studied the effect of changing pH (pH 3 to 10) on the antimicrobial activity of four different quaternary ammonium compounds (alkyldimethylethylbenzylammonium chloride, alkyldimethylbenzylammonium chloride, p-diisobutylphenoxyethoxydimethylbenzylammonium chloride, and methyldodecylbenzyltrimethylammonium chloride) against *Escherichia coli*, *Pseudomonas aeruginosa*, and *Micrococcus caseolyticus*. The activity of each of the quaternary ammonium compounds was similar (i.e., there was little variation attributable to the chemicals). There was, however, significant variation in activity, depending on the test organism. *E. coli* was the most susceptible at pH 6 to 7 and exhibited resistance below pH 6 and above pH 8. *M. caseolyticus* was resistant to the quaternary ammonium compounds at acidic pH levels, but quite susceptible to their action in the alkaline range. *P. aeruginosa* demonstrated just the opposite profile, with the greatest susceptibility at acidic pH levels. They concluded that because all four quaternary ammonium compounds had similar activity, the variation in results when the pH was changed was due to biological factors associated with the cell.

Tolerance to water hardness will vary depending on product formulation. In general, quaternary ammonium compounds are not known for good

TABLE 9 Effect of Carbon Chain Length on Hard-water[a] Tolerance of QACs[b]

| | R_1 | | | | | |
$R_1 + R_2$	6	7	8	9	10	11
19	700	800	500	550	—	—
20	500	900	1200	1400	1100	—
21	300	700	900	1300	1300	—
22	400	550	750	900	800	650

[a]Values in ppm as $CaCO_3$.
[b]AOAC Germicidal and Detergent Sanitizer Test (11), 30-s exposure to 200 ppm active QACs.
Source: Adapted from Ref. 88.

activity in the presence of hard-water ions. Hard-water tolerance is dependent on the type of quaternary ammonium compound and level of water hardness. Activity can be improved with chelators, such as EDTA, which sequester the hardness ions and negate some of the deleterious effects of these ions. The importance of quaternary ammonium compound chain length ($R_1 + R_2$) to hard-water tolerance has been demonstrated [88]. This work focused on structure and the maximum water hardness that would allow 99.999% kill in a 30-s exposure against *E. coli*. Compounds with R_1 and R_2 totaling 20 or 21 carbon atoms demonstrated the greatest tolerance to hard water, especially if R_1 equaled nine carbon atoms. The data are shown in Table 9.

Quaternary ammonium compounds, being cationic surfactants, are not compatible with anionic surfactants. This is important to keep in mind, as some cleaners may contain anionic surfactants, which have a negative charge and thus will neutralize the cation, and may form a precipitate.

This brief review of the physical and environmental factors (such as QAC concentration, solution pH and temperature, and presence of soil or hardness ions) that will alter cationic activity should help the reader appreciate that adherence to sound sanitation or manufacturing practices is primary to the performance of all chemicals used in sanitation, preservation, or disinfection and is not limited to the quaternary ammonium compounds.

VII. PRACTICAL APPLICATIONS

A. Health Care Applications

Quaternary ammonium compounds have a wide range of uses in the health care market, including applications in hospitals, nursing homes, dental offices, and veterinary facilities. Proper use of effective antimicrobial agents is a major factor in preventing nosocomial infections. Because of the antimicrobial as well as physical properties of the cationic surfactant, quaternary ammonium compounds are widely used for environmental disinfection in the health care setting, particularly for noncritical care items and housekeeping purposes. They are not recommended for disinfection of critical medical items or most semicritical items [89]. Support of this can be found in survey data where 35% of respondents used quaternary ammonium compounds for routine disinfection of noncritical patient care items, such as bedpans and electrocardiogram leads [90]. Quaternary ammonium compounds were used by 78% of the respondents in the same survey to disinfect floors, furniture, and other surfaces in patients' rooms.

However, quaternary ammonium compounds have previously seen some use in disinfection of critical care items. Previously, alkyldimethylbenzylammonium chloride was recommended for disinfection of surgical instruments [91], and others had reported on the use of QACs for such applications [92,93]. Because of the lack of activity against bacterial spores, mycobacteria, and hepatitis viruses, this practice has largely been discontinued.

For the most part the previously widespread use of benzalkonium chloride as a dental instrument disinfectant has also been discontinued [94], and the Council of Dental Therapeutics has withdrawn approval for its use [95]. On the other hand, 2% benzalkonium chloride in 70% alcohol is acceptable for use as a topical oral antiseptic in dental clinics [94].

Quaternary ammonium compounds have been evaluated as degerming agents for the skin. Benzalkonium chloride, alone and in various combinations with alcohol, has been reported as an antiseptic for the skin of both humans and animals [54–56]; however, its effectiveness as an antiseptic is diminished by the neutralizing effects of cotton or other materials [96]. A combination of alkyldimethylbenzylammonium chloride, cetyl alcohol, and isopropyl alcohol was described as a preoperative scrub for surgeons' hands [97]. Another study tested several materials and concluded that QACs had merit as skin degerming agents [98]. In a further trial, a cationic preparation was compared to 70% isopropanol and it was demonstrated that the cationic preparation yielded a higher reduction of *Staphylococcus epidermidis* inoculated on the skin of volunteers than did isopropanol [99].

The antibacterial activities of several antiseptics were evaluated in a mouse model as the assay method [100]. Investigators inoculated germ-free hairless mice by gastric intubation to produce a stable and reproducible skin flora. An antiseptic containing benzalkonium chloride produced log reductions ranging from approximately 0.8 log unit to approximately 1.8

log unit. Reportedly, this antiseptic preparation produced a 1.5- to 2-log-unit reduction in the skin flora of humans.

During the first half of this century, quaternary ammonium compounds were tested and subjected to clinical trials as wound irrigants and as urinary and gynecological lavages as well as antimicrobial ointments [76]. The majority of these applications are no longer in practice for the aforementioned reasons and because of the development of antibiotics and antibiotic preparations.

B. Food Service and Food-Processing Applications

Quaternary ammonium compounds are one of the most popular types of chemical sanitizers for food-contact surfaces, equipment, and premises [101]. Proper sanitation will reduce or eliminate potential spoilage and pathogenic organisms, creating a longer shelf life for perishable food products and safer food for consumption.

Trials to evaluate the sanitizing capabilities of alkyldimethylbenzylammonium chloride on artificially contaminated drinking glasses were successful [102]. Also, it was demonstrated that alkyldimethylbenzylammonium chloride could be used to reduce bacterial numbers on dairy equipment [103]. In a mechanical dishwashing operation, effective sanitation was achieved using a quaternary ammonium compound as a supplement to heat in the rinse cycle [104]. The quaternary ammonium compound was used at 200 ppm for 5 to 10 s at 120°F.

Quaternary ammonium compounds have been shown to be good sanitizers for the treatment of farm milk cans, dairy plant equipment, and environmental surfaces [105]. However, 5 to 10 ppm QAC was also reported to retard the growth of *Streptococcus lactis* used for cheese production, so that QACs have limited application in the dairy industry. Additionally, it was reported that quaternary ammonium compounds built up films on the inner surface of glass piping and gaskets after short periods of use as clean-in-place (CIP) sanitizers [106].

When used as a cow udder and teat wash, the quaternary ammonium compounds were less irritating to the skin of the udder and the hands of the milker than was chlorine, which had been the sanitizer of choice [107]. As an udder wash, 30 to 40 udders could be washed with a single solution of quaternary ammonium compound before there was a significant loss of antimicrobial activity [108].

More recently, the inhibitory effects of pure *n*-alkyldimethylbenzylammonium chloride on 13 different strains of dairy starter cultures have been evaluated [109]. Sensitivity, as measured by inhibition of acid production, varied from as little as 0.5 ppm QAC (for two strains of *Lactobacillus bulgaricus*) to 10 ppm QAC. Further, biological test procedures may be affected by the presence of QACs, some of which are listed in Table 10.

Mosley et al. developed a procedure to simulate the actual use pattern of sanitizers in the food and dairy industries and evaluated the effects of various sanitizers on food spoilage organisms [117]. They found that QAC should be at 200 ppm for effectiveness against organisms adhering to surfaces.

A variety of industrial applications have been suggested for QACs [118]. In the poultry industry it was reported that a quaternary ammonium compound was superior to chlorine for washing and disinfecting dirty shell eggs in the processing plant [119]. Quaternary ammonium compounds are used for sanitizing floors, walls, equipment, and furnishings in meat and poultry plants, and they are also recommended as terminal disinfectants for livestock and poultry trucks after each trip [118]. In the seafood-processing industry, QACs are recommended as sanitizers at 200 ppm for equipment and utensils, while 500 to 800 ppm is recommended for rough surfaces such as concrete floors and walls [118].

Products containing quaternary ammonium compounds were found to be effective against slime-producing bacteria, which shorten the shelf life of vacuum-packed meats and meat products [120]. The authors suggest that QACs be used to sanitize surfaces in the processing areas of these meat plants. Of various chemicals evaluated as sanitizers for food animal carcasses, quaternary ammonium compounds were found to reduce the total aerobic populations by less than 0.5 $\log_{10}$ cycles on carcasses [121]; thus QACs are not used routinely as carcass sanitizers. A patent recently was issued for a composition used to disinfect animal pens [122]. The

TABLE 10 Minimum Concentration[a] to Affect Bacteriological Assays

Assay	Concentration	Ref.
Standard plate count (partial inhibition)	20 (4 days' incubation)	111
	200	112
	500	113
Standard plate count (complete inhibition)	2000	114
	<20	111
Coliform count	20 (2 days' incubation)	111
Psychrophilic count	20 (4 days' incubation)	111
B. subtilis disk assay	100	115
B. stearothermophilus disk assay	100	116

[a]Quaternary ammonium compound concentration in ppm.
Source: Adapted from Refs. 110 to 116.

composition was comprised of dialkyl (C_{10}) dimethylammonium chloride and inorganic peroxides.

Quaternary ammonium compounds are listed among those sanitizers most applicable to fruit and vegetable processing plants [118]. They have also been used with success for sanitation in the brewing industry [118,123]. Quaternary ammonium compounds have been recommended for sanitation of used containers in wineries to control the growth of mold [124]. A patent was issued for the use of QACs in the sugar refining industry [125].

Tebbutt [126] conducted a microbiological evaluation of sanitization practices in food service establishments. While hypochlorites were frequently used, quaternary ammonium compounds were also used to sanitize cutting boards, work surfaces, and wiping cloths, and were found to help reduce the danger of cross-contamination.

Transfer of bacteria from human hands to raw or cooked food products has been implicated in outbreaks of foodborne disease [127,128]. Quaternary ammonium compounds could be used as a hand sanitizer by dipping hands in a solution containing 50 ppm QAC [118]. Sheena and Stiles [129] reported that a QAC dip solution produced a significant decrease in the number of bacteria on the skin.

C. Miscellaneous Applications

Quaternary ammonium compounds have been used successfully in industrial applications such as captive water treatment to control the growth of bacteria and algae [130] and in the pulp and paper industry to control the growth of slime-producing organisms [131].

A combination of alkyldimethylbenzylammonium chloride and alkyldimethylethylbenzylammonium chloride has been used to sanitize laundered fabric [132]. This preparation also provided residual bacteriostatic properties to the fabric.

Greenwald and Halsrud [133] disclosed a film-forming composition comprised of a quaternary ammonium compound and an anionic polymer. The resulting film is reported to be antimicrobial with the capability to withstand multiple water rinses for applications in the health care, food-processing, and food service industries, or in areas where prolonged control of microbial growth is needed.

VIII. SUMMARY

Quaternary ammonium compounds were initially developed to be used as simple disinfectants. As is obvious from this brief review, formulations containing QACs have found their way into many industries and have both myriad applications and limitations.

In 1683, Antonie von Leeuwenhoek, the Dutch lens grinder and scientist, reported observing small "animals" as he examined scrapings from his teeth under his microscope. Leeuwenhoek subsequently used "strong wine-vinegar" to gargle and to rinse his mouth. He observed that "an innumerable quantity of animals" remained when he examined a fresh scraping. Leeuwenhoek then mixed a small amount of the "wine-vinegar" with the scraping and observed that the "animals dyed". Leeuwenhoek concluded that while the "wine-vinegar" killed exposed "animals" within the scrapings, it could not penetrate the substance of the scrapings and kill "animals" within the scrapings. This may have been the first experiment on antimicrobials (S. S. Block, personal communication, 1987).

But over 300 years later, the findings of Leeuwenhoek provide us with more than a scientific experimental curiosity. Leeuwenhoek's discoveries demonstrate the need for precleaning prior to terminal sanitation, and they teach that an antimicrobial does have limitations. By gaining knowledge of those limitations, the user will be able to determine appropriate use applications to achieve the desired result.

REFERENCES

1. A. N. Petrocci, in *Disinfection, Sterilization and Preservation* (S. S. Block, ed.), Lea & Febiger, Philadelphia, 1983, pp. 309–329.
2. R. E. Dixon, R. A. Kaslow, D. C. Mackel, C. C. Fulkerson, and G. F. Mallison, *J. Am. Med. Assoc. 236*: 2415 (1976).
3. W. A. Jacobs and M. Heidelberger, *J. Biol. Chem. 20*: 659 (1915).
4. W. A. Jacobs and M. Heidelberger, *J. Biol. Chem. 20*: 685 (1915).
5. W. A. Jacobs and M. Heidelberger, *J. Biol. Chem. 21*: 103 (1915).
6. W. A. Jacobs and M. Heidelberger, *J. Biol. Chem. 21*: 145 (1915).
7. W. A. Jacobs and M. Heidelberger, *J. Biol. Chem. 21*: 403 (1915).
8. W. A. Jacobs and M. Heidelberger, *J. Biol. Chem. 21*: 439 (1915).
9. W. A. Jacobs and M. Heidelberger, *J. Biol. Chem. 21*: 455 (1915).
10. W. A. Jacobs and M. Heidelberger, *J. Biol. Chem. 21*: 465 (1915).
11. G. Domagk, *Dtsch. Med. Wochenschr. 61*: 829 (1935).
12. R. A. Cutler, E. B. Cimijotti, T. J. Okolowich, and W. F. Wetterau, *Proceedings of the 53rd Annual Chemical Specialties Manufacturers' Association Meeting*, 1966.
13. K. Jono, T. Takayama, M. Kuno, and E. Higashide, *Chem. Pharm. Bull. 34*: 4215 (1986).
14. R. Koch, *Mitt. Kaiserlichen Gesundheitsamt 1*: 234 (1881).
15. S. Rideal and J. T. A. Walker, *J. Roy. Sanit. Inst. 24*: 424 (1903).
16. T. Ikeda, H. Yamaguchi, and S. Tazuke, *J. Bioact. Compat. Polym. 5*: 31 (1990).
17. H. Chick and C. J. Martin, *J. Hyg. 8*: 654 (1908).

18. L. S. Stuart, L. F. Ortenzio, and J. L. Friedl, *J. Assoc. Off. Agric. Chem.* *36*: 466 (1953).
19. J. C. Kelsey and G. Sykes, *Pharm. J. 202*: 607 (1969).
20. American Public Health Association, *Am. J. Public Health 8*: 506 (1918).
21. K. Helrich, ed., *Official Methods of Analysis of the Association of Official Analytical Chemists*, Association of Official Analytical Chemists, Arlington, Va., 1990, pp. 135–137.
22. J. R. Rubino, J. M. Bauer, P. H. Clarke, B. B. Woodward, F. C. Porter, and H. G. Hinton, *J. Assoc. Off. Agric. Chem. 75*: 635 (1992).
23. C. W. Chambers, *J. Milk Food Technol. 19*: 183 (1956).
24. ASTM in *Annual Book of ASTM Standards,* American Society for Testing and Materials, Philadelphia, 1987, Vol. 11.04, pp. 850–855.
25. ASTM in *Annual Book of ASTM Standards,* American Society for Testing and Materials, Philadelphia, 1985, Vol. 11.04, pp. 643–645.
26. ASTM in *Annual Book of ASTM Standards,* American Society for Testing and Materials, Philadelphia, 1985, Vol. 11.04, pp. 646–648.
27. A. Beloian, A. Crémieux, J. Fleurette, M. K. Bruch, E. Asbury, J. H. S. Chen, T. A. Koski, T. J. Czerkowitz, and S. E. Leland, in *Disinfection, Sterilization and Preservation* (S. S. Block, ed.), Lea & Febiger, Philadelphia, 1983, pp. 885–1011.
28. Z. Baker, R. W. Harrison, and B. F. Miller, *J. Exp. Med. 73*: 249 (1941).
29. W. L. Wang, B. W. Powers, N. W. Luechtefeld, and M. J. Blaser, *Appl. Environ. Microbiol. 45*: 1202 (1983).
30. W. L. Wang, M. J. Blaser, J. Cravens, and M. A. Johnson, *Ann. Intern. Med. 90*: 614 (1979).
31. J. J. Miller, W. E. Brown, and V. J. Krieger, *Dev. Ind. Microbiol. 22*: 763 (1981).
32. A. Mustapha and M. B. Liewen, *J. Food Prot. 52*: 306 (1989).
33. M. Best, M. E. Kennedy, and F. Coates, *Appl. Environ. Microbiol. 56*: 377 (1990).
34. D. F. Greene and A. N. Petrocci, *Soap Cosmet. Chem. Spec. 56*: 33 (1980).
35. B. Terleckyj and D. A. Axler, *Antimicrol. Agents and Chemother. 31*: 794 (1987).
36. H. J. Hueck, D. M. M. Adema, and J. R. Wiegman, *Appl. Microbiol. 14*: 308 (1966).
37. M. Klein and A. Deforest, *Soap Chem. Spec. 39*: 70 (1963).
38. H. N. Prince, *Part. Microbiol. Control 2*: 55 (1983).
39. G. C. Lavelle, *Chem. Times Trends 10*: 45 (1987).
40. T. T. Brown, *Am. J. Vet. Res. 42*: 1033 (1981).
41. J. N. Mbithi, V. S. Springthorpe, and S. A. Sattar, *Appl. Environ. Microbiol. 56*: 3601 (1990).
42. E. L. Sing, P. R. Elliker, and W. E. Sandine, *J. Milk Food Technol. 27*: 101 (1964).
43. S. Watkins, H. A. Hays, and P. R. Elliker, *J. Milk Food Technol. 20*: 84 (1957).

44. M. L. Ancelin, H. J. Vial, and J. R. Philippot, *Biochem. Pharmacol. 34*: 4068 (1985).
45. B. R. Cords, in *Antimicrobials in Foods* (A. L. Branen and P. M. Davidson, eds.), Marcel Dekker, New York, 1983, pp. 257–298.
46. A. D. Russell, *Pharm. Int. 7*: 300 (1986).
47. C. E. Chaplin, *J. Bacteriol. 63*: 453 (1952).
48. K. Nishikawa, S. Oi, and T. Yamamoto, *Agric. Biol. Chem. 43*: 2473 (1979).
49. R. A. Kaslow, D. C. Mackel, and G. F. Mallison, *J. Am. Med. Assoc. 263*: 2407 (1976).
50. S. A. Plotkin and R. Austrian, *Am. J. Med. Sci. 235*: 621 (1958).
51. M. D. Shickman, L. B. Guze, and M. L. Pearce, *N. Engl. J. Med. 260*: 1164 (1959).
52. W. F. Malizia, E. J. Gangarosa, and A. F. Goley, *N. Engl. J. Med. 263*: 800 (1960).
53. J. C. Lee and P. J. Fialkow, *J. Am. Med. Assoc. 177*: 708 (1961).
54. J. G. Fox, C. M. Beaucage, C. A. Folta, and G. W. Thornton, *J. Clin. Microbiol. 14*: 157 (1981).
55. R. L. Sautter, L. H. Mattmann, and R. C. Legaspi, *Infect. Control 5*: 223 (1984).
56. A. K. Nakashima, M. A. McCarthy, W. J. Martone, and R. L. Anderson, *J. Clin. Microbiol. 25*: 1014 (1987).
57. N. J. Ehrenkranz, E. A. Bolyard, M. Wiener, and T. J. Cleary, *Lancet 2*: 1289 (1980).
58. G. Sundheim, T. Hagtvedt, and R. Dainty, *Food Microbiol. 9*: 161 (1992).
59. L. Neumayr, G. Heyderhoff, and J. Krämer, *Yeast 5*: S17 (1989).
60. A. D. Russell and G. W. Gould, *J. Appl. Bacteriol. 64*: 167S (1988).
61. Y. Sakagami, H. Yokoyama, H. Nishimura, Y. Ose, and T. Tashima, *Appl. Environ. Microbiol. 55*: 2036 (1989).
62. M. V. Jones, T. M. Herd, and H. J. Christie, *Microbios 58*: 49 (1989).
63. A. K. Nakasima, A. K. Highsmith, and W. J. Martone, *J. Clin. Microbiol. 25*: 1019 (1987).
64. D. S. Dhaliwal, J. L. Cordier, and L. J. Cox, *Lett. Appl. Microbiol. 15*: 217 (1992).
65. E. P. Krysinski, L. J. Brown, and T. J. Marchisello, *J. Food Prot. 55*: 246 (1992).
66. W. B. Hugo, in *Surface-Active Agents in Microbiology*, S. C. I. Monograph 19, Society of the Chemical Industry, London, 1965, pp. 67–82.
67. Z. Baker, R. W. Harrison, and B. F. Miller, *J. Exp. Med. 73*: 249 (1941).
68. R. D. Hotchkiss, *Ann. N.Y. Acad. Sci. 46*: 479 (1946).
69. M. T. Dyar and E. J. Ordal, *J. Bacteriol. 51*: 149 (1946).
70. K. McQuillen, *Biochim. Biophys. Acta 5*: 643 (1950).
71. R. D. Hotchkiss, *Adv. Enzymol. 4*: 153 (1944).
72. M. R. J. Salton, *J. Gen. Microbiol. 5*: 391 (1951).
73. Z. Baker, R. W. Harrison, and B. F. Miller, *J. Exp. Med. 74*: 661 (1941).
74. M. G. Sevag and O. A. Ross, *J. Bacteriol. 48*: 677 (1944).

75. W. McD. Armstrong, *Arch. Biochim, Biophys. 71*: 137 (1957).
76. C. A. Lawrence, in *Cationic Surfactants* (E. Jungermann, ed.), Marcel Dekker, New York, 1970, pp. 491–526.
77. C. A. Lawrence, *Surface-Active Quaternary Ammonium Germicides,* Academic Press, New York, 1950.
78. C. A. Lawrence, in *Disinfection, Sterilization and Preservation* (C. A. Lawrence and S. S. Block, eds.), Lea & Febiger, Philadelphia, 1968, pp. 430–452.
79. B. Lynn, in *Pharmaceutical Microbiology* (W. B. Hugo and A. Russell, eds.), J. B. Lippincott, Philadelphia, 1977, pp. 155–184.
80. P. Gelinas, J. Goulet, G. M. Tastayre, and G. A. Picard, *J. Food Prot. 47*: 841 (1984).
81. C. K. Johns, *Can. J. Technol. 32*: 71 (1954).
82. C. D. Freke and D. Haggie, *J. Food Prot. 44*: 699 (1981).
83. N. A. Miner, E. Whitmore, and M. L. McBee, *Dev. Ind. Microbiol. 16*: 23 (1975).
84. H. Hornung, *Z. Immunitaetsforch. 84*: 119 (1935).
85. G. Schneider, *Z. Immunitaetsforch. 85*: 194 (1935).
86. R. A. Quisno and M. J. Foter, *J. Bacteriol. 52*: 111 (1946).
87. K. F. Soike, D. D. Miller, and P. R. Elliker, *J. Dairy Sci. 35*: 764 (1952).
88. A. N. Petrocci, H. A. Green, J. J. Merianos, and B. Like, *Proceedings of the 60th Mid-year Meeting of the Chemical Specialties Manufacturers' Association,* 1974.
89. M. S. Favero, in *Disinfection, Sterilization and Preservation* (S. S. Block, ed.), Lea & Febiger, Philadelphia, 1983, pp. 469–492.
90. American Health Consultants, *Hosp. Infect. Control 13*: 56 (1986).
91. C. W. Walter, *Surg. Gynecol. Obstet. 67*: 683 (1938).
92. F. Caesar, *Fortschr. Ther. 4*: 249 (1935).
93. W. Schmidt, *Med. Welt 29*: 1043 (1935).
94. J. J. Crawford, in *Disinfection, Sterilization and Preservation* (S. S. Block, ed.), Lea & Febiger, Philadelphia, 1983, pp. 505–523
95. American Dental Association, *J. Am. Dent. Assoc. 97*: 855 (1978).
96. R. B. Kundsin and C. W. Walter, *AMA Arch. Surg. 75*: 1036 (1957).
97. C. W. Walter, *Am. J. Surg. 109*: 691 (1965).
98. P. E. Verdon, *J. Clin. Pathol. 14*: 91 (1961).
99. B. Christiansen, *Zentralbl. Bakteriol. Hyg. 186*: 368 (1988).
100. M. C. Barc, F. Tekaia, and P. Bourlioux, *Appl. Environ. Microbiol. 55*: 1911 (1989).
101. G. J. Banwart, *Basic Food Microbiology,* AVI, Westport, Conn., 1979.
102. A. J. Krog and C. G. Marshall, *Am. J. Public Health 30*: 341 (1940).
103. A. J. Krog and C. G. Marshall, *J. Milk Technol. 5*: 343 (1942).
104. W. L. Mallman and L. Zaikowski, *J. Milk Food Technol. 10*: 206 (1947).
105. P. R. Elliker, J. W. Keesling, D. D. Miller, and G. H. Wilster, *J. Milk Food Technol. 13*: 156 (1950).
106. F. F. Fleischman and R. F. Holland, *J. Milk Food Technol. 16*: 9 (1953).

107. E. M. Foster, *J. Milk Food Technol. 12*: 14 (1949).

108. W. C. Mueller, D. B. Seeley, and E. P. Larkin, *Soap Sanit. Chem. 23*: 123 (1947).

109. N. Guirguis and M. W. Hickey, *Aust. J. Dairy Technol. 42*: 11 (1987).

110. D. G. Dunsmore, *Residue Rev. 86*: 1 (1983).

111. H. V. Atherton and R. A. Johnson, *Sci. Tec. Caseavia 27*: 305 (1976).

112. L. E. Mull and E. L. Fouts, *J. Milk Food Technol. 10*: 102 (1947).

113. M. Sotlar, *Zb. Bioteh. Fak. Univ. Ljubl. 20*: 103 (1973).

114. A. S. Dubois and D. D. Diblee, *J. Milk Food Technol. 9*: 260 (1946).

115. J. Palmer, *J. Milk Food Technol. 27*: 311 (1964).

116. J. Richard and L. Kerherve, *Rev. Lait Fr. 306*: 127 (1973).

117. E. B. Mosley, P. R. Elliker, and H. Hays, *J. Milk Food Technol. 39*: 830 (1976).

118. N. G. Marriott, *Principles of Food Sanitation,* Van Nostrand Reinhold, New York, 1989.

119. V. J. Penniston and L. R. Hedrick, *U.S. Egg Poult. Mag. 50*: 26 (1944).

120. P. M. Mäkelä, H. J. Korkeala, and E. K. Sand, *J. Food Prot. 54*: 632 (1991).

121. M. E. Anderson, R. T. Marshall, W. C. Stringer, and H. D. Naumann, *J. Food Sci. 42*: 326 (1977).

122. S. Araki, M. Suzuki, T. Iwasaki, and Y. Hioki, Japanese patent 62,155,203, to Eisai Company, Kao Corp. (1987).

123. G. J. Lehn and R. L. Vignolo, *Brewers Dig. 21*: 41 (1946).

124. M. A. Amerine and M. A. Joslyn, *Table Wines: The Technology of Their Production,* University of California Press, Berkeley, Calif., 1970.

125. J. A. Casey, U. S. patent 3,694,262 (1972).

126. G. M. Tebbutt, *J. Hyg. 92*: 365 (1984).

127. D. O. Cliver and K. D. Kostenbader, Jr., *Int. J. Food Microbiol. 1*: 75 (1984).

128. D. Roberts, *J. Hyg. 89*: 491 (1982).

129. A. Z. Sheena and M. E. Stiles, *J. Food. Prot. 45*: 713 (1982).

130. D. J. Berenschot, E. G. King, R. K. Stubbs, and G. R. Babalik, U.S. patent 3,140,976 (1964).

131. B. F. Shema, U. S. patent 3,231,509 (1966).

132. E. G. Shay and A. N. Petrocci, *41st Annual Meeting of the Soap and Detergent Association,* 1968.

133. R. B. Greenwald and D. A. Halsrud, U.S. patent 5,158,766, to Ecolab Inc. (1992).

3

Current Topics on the Toxicity of Cationic Surfactants

HANS P. DROBECK   Consultant, Rensselaer, New York

I. INTRODUCTION

A comprehensive review of the toxicity of the cationic surfactants was provided in 1970 in a book entitled *Cationic Surfactants*, edited by Eric Jungermann [1]. To the author's knowledge, at the time it was the first treatise on this subject in the literature, and apparently, it has remained so. In this chapter we collect and review some of the toxicological information related to this group of compounds that has accumulated since Jungermann's monograph. It is rather surprising that despite an extensive literature search, the amount of new toxicologic data found on these cationic surfactants is relatively scarce. Two areas of investigation, however, that were poorly represented previously seem to dominate the new literature: topical toxicity (principally of the eye) and cytotoxicity. New acute and/or multiple-dose systemic toxicity studies are in the minority but are discussed as appropriate. As in the earlier treatise, the toxicologic literature of the 1970s and 1980s has been dominated by the quaternary ammonium compounds, which continue to play a significant role as disinfectants, sanitizers, and germicides. Thus most of our attention here is devoted to them.

II. SYSTEMIC TOXICITY

The term *systemic* in reference to a type of toxic effect or toxicity study has become common parlance. A typical dictionary definition is "pertaining specifically, in physiology, to affecting the entire bodily system." Toxicologists have cast the term to mean that the route of administration of a particular chemical presumably exposes the entire body to its potential adverse effects, which the toxicologist then hopes to be able to detect. It is in this sense that the following cationic surfactants are discussed.

In general, the acute (single-dose) toxicity of the quaternaries is characterized, at lethal doses, by peripheral paralysis and central nervous system (CNS) stimulant (convulsant)-like effects. In chronic (multiple-dose) studies, their toxic effects commonly consist of adverse effects on body weight or growth, reduced food consumption, dehydration, and increased mortality. The pathologic consequences of chronic oral administration, which are found principally in the gastrointestinal (GI) tract, are diarrhea, gastritis, emesis, and enteritis, and necrosis, hemorrhage, and erosion of the GI mucosa.

Perhaps the most widely investigated of the quaternaries is benzalkonium chloride (BAC), a member of the alkyldimethylbenzylammonium group of compounds in which the alkyl component is a mixture of C_8H_{17} to $C_{18}H_{37}$ groups. It is a common ingredient in cosmetics, serving as a foaming and cleansing agent, conditioner, and bactericide. It is also commonly used in ophthalmic medications and contact lens solutions, hand washes, antimicrobial soaps, skin wound cleanser, and preoperative skin preparation solutions. As a result of these widespread uses it comes into contact with almost all body surfaces and cavities and has the potential to be absorbed, inhaled, and accidentally ingested.

A. Systemic Effects by Inhalation

CD strain rats and Syrian golden hamsters were exposed to 0.1% BAC as part of an aerosolized hair conditioner, 4 h a day, 5 days a week for 14 weeks and subjected to a comprehensive toxicologic evaluation, including gross and microscopic pathology [2]. No adverse effects were found, either during the course of the study or at termination, which could be interpreted as being due to BAC inhalation. In humans, bronchoconstriction in asthmatic patients has been reported after inhalation of antiasthmatic respirator solutions containing BAC as preservative [4,5]. Such bronchoconstriction in asthmatics has been claimed to occur at a concentration of 0.2 μg/mL, as found in beclamethasone nebulizer solution. The mechanism is uncertain but may be due to enhancement of the release of histamine [6]. In a study of 28 asthmatic subjects [5] in which the patients inhaled histamine and BAC on alternate days, the dose that caused a 20% fall in initial expiratory volume ($PD_{20}FEV_1$) was measured. Subjects who responded to BAC were more sensitive to histamine than were subjects who did not respond. Among the BAC responders there was a significant correlation between $PD_{20}FEV_1$ values for histamine and benzalkonium chloride. These results showed that in asthmatic subjects with bronchial hyperresponsiveness, BAC caused a reproducible bronchoconstriction and enhanced response to histamine. However, those subjects who did not respond to BAC were less sensitive to histamine than were BAC responders, suggesting that the amount of histamine released was not enough to cause a response in these patients. Release of mediators such as leukotrienes or prostaglandins may also be involved in the BAC effect via its known cytolytic effect on mast cells.

It thus appears that there may be a segment of the population of asthmatics that may respond to BAC in an exaggerated way. It also raises the question of whether this effect is dose (concentration) dependent, and if so, whether lowering the concentration would limit or delete it.

B. Systemic Effects by Oral Ingestion

As described previously, the effects of oral administration of quaternaries are principally on the GI tract and resemble corrosive activity. It is generally known [7] that BAC in concentrations above 1% is corrosive to mucous membranes, skin, and eye. An example of this effect was seen in infant twins [8] who accidentally received applications of 5.6% in the mouth for treatment of oral candidiasis. The result was rapid appearance of chemical burns of the mouth and pharynx associated with fever, anorexia, and dehydration in both twins and pneumonitis in one of the two. Both recovered without serious systemic effects, presumably because the amounts actually ingested or inhaled were either nil or limited.

Similar findings were made with a germicide of slightly different alkyl distribution than BAC in large flocks of white Leghorn chickens that accidentally ingested high concentrations in drinking water as a germicide [9]. These flocks had high mortality, dehydration, and poor egg production. Pathological scrutiny revealed lesions on the oral and pharyngeal mucosa and tongue consisting of ulceration and necrosis. In some cases the upper esophagus was also involved.

In a prior study [10], turkeys were given the same product in drinking water for 5 to 6 weeks at several concentrations to determine the no-effect and toxic levels. The highest no-effect level was 200 ppm (0.02%), whereas 500 ppm was toxic, causing mortality, dehydration, stunting of growth, and ulceration and hyperkeratosis of the squamous epithelium of the tongue.

Thus use of the quaternaries as germicides or fungicides where they come in contact with the mouth must be done with knowledge that significant adverse effects can accompany overdose or concentrations accidentally higher than recommended.

Studies to characterize the toxicity of one of the alkyltrimethylammonium compounds, cetyltrimethylammonium bromide (CTAB), in rats have revealed that its profile of effect is similar to that for BAC [16,17]. The acute intravenous LD_{50} of CTAB in mice and rats was reported as 32 and 44 mg/kg, respectively, and when given in the drinking water at doses of 10, 20, and 45 mg/kg for 1 year resulted in reduced efficiency in food conversion and reduction in growth rate at the highest dose level. When perfused in rat and rabbit intestines in conjunction with glucose, methionine, and sodium butyrate, the absorption of these nutritional compounds was depressed, indicating that the adverse effect on growth in the absence of histopathologic changes in the GI tract may be due to this effect on intestinal absorption [19].

One of the dialkyl quaternary ammonium compounds, known as Quaternium-18, Arquad 2HT, or Aliquot H226, has been subjected to thorough review by the Cosmetic Ingredient Review Expert Panel [18]. The review indicates that Quaternium-18 is used in 20 cosmeticlike products in concentrations of 0.1 to 10%. The acute oral LD_{50} in rats was reported as approximately 5250 mg/kg. When fed to guinea pigs for 12 days the lowest toxic dose level was 1 g/kg. When fed to dogs and rats for 90 days at 2800 ppm (0.28% of the diet), no toxic effects were found. These data suggest that Quaternium-18 has a very low order of systemic toxicity.

C. Systemic Effects on Reproductive Function

There appears to have been very little research interest in the potential for the quaternary ammonium compounds to influence the reproductive process or the development of fetuses. However, BAC is an effective spermicidal agent and vaginal contraceptive and, as such, is available over the counter in Canada. This has led to a question as to whether spontaneous abortions and/or congenital defects result from the use of such agents. In a study to investigate this, BAC aqueous solutions were instilled into the vaginas of female Wistar rats once on day 1 of gestation in dose levels of 25, 50, 100, or 200 mg/kg [11]. When the dams were sacrificed on day 21 of gestation, no congenital malformations were found, but dose-related decrease in pregnancy rate, live litter size, and litter weight were seen. A significant fetocidal effect was thus found. In addition, there was a significant reduction in the number of implantation sites in the highest dose level, indicating that there was also an embryocidal effect. The mechanism of these changes was not defined in this study, but in an in vitro experiment with BAC in which mouse embryos were exposed to concentrations of 0.1 to 100 ng/mL of culture medium, cell cleavage, and hatchability were suppressed at as low a level as 10 ng/mL [12]. These data strongly suggested at least that the embryocidal–fetocidal effects seen in the prior study were the result of the presence of BAC and not an undefined toxicity to the maternal rats caused by the effects of the compound in the vagina. When given orally by gavage to mice in dose levels ranging from 1 μg/kg to 30 mg/kg during either the first 6 days of gestation or the complete 18 days, a dose-related effect was seen [13]. It was found that 3 mg/kg was the highest no-effect dose and that 10 mg/kg during the first 6 days of gestation resulted in a slight decrease in pregnancy rate. The high dose of 30 mg/kg was also associated with an increase in resorption rate and number of dead fetuses. Thus, a dose of 30 mg/kg was interpreted as an abortive oral dose in the mouse.

In the foregoing three experiments, BAC was found embryocidal or fetocidal in an in vitro culture system, by vaginal exposure and by oral administration, clearly suggesting that accidental or intentional use in humans may produce the same effect. However, none of these studies revealed congenital malformations.

Further evidence of the lack of developmental effects of the quaternaries on rats was determined in a study in which three different compounds were applied topically during days 6 to 15 of gestation [14]. The three compounds—benzyldimethylstearylammonium chloride, trimethylstearylam-monium chloride, and dimethyldistearylammonium chloride—were each applied at three concentrations that produced dose-related local skin re-actions. In addition to the lack of effect on development of the fetuses, there were also no adverse effects on litter size, postimplantation loss, or litter or fetal weights. Also, when Wistar rats were administered various quaternaries at 5, 15, and 50 mg/kg orally during the organogenesis period [15], once again no teratogenic activity was seen.

D. Systemic Effects on the Ear

Histopathologic examination of the ears of guinea pigs (organ of Corti, vestibulum, tympanic cavity, cochlea) after exposure of the round window to 0.1% BAC solutions in alcohol or distilled water revealed that in both vehicles it caused significant pathology [3]. Changes included hair cell loss from the organ Corti, fibrosis of the tympanic cavity and cochlea, and damage to the neuroepithelium of the vestibulum. Exposure periods of 10, 30, and 60 min demonstrated that the longer the exposure, the greater the damage to these organs, but that there were only minor differences between the solutions. No attempts to titrate the concentration of BAC to a no-effect level were made.

III. OCULAR TOXICITY

Interest in the ocular irritation of cationic surfactants dates back into the 1950s, when investigators became aware of the testing procedure devised by Draize et al. in 1944 [20]. This new procedure provided a way to combine the reactions seen in the conjunctiva, cornea, and iris of rabbits via de-scriptive and numerical scores, and arrive at an assessment of the degree of irritancy. Draize's method was reliable enough and consistent enough to allow for comparative testing of chemicals, and eventually it was made the official U.S. Food and Drug Administration test method. In 1950, Hoppe et al. [21] introduced an interesting sophistication of the Draize procedure: the concept of the percent concentration of a chemical required to result in minimal irritancy, the threshold. By using graded concentrations it was possible to find this threshold. Hoppe used this procedure to compare the irritancy in the rabbit eye of three dialkyl-, two alkyldimethylbenzyl-, and one alkyldimethyl-substituted benzylammonium chloride quaternary ammonium compound. In this way it was possible to determine that of the six compounds in three different chemical classes, all had threshold irri-tation concentrations (TICs) of 0.063% except one that had a TIC of 0.125%, presumably slightly less irritating.

In 1952, Draize and Kelley [22] declared that cosmetic formulations containing surface-active agents should be tested for eye irritation if the cosmetic could gain access to the eye during routine use. Utilizing his rodent eye test procedure, he categorized 23 surface-active agents according to the maximal tolerated concentrations and found that the order of eye toxicity was cationic > anionic > nonionic.

Subsequently, Hazelton [23] used the Draize scoring system to evaluate the ocular irritancy of 28 nonionic, anionic, and cationic surfactants. Using the rationale that the protein precipitating properties of the cationics would make them more likely to irritate the eye than would nonprotein precipitating nonionics, the Draize test again demonstrated a rank of irritation of cationic > anionic > nonionic. Other characteristics, such as surface activity, pH, wetting power, foaming power, combined wetting and foaming, capacity to produce opacity, and erythrocyte compatibility, were less reliable indices of irritancy.

In another example of how quaternary ammonium compounds have been compared for eye irritancy by the Draize method, a group of six alkyltrimethylammonium compounds—Arquad 12, Arquad 16, Arquad T (tallow compound), Arquad 18, Arquad C (coco compound), and Arquad S (soya compound)— were all tested using 0.1 mL of 10% aqueous solutions [24]. The first three were serious irritants; the latter three were only slightly irritating. With data such as these, investigators could begin to make decisions regarding mechanism of action and marketing choices regarding the eye safety of quaternary products.

The Draize test continued to play a dominant role in eye irritancy testing during the 1970s and 1980s, particularly in the cosmetic industry and, in the United States, as a regulatory tool. The dialkyl quaternary Quaternium-18, which is widely used as a thixotropic agent in cosmetics, was studied by the Draize method at concentrations of 4, 5, and 7.5% and was determined to be a minimal irritant [18]. Matsuura et al. [25] tested a series of surfactants at 10% in both rabbit and guinea pig eyes and in essential agreement with Draize and Hazelton, found that in both species the irritancy was in the order cationic > amphoteric > anionic > nonionic (Table 1). BAC was found to cause desquamation and necrosis of the corneal epithelium at a concentration as low as 0.5%. It was also determined that the rabbit was a more sensitive test species than the guinea pig for these compounds. Further evaluation of BAC continued to confirm its irritancy potential for the eye. Although generally recognized as relatively safe for the human eye, its accidental introduction into the eye at higher than recommended concentrations or volumes is known to be associated with significant irritancy. The volume factor was investigated in the rabbit eye to determine which volume would be most predictable for the human

TABLE 1 Ocular Irritation and Toxicity of Benzalkonium Chloride

Animals tested	Test substance	Methodology	Results	Ref.
Rabbits	2.0, 1.0, 0.5, 0.1, and 0.01% benzalkonium chloride	All five solutions instilled into both eyes twice daily for 7 days; 0.1% solution instilled into eyes five times daily for 1 week	2.0% solution caused conjunctival necrosis, ulceration and haziness of the cornea, and massive iritis; the 1.0 and 0.5% solutions caused damage to the cornea and bulbar and palpebral conjunctivae; the 0.1 and 0.01% solutions (instilled twice daily) did not cause ocular damage; 0.1% solution (instilled five times daily) damaged corneal endothelium	28
New Zealand albino rabbits	0.1% benzalkonium chloride	Instilled into one eye	No ocular irritation	27
Rabbits	10.0, 1.0, 0.5, and 0.1% benzalkonium chloride	Single instillation; corneas examined at 4 h and 1 and 7 days postinstillation	Desquamation and necrosis of corneal epithelium (10.0, 1.0, 0.5, and 0.1% concentrations)	25

Species	Test substance	Method	Findings	Ref.
New Zealand albino rabbits	0.01% benzalkonium chloride	Drops instilled into both eyes; corneas examined at 5, 15, and 30 min and 1, 3, and 6 h after instillation	Desquamation of epithelial cells and extensive disruption of plasma membranes	34
New Zealand white rabbits	0.01% benzalkonium chloride	Excised corneas flooded with solution for 2 min	Ruptured endothelial cells with exposed nuclei; severe cellular edema	40
New Zealand white rabbits	0.01% benzalkonium chloride	Aqueous humor removed from each eye and replaced with 0.01 mL of test substance	Swelling of corneal endothelium, ruptured plasma membranes, dilation of endoplasmic reticulum, and mitochondrial swelling	40
Albino and pigmented rabbits (two albinos and two pigmented/group)	0.01 and 0.007% benzalkonium chloride	Subconjunctival injection of both solutions (0.1 mL) once daily for 2 weeks; eyes examined via ophthalmoscopy (daily) and electron microscopy	Extensive elevation of retina and retinal detachment in both treatment groups	36
Albino rabbits	6.5×10^{-6} to 6.5×10^{-3}% benzalkonium chloride	Corneal endothelium perfused for 3 h	Severe endothelial cellular damage	38
Guinea pigs	10.0, 1.0, 0.5, and 0.1% benzalkonium chloride	Single instillation; corneas examined at 4 h and 1 and 7 days postinstillation	Desquamation and necrosis of corneal epithelium (10.0, 1.0, and 0.5% concentrations)	25

response [27]. BAC, one of 21 chemicals used, was tested at 0.1%, which is known to be mildly irritating in the human eye. When volumes of 0.01, 0.03, and 0.10 mL were instilled into the rabbit eye and scored via the Draize procedure, 0.01 mL resulted in the degree of irritancy most consistent with the known human response. This also appeared to be true for most of the chemicals tested and therefore suggested that the 0.1 mL volume stipulated in the Draize test should be reduced to translate the rabbit eye data more realistically to humans. It also suggested that the common criticism that the rabbit eye was too sensitive was not necessarily true, but rather, that a volume correction in the procedure would make the rabbit more predictable. In a design to test concentration range effects, concentrations of 0.01, 0.1, 0.5, 1.0, and 2.0% aqueous solutions of BAC were instilled into both eyes of rabbits twice a day for 7 days [28]. The 0.01 and 0.1% concentrations did not cause any ocular damage, whereas the higher concentrations resulted in concentration-related effects to the conjunctiva, cornea, and iris. However, when the 0.1% level was instilled five times a day for 7 days, corneal endothelial damage occurred. The latter two studies of topical irritation were good examples of the toxicologist's axiom of dose response, one case based on concentration, the other on dose.

One common and highly effective use of BAC is as a preservative in topical ophthalmic solutions at concentrations ranging from 0.004 to 0.02%. The usual clinical concentration is 0.01%. Solutions used to sterilize contact lenses, usually overnight, must contain a compatible and effective preservative, and one of the most prevalent is benzalkonium chloride. Hydrophilic (soft) contact lenses were stored overnight in 4 mL of aqueous BAC (0.0025%) plus ethylenediaminetetraacetic acid (EDTA) (0.01%), then fitted to rabbits and worn 6 h daily [29]. Severe irritation occurred after a few days of this regimen, preventing continuation. This was a confirmation of an earlier finding that hydrophilic lenses absorb and concentrate bactericides and increase their contact time with the cornea, resulting in increased irritation [30]. In this case the lens/animal model was instrumental in revealing the enhanced toxicity that would not have been seen in the routine rabbit eye test model. Another example of this occurred accidentally in humans [31]. A female subject wearing a corneal bandage soft contact lens after surgery ran out of her supply of irrigating solution, which contained 0.001% thimerosal. She then began using a different solution, which contained BAC as preservative, and after 3 days developed severe irritation in the eye. After soaking of the lens in the new irrigating solution was discontinued, the eye returned to normal. The investigator then used an agar plate diffusion method to determine the concentration of BAC in the patient's soft lens. Soaking soft lenses for 12 h in 0.005,

0.01, 0.02, or 0.04% BAC and placing a trephined sample of each on solid blood agar medium seeded with *Staphylococcus aureus* demonstrated that the concentration of 0.02% was identical to that of the patient's soft lens, which should have been at 0.01%. These studies appear to have demonstrated that soft lens solutions cannot be reliably tested for safety by the standard method of applying drops on rabbit corneas, whereas nonabsorbing hard lenses may be reliably so tested.

During the last two decades the standard rabbit eye test has come under close review in regard to its accuracy and reproducibility [26]. In addition, the animal welfare movement has questioned the humaneness of the procedure. These developments have generated a widespread effort toward the development of new methods that would reduce the use of animals for this type of testing. The result has been a multiplicity of diverse alternative method developments intended either to make in vivo tests more accurate and/or to substitute for the animal altogether. One of the interesting findings is that of the large number of cationic surfactants available for research tools for such investigations, the great majority of studies have utilized BAC, a mixture of alkyldimethylbenzylammonium chlorides. Presumably the reason for this is the almost ubiquitous use of the compound in the cosmetic formulary. According to a Cosmetic Ingredient Review report [2], in 1986 it was an ingredient in 83 cosmetic formulations as well as a large number of noncosmetic products, principally as an antimicrobial preservative. Some of the newer approaches in which BAC was utilized as a test agent follow.

Alteration in the permeability of epithelial surfaces as the result of damage by chemical agents is well known. In an attempt to utilize this principle, Maurice and Singh [32] exposed the corneas of mice to BAC solutions ranging in concentration from 0.001 to 0.01% for 1 min. After washing, a drop of a fluorescent dye solution, sulforhodamine, was placed on the cornea for 1 min. The fluorescence of the corneas was then measured with a slit-lamp fluorophotometer. Using the ratio of fluorescent measurement from control to test corneas gave an index of permeability of the fluorescent solution. BAC gave a graded response, with the maximum increase being a factor of 15 at the 0.1% concentration. Sodium chloride at 0.1 N concentration gave similar results. Both compounds are known as irritants to the eye at these concentrations. It was suggested that validation of this procedure could potentially provide a relatively cheap, rapid screen for acute ocular toxicity.

To improve the level of sophistication regarding the damage to ocular tissue by BAC, light microscopy, slit-lamp ophthalmoscopy, and scanning electron microscopy have been applied in more recent years. Slit lamp combined with light microscopic evaluation in one instance revealed, in rabbits, that a 0.01% concentration was free of damage [33]. However, when BAC 0.01% was combined with 0.5% hydroxyethyl cellulose, a viscosity-enhancing agent, sloughing of the superficial cell layer and loss of polarity of the basal cells were seen. Thus the use of modern microscopy revealed the enhanced irritant effect of 0.01% BAC, presumably due to the prolonged contact time caused by the increased viscosity.

Scanning electron microscopy has added still further to understanding of the cytotoxic effects at the cellular level. Its high resolution and depth of focus, which permit resolution of minute cellular details, made it possible to distinguish between effects on the corneal epithelium of rabbits of 0.01 and 0.05% concentrations at an exposure time of 30 min [35]. At 0.01%, only the top cell layer was affected, with loss of cellular microvilli and necrosis of cells. At 0.05% the foregoing two effects plus desquamation of the cells in this layer were seen and all three effects were found in the second-layer cells as well. Similar effects and difference in severity between generally used concentrations and fivefold higher concentrations were seen with four other preservatives: hydroxybenzoic acid ethyl ester, chlorhexidine, formaldehyde, and hexachlorophene. Hexachlorophene appeared to be the most damaging, causing significant effects on the third-layer cells. In another very comprehensive scanning electron microscopic study [34], corneal surface effects of a series of drugs, vehicles, and preservatives, including BAC, were evaluated and plasma membrane effects graded as absent, moderate, or significant. At 0.01%, BAC, together with two drug preparations (pilocarpine and gentamicin) containing BAC, were classified as causing significant plasma membrane injury and cell death when an exposure time of 30 min was used.

BAC is used as a preservative in β-blocker solutions for treatment of glaucoma. In two such products the concentrations are 0.007% or 0.01%. In a study in which these products with or without BAC were injected subconjunctivally daily for 2 weeks into albino and pigmented rabbits, ophthalmoscopic examination of the fundus, electroretinogram (ERG), light microscopic examination, and electron microscopic examination were all brought to bear on evaluation of potential effects on the retina and choroid [36]. Albino rabbits were unaffected by the treatment. Pigmented rabbits revealed significant effects in the microscopic studies of the retina and the ERG with both complete drug solutions and with an ophthalmic base solution containing only BAC, whereas the two drug solutions without BAC were free of effects. These results strongly imply that absorption and concentration of benzalkonium chloride by the melanin pigment of the retina could be damaging to these tissues in pigmented eyes, such as those of the pigmented rabbit. The lack of deleterious effects in albino rabbits after subconjunctival injection for 2 weeks was interpreted as an indication that the normal use of β-blocker solutions containing BAC in the form of eyedrops is safe for the posterior segment of the eye.

BAC is also used as a preservative in solutions injected into the anterior chamber of the eye during surgery that exposes the endothelium. Graded concentrations injected into the anterior chamber of albino rabbits revealed a threshold for irritation of 0.03% in one study [37], the highest nonirritating concentration being 0.01%. Higher concentrations caused conjunctival congestion, swelling and discharge, flare, iritis, corneal clouding, fluorescein staining, and increased corneal thickness. Evaluation was done by Draize scoring, slit lamp, and pachometer.

The response of the corneal endothelium following perfusion under the specular microscope, injection into the aqueous, or exposure of excised corneas from enucleated rabbit eyes has revealed that the endothelium may be more sensitive than the epithelium to many drugs and preservatives [38–40]. The threshold for physiological and ultrastructural adverse effects of BAC was found to be as low as 0.0001% or less and became apparent as early as 15 min after exposure. This was demonstrated dramatically in a case of a man with keratoconjunctivitis [41] in whom prolonged treatment with topical formulations containing BAC developed such profound pathology of the cornea, including the endothelium, that corneal transplantation was required. Not until discontinuation of the topical medication were the symptoms relieved. The case prompted caution in the use of preservative medications for prolonged periods in patients with ocular surface disease. Similar profound effects were reported following cataract extraction in 10 patients who received a variety of treatment formulations containing BAC at concentrations ranging from 0.004 to 0.02% [42]. The authors speculated that the preservative may have contributed to the adverse topical effects by virtue of its ability to enhance drug penetration because of its surface-active properties, its binding to cell membranes, and the hastening of corneal drying. This corneal drying effect has also been described by Wilson et al. [59], who demonstrated a dose (concentration)-related effect in rabbits and a significant shortening of the drying time in humans when a 0.01% BAC solution was compared to saline. An explanation for this is that BAC may disrupt the oily layer secreted by the meibomian glands which overlies the watery lacrymal film lying on the surface of the cornea.

IV. IN VITRO CYTOTOXIC EFFECTS

Over the years since its inception, the in vivo Draize test has become involved in controversy regarding its subjectivity, between-laboratory reliability, apparent ultrasensitivity, and the ethical considerations of subjecting the eyes of live animals to irritants. These considerations have motivated the development of a diversity of new procedures to circumvent these issues. A major effort has gone into finding in vitro tests that would be predictable for the response in the human eye, be inexpensive, and serve as reliable models for the many classes of chemicals that contact the eye.

BAC has been tested in a number of potentially useful in vitro assays. An extensive review and evaluation of many of these alternative methods was recently presented by the Pharmaceutical Manufacturers Association Drug Safety Subsection In Vitro Toxicology Task Force [45]. The Task Force concluded that "no single in vitro test system or battery of tests is currently capable of replacing the Draize Eye Test for ocular toxicity." The Task Force's classification of these tests into (1) those showing promise, (2) those of questionable relevance, and (3) those that were inadequate revealed that the various methods in which BAC had been tested were distributed more or less throughout these three classifications.

Regardless of their classification and eventual utility, the various in vitro studies that have resulted have contributed significantly to the understanding of the effect of BAC (plus other cationic surfactants) on the cytotoxicity of ocular and other cells and tissues. In general, in vitro studies have attempted to involve one or more of the numerous endpoints of cytotoxicity, such as viability of the cell line, cell detachment, cloning efficiency, membrane integrity, and growth inhibition. Such numerous effects are reflections of the many responses of the eye to injury and make it unlikely that one or even two in vitro tests would be adequate to cover them all, as well as the multiple responses to varied classes of chemicals.

A. Cell Proliferation Tests

BAC has been evaluated in a cell proliferation study with BALB/c3T3 mouse fibroblasts [43] (Table 2). In this BALB/c3T3 assay cytotoxicity was based on exposing the cells to four concentrations of the test material for 30 min and calculating the concentration resulting in a 50% inhibition of growth (GI_{50}). The surfactants tested were the most cytotoxic agents in this study, which included aromatic compounds, alcohols, ketones, and acetates. To test the in vitro data for correlation with in vivo, the results were calibrated against in vivo data from the percent of swelling of rabbit corneas tested with the same agent. Cytotoxicity correlation between the two methods was good for the surfactant compounds and the alcohols. Cetylpyridinium bromide was rated as the most potent and BAC as the next most potent cytotoxin of the six surfactants tested. The successful correlation of in vitro and in vivo data for the surfactants would suggest that for this class of chemicals, this assay may have promise as an alternative for the Draize test even though a mouse fibroblast (nonocular) cell line was used.

Inhibition of cell proliferation was again the criterion in a study in which SIRC rabbit corneal cells were used [46] and the cytotoxicity expressed as a percentage of the proliferation rate of the control cell culture (ID_{50}). The inhibition dose 50 (ID_{50}) of BAC was 1.4 µg/mL (0.00014%), compared to that of thimersol of 0.04 µg/mL, suggesting that thimersol was the more cytotoxic or cytostatic. The results also indicated that this cell system was very sensitive since the customary use concentration of BAC in ophthalmic solutions is 0.01%. As a comparison, human skin fibroblasts in culture were exposed to cetyltrimethylammonium bromide (CTAB) at concentrations ranging from 0.1 to 1.0 µg/mL for 1, 3, or 6 days and the effect on cell growth determined [47]. Depressed cell counts were produced by concentrations of 0.4, 0.6, and 1.0 µg/mL and were exposure-time dependent. The only questionable effect occurred at the 0.2 µg/mL level after 6 days of exposure.

BAC has been demonstrated to have a growth depressant effect on a rarely studied cell, the human ocular trabecular meshwork cell [54]. These cells, named HTMW-2 cells, were found to be very sensitive, in that BAC caused significant reduction in their growth at $2 \times 10^{-5}\%$ or $2 \times 10^{-6}\%$ (0.00002% or 0.000002%). The significance of the finding of such a high level of sensitivity by a cell type that would probably rarely be contacted by ocular preservatives is unknown.

B. Cell Membrane Integrity Tests

Study designs dependent on criteria dealing with cell membrane integrity have been used to assess the effect of cationic surfactants, including BAC. The use of human ocular endothelial cells in culture has been shown to be practical, reproducible, relevant to in vivo eye irritancy pathology, and avoids the problem of species extrapolation [44]. Using such a cell culture, Douglas and Spilman devised a method of ^{51}Cr release to derive concentrations resulting in 50% maximal toxicity for test substances (ED_{50}). BAC resulted in a ^{51}Cr-retention ED_{50} value of $3.3 \times 10^{-5} M$, while cetylpyridinium chloride resulted in an ED_{50} value of $1.1 \times 10^{-2} M$.

Leakage of lactic dehydrogenase (LDH) from cells and incorporation of [^3H]leucine as a measure of de novo protein synthesis using a murine epidermal cell line (HEL/30) have been found to be sensitive models for determining surfactants toxicity [49]. When surfactants were compared for toxic potency the classic orientation was found: cationic > anionic > nonionic > amphoteric, as with many other toxicity models. Beginning leakage

TABLE 2 Outline of Selected In Vitro Cytotoxicity Data

Compound[a]	Cell line	Criteria	Data	Ref.
BAC	BALB/c3T3 mouse fibroblasts	Growth inhibition (GI_{50})	GI_{50} = 0.024 µL 10% BAC/mL	43
CPB	BALB/c3T3 mouse fibroblasts	Growth inhibition (GI_{50})	GI_{50} = 0.013 µL 10% CPB/mL	43
BAC	SIRC rabbit corneal cells	Inhibition dose 50 (ID_{50})	ID_{50} = 1.4 µg/mL	46
CTAB	Human skin fibroblasts	Depression in cell count	At 3 and 6 days of exposure concentrations of 0.4, 0.6, and 1.0 µg/mL CTAB depressed cell counts	47
BAC	Human corneal trabecular meshwork (HTMW-2) cells	Reduction in growth	Significant reduction in growth at 2×10^{-5}% and 2×10^{-6}%	54
BAC	Corneal endothelial cells	^{51}Cr release	ED_{50} = $3.3 \times 10^{-5}M$	44
CPB	Corneal endothelial cells	^{51}Cr release	ED_{50} = $1.1 \times 10^{-2}M$	44
BAC	Murine epidermal cells (HEL/30)	[^{3}H]leucine incorporation	IC_{50} (50% inhibition) = 2.5 µg/mL	49
Benzethonium chloride	Murine epidermal cells (HEL/30)	[^{3}H]leucine incorporation	IC_{50} (50% inhibition) = 17.4 µg/mL	49
Alkyltrimethylammonium bromide	Murine epidermal cells (HEL/30)	[^{3}H]leucine incorporation	IC_{50} (50% inhibition) = 18.6 µg/mL	49
CPB	Murine epidermal cells (HEL/30)	[^{3}H]leucine incorporation	IC_{50} (50% inhibition) = 5.2 µg/mL	49
BAC	Rabbit corneal epithelial cells	Inhibition of [^{3}H]thymidine incorporation	5-min exposure = 0.02%; 30-min exposure = 0.0075%; 60-min exposure = 0.0004%	50

CTAB	HEp2 cells	Concentration required to damage 50% of the cell population as determined by loss of fluorochromasia	In decreasing cytotoxic potency: CTAB > MTAB > LTAB	48
LTAB	HEp2 cells	Concentration required to damage 50% of the cell population as determined by loss of fluorochromasia	In decreasing cytotoxic potency: CTAB > MTAB > LTAB	48
MTAB	HEp2 cells	Concentration required to damage 50% of the cell population as determined by loss of fluorochromasia	In decreasing cytotoxic potency: CTAB > MTAB > LTAB	48
BAC	Rat peritoneal mast cells	Inhibition of histamine release from cells treated with 48/80; release of histamine from mast cells	81% at 3 µg/mL; effective at 10–30 µg/mL	51
BAC	Rat peritoneal mast cells	Inhibition of 5-hydroxytryptamine secretion induced by polylysine, corticotropin, and IgE decapeptide	Concentration-dependent inhibition at 1 and 5 µg/mL	52
BAC	Rabbit erythrocytes	10, 50, and 100% hemolysis	10%: $2.2 \times 10^{-5}M$; 50%: $3.3 \times 10^{-5}M$; 100%: $4.2 \times 10^{-5}M$	53

TABLE 2 (*Continued*)

Compound[a]	Cell line	Criteria	Data	Ref.
BAC	Murine tumor cell line P815	^{51}Cr-release cytotoxic dose 50 (CD_{50})	$\frac{1}{2}$-h exposure: 7.2 ppm 2-h exposure: 3.5 ppm 4-h exposure: 0.6 ppm	55
BAC 0.01%	Human corneal epithelial cell line	Morphological criteria	Cell degeneration within 2 h	56
Chlorobutanol 0.5%	Human corneal epithelial cell line (comparator)	Morphological criteria	Cell degeneration within 8 h	56
BAC	L5178Y mouse lymphoma cells	Concentration required to kill 50% of the cells in 1 h	15 μg/mL	57
BAC	Chang's human conjunctival cells	Time of exposure causing 50% cell damage (CDT_{50})	CDT_{50} of 0.007% BAC in 90–100 s	58

[a]BAC, benzalkonium chloride; CPB, cetylpyridinium bromide; CTAB, cetyltrimethylammonium bromide; LTAB, lauryltrimethylammonium bromide; MTAB, myristyltrimethylammonium bromide.

of LDH occurred with all four cationics tested (BAC, alkyltrimethylammonium bromide, cetylpyridinium bromide, and benzethonium chloride) at levels as low as 10 μg/mL after 2 h of exposure. BAC appeared to be the most toxic. In the [³H]leucine incorporation evaluation (de novo protein synthesis), which may be a measure of toxicity on the endoplasmic reticulum (an intracellular target), BAC and all three of the other cationics affected protein synthesis at a concentration as low as 1 μg/mL. [³H]thymidine incorporation in confluent cultures of rabbit corneal epithelium was also inhibited by BAC in a concentration- and time-dependent fashion in a study using concentrations ranging from 0.0004 to 0.02% [50]. Similar effects were found with two other topical ophthalmic preservatives, thimersol and chlorobutanol, in concentration ranges used clinically. BAC was the most potent of the three compounds, requiring 20- to 550-fold lower concentrations to impair [³H]thymidine incorporation, depending on length of exposure.

Another assessment of detergent toxicity via cell membrane integrity has been carried out with a combination of HEp2 adherent cells using fluorescein diacetate/ethidium bromide and HeLa cells for alkaline phosphatase release [48]. The concentrations of detergent to damage 50% of the HEp2 cells (CD_{50}) as measured by loss of fluorescein, and to release 50% of the cellular content of alkaline phosphatase (ED_{50}) of the HeLa cells were determined to rank the compounds. The three cationics studied via the fluorescein HEp2 cells were ranked in accordance with decreasing toxicity: cetyltrimethylammonium bromide, myristyltrimethylammonium bromide, and lauryltrimethylammonium bromide. In combination, the procedures used in this study once again demonstrated the classic decreasing order of toxicity: cationic > anionic > nonionic.

Attempts have been made to determine what the activity of BAC is at the level of the cell membrane since it is generally regarded as a cell membrane–active compound. The question as to whether its effects are due to nonspecific detergent activity or to a specific receptor effect was addressed by testing it against the potent polyamine histamine releaser 48/80 in rat peritoneal mast cells [51]. BAC was determined to be an inhibitor of 48/80 histamine release in the concentration range 1 to 3 μg/mL. At 3 μg/mL it inhibited the releasing effect by 81%. When BAC was tested at 10 and 30 μg/mL it was a histamine releaser itself. Since mast cells inactivated by 50°C for 15 min do not respond to 48/80, but when exposed to BAC released histamine on a response curve similar to that of noninactivated cells, it appeared that the effect of BAC was due to a nonspecific detergent action on the membrane.

In another study using rat peritoneal mast cells, BAC inhibited 5-hydroxytryptamine release induced by polylysine, corticotropin, and IgE decapeptide [52]. The similarity of the antagonism of these three peptides with that of the histamine releaser 48/80 described above by BAC indicated that the mode of action of these compounds was similar. The hemolytic effect of BAC on rabbit erythrocytes in vitro in the presence of 0.6% NaCl was demonstrated by Ansel and Cabre [53]. Molar concentrations that produced 10, 50, and 100% hemolysis were 2.2×10^{-5}, 3.3×10^{-5}, and 4.2×10^{-5}. In the same system the values for phenol were 4.3×10^{-2}, 4.6×10^{-2}, and 5.0×10^{-2} and for thimersol were 4.8×10^{-5}, 6.1×10^{-5}, and 6.8×10^{-5}. These data and those of several other preservatives were used to select test concentrations in a study to determine the effect of dimethyl sulfoxide (DMSO) on hemolysis. DMSO was an effective inhibitor of hemolysis, apparently via a cellular-based mechanism rather than via an extracellular chemical reaction.

C. Cell Viability Tests

Cell viability (mortality) and associated morphological changes have been the subject of studies on the effects on in vitro cell lines of BAC and other preservatives commonly used in ophthalmic preparations. Acute cell injury, cytolysis, and cell death are the target of such studies. Determinations of cytotoxic dose (CD_{50}) were made in three ^{51}Cr-radiolabeled cell lines, murine P815 and YAC-1 and rabbit corneal epithelial cells (SIRCs). BAC was compared to sodium dodecyl sulfate, alkylethoxylated sulfates (AE_1S and AE_3S), polysorbate 20, and coconut soap [55]. BAC consistently ranked the most toxic in all three cell lines. Morphological criteria were used to assess the effects of BAC in a study on human corneal epithelium cultures in which it was compared to chlorobutanol in their typical use concentrations of 0.01% and 0.5%, respectively [56]. Both compounds began to elicit toxic changes 2 to 3 h after exposure began. Cessation of cytokinesis and mitotic activity occurred very quickly, and vacuolation and pyknosis of the nuclei were also observed. The degenerative effects caused by chlorobutanol appeared to be less severe and occurred over a longer time frame than those of BAC.

Cytotoxicity has been determined for BAC in two other cell lines: the mouse lymphoma L5178Y line and Chang's conjunctival cells [57,58]. When the L5178Y cells were exposed to BAC, chlorohexidine, or thimersol for 1 h it was found that the concentrations to kill 50% of the cells were, respectively, 15, 20, and 3 μg/mL. Thus BAC and thimersol were toxic in this assay at their use concentrations. In Chang's conjunctival cell cultures the BAC LD_{50} on cell population caused the investigator to suggest that it should be used in ophthalmic solutions at concentrations of less than 0.005%. Thus the question of extrapolating in vitro data directly to the in vivo condition emerged since BAC has been used successfully at 0.01% for many years, although not without occasional adverse effects.

V. IRRITANT DERMATITIS

At the outset of this chapter it was indicated that a prior review of published data up to the late 1960s had been made in *Cationic Surfactants*, published in 1970 [1]. In a discussion of the irritancy of cationic surfactants it was pointed out that the majority of the data were on the quaternary ammonium compounds and that, in this group, benzalkonium chloride was the clear leader. In the current review, covering about 20 years, this still appears to be true. It is apparent that the reason is still the same: its very widespread use in so many antibacterially oriented preparations. Nevertheless, we will describe the irritant properties and comparative irritancy of a few other cationics on which adequate literature has accumulated during the last two decades.

As emphasized by Mathias in *Dermatotoxicology*, 3rd ed. [60], a discussion of cutaneous or dermal irritation should first define irritation. The most fitting definition to describe the proposed effects of the cationic surfactants on skin is that, as primary irritants, they may result in an inflammatory response characterized by erythema and edema and/or corrosion after a single or multiple contacts with the local site. It is also pertinent to point out that penetration of the skin barrier, the stratum corneum, must occur for a chemical's irritant potential to be expressed. This is dependent on its ability to interact with the skin barrier, which, in turn, is influenced by many extrinsic and intrinsic factors. Thus the primary irritancy response is an extremely complex event dependent on the nature of the chemical agent, the condition of the skin, the status of the local ambient environment, and the responsiveness of the host. In addition, when predictive experimental models of irritancy are used, such as the Draize test in rabbits [20] or the various patch tests in humans, other highly critical factors that must be taken into account are the vehicle, which can dramatically influence the outcome, as well as the concentration of the test agent. As expected, cationic surfactant irritancy responds to these factors. Comparing various members of this large group of compounds in terms of their irritant potency would therefore be difficult unless these diverse factors were standardized.

A. Mechanisms

With a phenomenon as complex as skin irritation, which potentially is influenced by so many factors, it would also seem logical to expect that its induction could be the result of multiple mechanisms. Patrick et al. [61]

studied this in a topical application—a mouse ear thickness model, in which rate of absorption, differences in inflammatory mechanisms, differences in repair mechanisms, or combinations of these may all have played a role. Among the chemicals studied was the cationic surfactant benzalkonium chloride (BAC). Components of the inflammatory response, such as the number of blood vessels seen (measure of dilation), degree of edema, type of cellular infiltrate, and condition of the epidermis, dermis, and musculature, were determined histologically. Permeability of the site of application was measured by injecting trypan blue intraveneously and examining for blue staining. Quantification of vascular permeability effects was determined by injecting ^{125}I-labeled bovine serum albumin and counting the radioactivity of the ears (cpm). Difference in temperature of the ears was used to determine blood flow change. This spectrum of tests could be applied at various points during the changes in thickness of the ears. In addition, dose response based on different concentrations was studied. This study, using five known skin irritants of different chemical structures, demonstrated that they induced skin irritation by multiple mechanisms and that differences in the components of the inflammatory response are common. Both BAC and phenol resulted in permanent thickening of mouse ears after one application, using the criterion that the ears were still thicker than the solvent control ear at 6 weeks. Since this long-lasting effect did not correlate with the intensity of the inflammatory response of these two chemicals, noninflammatory mechanisms, not identified in this experiment, must have been responsible.

As mentioned earlier, a sine qua non of irritation is that penetration of the stratum corneum must occur. Since cationic surfactants are recognized as topical irritants [1], it is evident that they have penetrating ability which may be attributed to their surface-tension-reducing capability, removal of lipids and/or amino acids, and perhaps more direct barrier-destroying activity, such as protein denaturing. The cationic surfactant benzethonium chloride has been shown to be effective in enhancing absorption of Sarin, an organophosphorous cholinesterase inhibitor, via the skin of guinea pigs [62]. Exposure of the belly skin of guinea pigs to benzethonium chloride for 1, 5, or 30 min decreased the time to respiratory arrest of a 25-μL standard dose of Sarin in a time-related fashion. The time to Sarin respiratory arrest was approximately one-third that of guinea pigs pretreated with distilled water after a 30-min period and was similar to the nonionic surfactant alkyl ether–sulfate. With both compounds the barrier-penetrating effect appeared slowly in comparison with a solvent such as chloroform, which reached its maximum effect with the 1-min exposure, suggesting a different mechanism.

Since surfactants in general are recognized as skin barrier penetrants, indicating that they interact with the stratum corneum, it follows, as in the case with Sarin above, that they might enhance the permeation of drugs through the skin in a beneficial way for transdermal drug delivery. However, as is commonly found, they are also irritants, which limits their usefulness in this regard. In an attempt to study this, Aoyagi et al. [63] prepared polymers of a BAC monomer which were of increasing molecular weight. They were then tested in an in vitro rabbit abdominal skin model that measured the permeation of 5-fluorouracil (5-FU) into a diffusion cell chamber. In addition, several BAC monomers of different alkyl groups were tested to determine the effect of alkyl chain length on absorption of 5-FU. The most effective enhancer of absorption was the one with a hexadecyl group, and it was chosen to have the various polymers made of it. It was predicted that polymerization of this surfactant monomer would cause lower surfactant activity, less penetration, and less mobility, and thus have reduced irritancy as well. These studies, when coupled with determination of differential scanning calorimetry (thermal behavior of the rabbit whole skin and stratum corneum), demonstrated that the enhancing activity of the polymers was related to their molecular weight, with low-molecular-weight polymers having a high enhancing effect on drug absorption while having a low tendency to be irritants. The latter conclusions were based largely on the differential scanning calorimetry studies, which showed that whereas the monomeric compound penetrated to the dermis and was thus more likely to cause irritation, the polymeric derivative, because of its high molecular weight, did not. These studies presumably showed that active monomeric penetrants of the skin could be made less irritating, while maintaining their skin-penetrating quality, by synthesizing them into relatively low-molecular-weight polymers.

BAC has played a key role in numerous investigations concerning the basic factors in skin irritancy because it has a record of producing a dependable response at established concentrations, which, however, are well above its usual use concentration. Evidence for a significant role of prostaglandins in inflammation has been well documented [64,65] and it is accepted that they are important in the development of a sustained inflammatory response. In one example 10% aqueous BAC was used to produce a reliable primary irritant response on the thighs of human patients, and punch biopsies were then assayed for PGE activity [66]. It was clearly demonstrated that the prostaglandin system of skin is activated during inflammatory states and the authors suggested that it may have provided a basis for future therapy of dermatitis. Prostaglandins were the subject of another study in which 10% aqueous BAC was also used as the irritant on the abdominal skin of 12 human volunteers [67]. Using suction cups at 300 mmHg below atmospheric pressure, blister fluid was obtained from normal skin sites and compared with that of the BAC-caused blisters. Analyses for prostaglandin E_1 activity demonstrated that in the inflamed tissues and blister fluid, levels were approximately twice those from the uninvolved normal control skin of the same patients and that the highest level was in the epidermis. It was concluded that epidermal synthesis during inflammation was the principal source of prostaglandin activity and that dermal cells were minor contributors.

Another way to measure the effect of irritants on the barrier function of the skin is skin water vapor loss. It has been demonstrated in humans that some irritants exert an effect on water loss at concentrations well below their irritant concentration [68,69]. This suggests that significant effects on the stratum corneum barrier may occur even though no visible signs of inflammation are observed. In one instance BAC at 1% resulted in minimal water loss compared to the control, and in another case, when tested at 0.2%, it did not significantly influence water loss at all. Concentrations of BAC greater than 1% are known to cause a significant irritant response in the skin of humans under certain conditions. Water loss, or transepidermal water loss (TEWL) as it is usually referred to, may thus provide an additional key to the understanding of the mechanism of chemical irritancy and make it possible to establish a more accurate nonirritating concentration.

Enhanced mitosis and, consequently, epidermal hyperplasia are another of the broad spectrum of mechanisms involved in topical irritancy induced by chemicals. In human males, various irritants applied to the skin of the back in a standard model demonstrated that increases in mitotic index, determined microscopically from punch biopsies, were not necessarily related to the degree of clinically observed inflammation [70]. Whereas 5% BAC causes the most intense inflammatory response, the greatest increase in mitotic index occurred with 1% sodium lauryl sulfate. When measured over three 24-h periods, increase in mitosis lagged at 24 h, peaked at 48 h, and tended to fall at 72 h, a pattern that matched other types of skin injury responses, suggesting that a mitotic response is intimately involved in the normal response to irritants. A concentration gradient is also very important in that when concentrations are too low, penetration is not adequate, and when it is too high, cell killing prevents the mitotic increase from occurring.

Among the multitude of antimicrobial uses for cationic surfactants is their use as topical antiseptics for wound cleansing, disinfection of infected wounds, or bathing of surgical wounds. Interference with the normal regeneration of vascular structures and laying down of fibrous connective tissue, collagen, and many other factors impairs the normal healing process, resulting in delay of healing time and/or inadequate healing and strength of wound incisions. The effects on wound healing by dequalinium chloride, a heterocycle quaternary ammonium compound, used to treat skin infections and mucosal lesions, was investigated in rats by Mobacken et al. [71]. Surgical incisions were exposed to distilled water or aqueous solutions of 0.0025%, 0.01%, 0.1%, and 0.4% of dequalinium chloride for 5 min, and the breaking load of the incisions was measured with a tensiometer after 1 or 2 weeks of healing time. Concentration-related reduction of breaking load, compared to control, was found at 7 days with all four concentrations, and at 14 days with the 0.025%, 0.1%, and 0.4% concentrations. These findings, which measured the strength of the incisional wound, were an indication that the quality and/or amount of collagen being produced was being affected adversely and that potential irritants in surgical procedures should be used with caution.

B. Irritant Dermatitis of Selected Cationic Surfactants

Three of the most widely used cationic surfactants in commerce today, particularly in cosmetic, germicidal, and sanitizer products, are Quaternium-18 [18], benzethonium chloride [72], and benzalkonium chloride [2] (Table 3). These compounds are found in hundreds of products, especially cosmetics, at various concentrations and commonly come into contact with the skin continuously or several times each day for long periods of time. Their successful use for many years attests to their relative safety and low potential for sensitization. However, they do exhibit a graded skin irritation response dependent on concentration. These three compounds have been the subject of extensive reviews of their safety by the Cosmetic Ingredient Review Expert Panel, supported by the Cosmetic, Toiletry and Fragrance Association, and have been declared safe either as currently used or at some maximum recommended concentration.

1. Quaternium-18

Quaternium-18, a dialkyl quaternary ammonium chloride that is also used in the form of bentonite or hectorite clay is found in some cosmetic preparations in concentrations as high at 10%, but more commonly at 1 and 5%. In a group of skin irritancy studies in rabbits reported in the *Cosmetic Ingredient Review* [18], concentrations as high as 75% in the Draize test and 7.5% under patches allowed to remain on the skin for 21 days showed that it was only a mild irritant. The bentonite and hectorite clays were also found to be nonirritating in rabbit models, and in fact, the bentonite was

TABLE 3 Irritant Dermatitis of Selected Cationic Surfactants

Test substance	Animal tested	Methodology	Results	Ref.
Quaternium-18	Rabbit	0.5 mL of 4 or 5% aqueous dispersion under patch for 24 h	No reaction	18
Quaternium-18	Rabbit	7.5% concentration via the Draize method	Delayed erythema at 72 h; score of 1.92 out of possible 8	18
Quaternium-18	Rabbit	0.5 mL of aqueous dispersions of 1.5, 3.7, and 7.5% under patches for 21 days	Graded as mild irritant	18
Quaternium-18 hectorite	Rabbit	FHSA test; 0.5 g of 50% suspension on aqueous intact or abraded skin	No irritation	18
Quaternium-18 bentonite	Rabbit	Undiluted to intact and abraded skin 6 h/day for 5 days + 10 days' rest and 5 additional exposures	No irritation	18
Quaternium-18	Human, both sexes	7.5% concentration, 15 alternating 24-h patch tests	Mean primary irritation index 0.21 out of a maximum of 8	18
Quaternium-18 hectorite	Human, both sexes	100% concentration, 15 consecutive 24-h patch tests	No irritation	18
Quaternium-18 bentonite	Human, both sexes	4% concentration in repeated insult patch test	No irritation	18
Benzethonium chloride	Rabbit	2 mL of a 0.1% solution applied five times a week for 4 weeks	No irritation	73

Substance	Species	Test conditions	Result	Reference
Methylbenzethonium chloride	Rabbit	2 mL of a 0.1% solution applied five times a week for 4 weeks	No irritation	73
Benzethonium chloride	Human	Concentrations of 0.1, 0.12, and 0.2% in various topical procedures	No irritation	72
Methylbenzethonium chloride	Human	0.5% in skin cleanser, 24-h patch test	No irritation	72
Methylbenzethonium chloride	Human	0.5% in skin cleanser, repeated insult patch test	7% incidence of mild to barely perceptible irritation	72
Benzethonium chloride	Human	5% aqueous solution, 48-h occlusive patch test	51 reactions among 100 subjects; classed as irritant	72
Benzalkonium chloride	Rabbit	0.1, 1, 5, and 10% aqueous solutions, occlusive patches, 24 h	Induration of all test sites at all concentrations	2
Benzalkonium chloride	Rabbit	0.3% solution, 24-h occlusive patch	No irritation	2
Benzalkonium chloride	Guinea pig	0.1, 1.0, and 5%, patches, 24 h	Necrosis at 5% and 1%	2
Benzalkonium chloride	Rat	1% solution over a 3-month period	Hyperemia and necrosis	2
Benzalkonium chloride	Human	0.1% in a cream, closed patch, daily application for 21 days	Essentially no evidence of cumulative irritation	2
Benzalkonium chloride	Human	0.5% aqueous solution, 12-h exposure	No irritation	2
		2.5 or 5% aqueous solution, 12-h exposure	Confluent erythema, edema, and papules	
Benzalkonium chloride	Human	10% aqueous solution, 24-h patch test	Erythema, infiltration, and vesicle formation	67

declared inert. Among 50 human subjects in a repeated insult occluded patch test given 15 applications each of a 7.5% concentration, only 13 reactions occurred out of a total of 750 exposures, and these were mild. The primary skin irritation index was 0.26 out of a total of 8 and the result was termed practically nonirritating. Similar nonirritating results were reported for the 100% concentrations of Quaternium-18 hectorite and 4% of the bentonite when studied in 15 days of consecutive patch testing and in actual cosmetic formulation-use tests designed to demonstrate fatiguing and sensitization. These products therefore have apparently relatively low irritancy potential for the skin of animals or humans.

2. Benzethonium and Methylbenzethonium Chloride

Benzethonium and methylbenzethonium chlorides are quaternary ammonium compounds also known in commerce as Hyamine 1622 and Hyamine 10x and are used commercially in concentrations usually ranging from 0.1 to 1%. When benzethonium chloride was studied topically in human subjects, concentrations of 0.1, 0.12, and 0.2% produced no indications of irritation, whereas a 5% aqueous solution in a 48-h occlusive patch test resulted in 51 irritant reactions in 100 subjects tested. Methylbenzethonium chloride, tested in a single-exposure patch test at 0.5%, was nonirritating, while when studied in a repeated insult patch test resulted in only a 7% incidence of mild to barely perceptible irritant reactions. Thus, in humans, irritancy of these compounds began to occur at 0.5% and was clearly demonstrated at 5% [72]. In a chronic study in rabbits 5 days/week for 4 weeks, neither compound displayed any irritancy when 2 mL of a 0.1% solution was applied five times a week [73]. As with any chemical, however, unusual cases of irritancy may occur. Such a case was reported by Maibach and Mathias [74] in a female patient with severe, long-standing labiitis and vulvitis caused by the use of an ointment that contained an unknown concentration of benzethonium chloride. Discontinuation of use cleared the condition and a patch test was negative, indicating that it was not a case of allergic contact dermatitis. It thus appears that benzethonium chloride and methylbenzethonium chloride are essentially safe to use on the skin for cosmetic purposes up to a concentration of 0.5%.

3. Benzalkonium Chloride

Benzalkonium chloride (BAC), one of the alkyldimethylbenzylammonium compounds, is commonly used in dermal preparations at a concentration of 0.1% or less but occasionally at 1 to 5%. In general, skin irritation is found in both animals and humans when it is tested at greater than 0.1%. Results of topical studies in rabbits, guinea pigs, and rats were highly variable [2], and duration of exposure and concentration were the principal factors in the expression of irritancy. A high concentration such as 5%, for example, resulted in necrosis in both guinea pigs and rabbits when left in contact with the skin for 24 h, while 0.1% was much less likely to result in irritation. Similar concentration-related skin irritation has been observed in human subjects tested by various means of topical application. At 0.1% the occurrence of reactions is minimal. Even with 23-h exposure each day for 21 days under a daily patch, 0.1% did not induce irritant reactions in a group of 10 patients. If the concentration is increased to 0.5%, however, significant effects such as pustules and bullae (blisters) may occur even if exposure is limited to 48 h. This is also true with 1% and 2% concentrations [75]. At an even higher concentration of 10% under patch test exposure for 24 h, erythema, infiltration, and vesicle formation were observed [67]. A 0.1% concentration or less appears to be ideal from the standpoint of topical tolerance. BAC is also effective for most of its intended uses at this or lower concentrations. As reported previously [74] with regard to methylbenzethonium chloride, a severe case of irritant dermatitis of the vulva was described, which raised the question of potentially different reactivities at different skin sites. Using 17% BAC and 20% maleic acid, Britz and Maibach [76] compared reactions of the volar forearm and labia majora following topical application of 10 μL to 1-cm test sites. A significantly greater response to these irritants was found on the vulvar skin sites, suggesting that the vulvar stratum corneum is a less efficient barrier to irritants than is that of the forearm. This was confirmed by the greater degree and incidence of burning and stinging sensations. This demonstrated again that chemical irritancy is not the same on all skin locations. In addition, the microscopic changes that occur in contact dermatitis caused by BAC and other chemical irritants has been investigated in humans. To establish a concentration that would reproducibly induce mild to moderate reactions in a high incidence, aqueous BAC, at 0.5% and 1% concentrations, was patch tested for 48 h on healthy volunteers. Although neither concentration produced the high response rate sought, there was a distinct dose response, in that 1% did result in more intense reactions [77]. The 0.5% concentration appeared to be the optimum concentration for evaluating the histomorphologic and/or electron microscopic changes that occur in BAC irritant dermatitis. Light microscopic examination of biopsies after 48-h exposure revealed mild spongiosis, exocytosis of mononuclear cells, loci of necrotic damage in the stratum spinosum, and rarely, lipid accumulation in keratinocytes. The necrotic changes were found even in patch test areas rated as mild clinically, suggesting a different mechanism of irritation than those of the anionic detergent sodium lauryl sulfate, which caused more profound changes in keratinocytes, indicating an enhanced effect on epidermal cell turnover [78]. Electron microscopic study showed that the Langerhans cells were metabolically activated by BAC irritation, were in some cases damaged, and occasionally were placed into opposition to lymphocytes. Metabolic activation of the Langerhans cells was not seen with several other irritants studied, including sodium lauryl sulfate. These and other differences in reaction to clinically different irritants suggest that there may be specificity in the way in which topical irritants affect the skin and the way the skin reacts to them [79]. In another microscopic study with BAC, Willis et al. [80] evaluated both the temporal and cellular response to BAC at 0.5% after 48-h patch exposure. The cellular response was determined to be predominantly T-cell infiltrate. The earliest changes occurred at 6 h after 48-h patch testing, at which time increased numbers of lymphocytes and Langerhans cells were evident. At 12 h epidermal spongiosis and edema were also present. The peak response was found to be highly variable (between 24 and 96 h) and repair was under way by 168 h. Although one would not necessarily predict that the cellular response of laboratory animals to topical irritants would be different from that in humans, such differences have been reported. Whereas polymorphonuclear leucocytes are uncommon in the human inflammatory response to topical irritants, they are common in animals, the guinea pig being one example [81]. A concentration of BAC of only 0.1% applied twice daily for 3 weeks into the external auditory meatus of guinea pigs resulted in a complex of changes, such as acute and chronic inflammation, acanthosis, parakeratosis, hyperkeratosis, spongiosis, intracellular edema, migration of lymphocytes and neutrophils into the dermis, thickening of the stratum spinosum and stratum granulosum, and significant increase in the mitotic index of the basal epidermal layer. Thus commonly used aural preparations containing BAC, even in this relatively low and usually well tolerated concentration, should be used and observed with caution.

Although it may seem appropriate that there be species differences in response to the same chemical irritant based on differences in skin structure, basic tissue responses are found across classes of animals. One example is when rainbow trout are exposed to zephiran chloride (BAC), which has been found useful in water treatment for control of bacterial gill diseases. In a study by Byrne et al. [82] it was found that concentrations as low as 1 ppm for 1 h caused spongiosis of epithelium of the gills, and following 2- or 3-h exposures, cell necrosis and hydropic degeneration occurred. In higher concentrations, inflammatory cell infiltration, exfoliation, and vesiculation were found. Thus typical irritation-induced, dose-related effects based on duration and concentration were observed in the gill epithelium of fish, and in this case were interpreted to suggest that levels recommended for control of disease may be too high for routine use with rainbow trout.

Differences between individuals in cutaneous sensitivity to the same irritant have puzzled investigators for many years. The basic factors that govern these differences are essentially unknown. Genetic constitution is one influence that must be considered. BAC at a aqueous concentration of 0.5% was utilized in one attempt to answer this question [83]. Groups of monozygous and dizygous twins were patch tested for 24 h and the intrapair reaction strength of irritant reactions compared for degree of concordance between twins. This study showed that there was a significant tendency to greater intrapair likeness among the monozygous twins than among the dizygous twins or control group twins, indicating that a genetic influence may be possible. This suggests that genetic constitution may be a fruitful line of research to pursue in chemical irritant dermatitis.

The chronic topical toxicity of cationic surfactants has not been the subject of much significant published research. BAC, however, has been utilized at 8.5 and 17% in one study in which both mice and rabbits were studied for their lifetimes [84]. Mice received 0.02 mL on the dorsal skin and rabbits 0.02 mL into the interior left ear, each twice per week. The principal target of these studies was the carcinogenic response, although local and systemic effects were also evaluated. BAC showed no evidence of carcinogenicity in either species, although local topical effects such as inflammation, fibrosis, and ulceration were seen, which was not surprising considering the very high concentrations used.

REFERENCES

1. R. A. Cutler and H. P. Drobeck, in *Cationic Surfactants* (E. Jungermann, ed.) Marcel Dekker, New York, 1970, p. 527.
2. Anon. *J. Am. Coll. Toxicol.* 8(4): 589 (1989).
3. J. Aurnes, *Acta Otolaryngol.* 93: 421 (1982).
4. R. Beasley, *Lancet* 2(8517: 1227 (Nov. 22, 1986).
5. Y. G. Zhang, *Am. Rev. Respir. Dis.* 141(6): 1405 (1990).
6. J. W. Coleman, *Biochem. J.* 198: 615 (1981).
7. M. N. Gleason, R. E. Gosselin, H. C. Hodge, et al, in *Clinical Toxicology of Commercial Products* (3rd ed.) Williams & Wilkins, Baltimore, 1969, p. 167.
8. J. T. Wilson, *Am. J. Dis. Child.* 129: 1208 (1975).
9. A. S. Dhillon, R. W. Winterfield, and H. L. Thacker, *Avian Dis.* 26: 928 (1982).
10. H. W. Reuber, T. A. Rude, and T. A. Jorgenson, *Avian Dis.* 14: 211 (1970).
11. H. S. Buttar, *J. Appl. Toxicol.* 5: 398 (1985).
12. H. S. Buttar, C. Bura, and J. R. Moffatt, *Teratology* 33(3): 64c (1986).
13. J. Momma, K. Takada, Y. Aida, A. Takagi, H. Yoshimoto, Y. Suzuki, Y. Nakaji, Y. Kurokawa, and M. Tobe, *Eisei Shikenjo Hokoko 105*: 20 (1987).

14. A. K. Palmer, A. M. Bottomley, J. A. Edwards, and R. Clark, *Toxicology* *26*: 313 (1983).
15. K. R. Stevens, M. Knickerbocker, and M. A. Gallo, *Toxicol. Appl. Pharmacol. 45*: 346 (1978).
16. B. Isoma and K. Bjondahl, *Acta Pharmacol. Toxicol. 47*: 17 (1980).
17. B. Isoma, J. Reuter, and B. M. Djupsund, *Arch. Toxicol. 35*: 91 (1976).
18. Anon., *J. Am. Coll. Toxicol. 1*(2): 71 (1982).
19. J. A. Nissim, *Nature 187*: 308 (1960).
20. J. H. Draize, G. Woodard, and H. O. Calvery, *J. Pharmacol. Exp. Ther. 82*: 377 (1944).
21. J. O. Hoppe, E. B. Alexander, and L. C. Miller, *J. Am. Pharm. Assoc. Sci Ed. 39*: 147 (1950).
22. J. H. Draize and E. A. Kelley, *Proc. Sci. Sect. Toilet Goods Assoc. 17*: 1 (1952).
23. L. W. Hazelton, *Proc. Sci. Sect. Toilet Goods Assoc. 17*: 5 (1952).
24. *Products Data Bulletin 67-14*, Armour Industrial Chemical Co., Chicago, IL, (1967).
25. T. Matsuura, T. Tsuyosi, and Y. Masamoto, *J. Soc. Cosmet. Chem Jpn. 17*(2): 153 (1983).
26. C. S. Weil and R. A. Scala, *Toxicol. Appl. Pharmacol. 19*: 276 (1971).
27. J. F. Griffith, G. A. Nixon, R. D. Bruce, P. J. Reer, and E. A. Bannan, *Toxicol. Appl. Pharmacol. 55*: 501 (1980).
28. A. R. Gassett, Y. Ishii, H. E. Kaufman, and T. Miller, *Am. J. Ophthalmol. 78*(1): 98 (1974).
29. M. Davies, *J. Pharm. Pharmacol. Proc. 25*(Suppl.): 134P (1973).
30. B. A. Holden and A. J. Markides, *Aust. J. Optom.* p.325 (Oct 1971).
31. A. R. Gasset, *Am J. Ophthalmol. 84*(2): 169 (1977).
32. D. Maurice and T. Singh, *Toxicol. Lett. 31*: 125 (1986).
33. G. Durand-Cavagna, P. Delort, P. Duprat, Y. Bailly, B. Plazonnet, and L. R. Gordon, *Fundam. Appl. Toxicol. 13*(3): 500 (1989).
34. R. R. Pfister and N. B. Burstein, *Invest. Ophthalmol. 15*(4): 246 (1976).
35. J. A. M. A. Dormans and M. J. vanLogten, *Toxicol. Appl. Pharmacol. 62*(2): 251 (1982).
36. A. Chou, S. Hori, and M. Takase, *Jpn. J. Ophthalmol. 29*: 13 (1985).
37. B. Britton, R. Hervey, K. Kasten, S. Gregg, and T. McDonald, *Ophthalmic Surg. 7*(3): 46 (1976).
38. K. Green, D. S. Hull, E. D. Vaughn, A. M. Anthony, and K. Bowman, *Arch. Ophthalmol. 95*: 2218 (1977).
39. D. S. Hull, *South. Med. J. 72*(11): 1380 (1979).
40. J. B. Lavine, P. S. Binder, and M. G. Wickham, *Ann. Ophthalmol. 11*(10): 1517 (1979).
41. M. A. Lemp and L. E. Zimmerman, *Am. J. Ophthalmol. 105*: 670 (1988).
42. R. W. Zabel, G. Mintsioulis, I. M. MacDonald, J. Valberg, and S. J. Tuft, *Can. J. Ophthalmol. 24*(7): 311 (1989).
43. H. E. Kennah II, D. Albulescu, S. Hignet, and C. S. Barrow, *Fundam. Appl. Toxicol. 12*: 281 (1989).

44. W. H. J. Douglas and S. D. Spilman, *Alternative Methods in Toxicology, Vol. I Product Safety Evaluation*; Mary Ann Liebert Publishers, New York, 1983, p.207.
45. D. E. Wierenga, *Evaluation of Alternatives to the Draize Test*, Pharmaceutical Manufacturers' Association, Washington, D.C., 1990.
46. I. M. Clarke, G. C. Jefferson, R. J. McBride, and G. C. Priestley, *J. Pharm. Pharmacol. 37*(Suppl.): 63P (1985).
47. J. Benoit, M. Cormier, and J. Wepierre, *Cell Biol. Toxicol. 4*(1): 111 (1988).
48. M. C. Scaife, *Food Chem. Toxicol. 23*(2): 253 (1985).
49. M. Marinovich, E. Tragani, A. Corsini, and C. L. Galli, *J. Toxicol. Cutaneous Ocul. Toxicol. 9*(3): 169 (1990).
50. P. Imperia, H. M. Lazarus, R. E. Botti, and J. H. Lass, *J. Toxicol Cutaneous Ocul. Toxicol. 5*(4): 309 (1986).
51. G. W. Read and E. F. Kiefer, *J. Pharmacol. Exp. Ther. 211*(3): 711 (1979).
52. J. W. Coleman, *Immunol. Lett. 5*(4): 197 (1982).
53. H. C. Ansel and G. E. Cabre, *J. Pharm. Sci. 59*(4): 478 (1970).
54. J. R. Samples, P. S. Binder, and S. Nayak, *Exp. Eye Res. 49*(1): 1 (1989).
55. J. A. Shadduck, J. Everitt, and R. A. Meccoli, *Soap Cosmet. Chem. Spec. 61*(11): 36, 56 (1985).
56. B. J. Tripathi and R. C. Tripathi, *Lens Eye Toxic Res. 6*(3): 395 (1989).
57. T. J. Withrow, N. T. Brown, V. M. Hitchens, and A. G. Strickland, *Photochem. Photobiol. 50*(3): 385 (1989).
58. N. Takahashi, *Acta Soc. Ophthalmol. Jpn. 84*(9): 1171 (1980).
59. W. S. Wilson, A. J. Duncan, and J. L. Jay, *Brt. J. Ophthalmol. 59*: 667 (1975).
60. C. G. Toby Mathias, *Dermatotoxicology*, 3rd ed., Vol 7, *Clinical and Experimental Aspects of Cutaneous Irritation*; Hemisphere, Washington, D.C., 1987, p. 173.
61. E. Patrick, H. I. Maibach, and A. Burkhalter, *Toxicol. Appl. Pharmacol. 81*: 476 (1985).
62. T. Fredriksson, *Acta Dermato-Venerol. 49*: 55 (1969).
63. T. Aoyagi, O. Terashima, N. Suzuki, K. Matsui, and Y. Nagase, *J. Controlled Release 13*: 63 (1990).
64. M. W. Greaves, J. Sondergaard, and W. McDonald-Gibson, *Br. Med. J. 2*: 258 (1971).
65. G. P. Velo, C. J. Dunn, J. P. Giroud, J. Timsit, and D. A. Willoughby, *J. Pathol. 111*(3): 149 (1973).
66. H. P. Jorgensen and J. Sondergaard, *Acta Dermatol. (Stockholm) 56*: 11 (1976).
67. V. Kassis, T. Mortensen, and J. Sondergaard, *Acta Dermatol. (Stockholm) 61*: 429 (1981).
68. P. G. M. van der Valk, J. P. Nater, and E. Bleumink, *Derm. Beruf. Vmwelt. 33*(3): 89 (1985).
69. E. Berandesca, D. Fideli, P. Gabba, M. Aspa, G. Rabbiosi, and H. I. Maibach, *Contact Dermatitis 23*: 1 (1990).
70. L. B. Fisher and H. I. Maibach, *Contact Dermatitis 1*: 273 (1975).

71. H. Mobacken, M. Romanus, and C. Wengstrom, *Dermatologica 148*: 154 (1974).

72. Anon., *J. Am. Coll. Toxicol. 4*(5): 65 (1985).

73. J. K. Finnegan and J. B. Dienna, *Proc. Sci. Sect. Toilet Goods Assoc. 20*: 16 (1953).

74. H. I. Maibach and C. T. Mathias, *Contact Dermatitis 13*(5): 340 (1985)

75. J. E. Wahlberg, K. Wrangsjo, and A. Hietasalo, *Contact Dermatitis 13*: 266 (1985).

76. M. B. Britz and H. I. Maibach, *Contact Dermatitis 5*: 375 (1979).

77. C. M. Willis, C. J. M. Stephens, and J. D. Wilkinson, *Contact Dermatitis 18*: 20 (1988).

78. C. M. Willis, C. J. M. Stephens, and J. D. Wilkinson, *J. Invest. Dermatol. 93*(5): 695 (1989).

79. C. M. Willis, C. J. M. Stephens, and J. D. Wilkinson, *J. Invest. Dermatol. 95*(6): 711 (1990).

80. C. M. Willis, E. Young, D. R. Brandon, and J. D. Wilkinson, *Br. J. Dermatol. 115*: 305 (1985).

81. W. S. Monkhouse, P. Moran, and A. Freedman, *Clin. Otolaryngol. 13*(2): 121 (1988).

82. P. Byrne, D. Spearl, and H. W. Ferguson, *Dis. Aquat. Org. 6*(3): 185 (1989).

83. R. Holst and H. Moller, *Br. J. Dermatol. 93*: 145 (1975).

84. F. Steuback, *Acta. Pharmacol. Toxicol. 41*: 417 (1977).

4

Environmental Aspects of Cationic Surfactants

ROBERT S. BOETHLING   Office of Pollution Prevention & Toxics, U.S. Environmental Protection Agency, Washington, D.C.

The views expressed in this article are those of the author, and do not necessarily reflect policy of the U.S. Environmental Protection Agency.

I. PHYSICAL/CHEMICAL PROPERTIES

Table 1 contains a list of chemical structures and abbreviations for selected cationic surfactants. Measured values of environmentally important physical/chemical properties are summarized in Table 2. As the table shows, available data are sparse. Despite having a positive charge, cationics as a class have rather low water solubility. This is especially true for fabric-softening dialkyl quaternaries such as DTDMAC, which have solubilities in the low milligram per liter range. Vapor pressures should be extremely low, however, so that cationics are not expected to volatilize significantly from soil or water.

The most important property of cationic surfactants from an environmental perspective is that they are strongly sorbed by a wide variety of

TABLE 1 Structures and Abbreviations for Selected Quaternary Ammonium Compounds

General Structure	Abbreviation	Quaternary Ammonium Compound
pyridinium ring, R–N^{+}, X^{-}	A_nPB A_nPC	alkyl (C = n) pyridinium bromide alkyl (C = n) pyridinium chloride
$R-\overset{\overset{CH_3}{\|}}{\underset{\underset{CH_3}{\|}}{N^{+}}}-CH_3\;\; X^{-}$	TTMAC A_nTMAB A_nTMAC	tallowtrimethylammonium chloride alkyl (C = n) trimethylammonium bromide alkyl (C = n) trimethylammonium chloride
$\langle\text{C}_6\text{H}_5\rangle-CH_2-\overset{\overset{CH_3}{\|}}{\underset{\underset{CH_3}{\|}}{N^{+}}}-R\;\; Cl^{-}$	A_nDMBAC	alkyl (C = n) dimethylbenzylammonium chloride
$R-\overset{\overset{CH_3}{\|}}{\underset{\underset{CH_3}{\|}}{N^{+}}}-R\;\; X^{-}$	DTDMAC DTDMAMS DA_nDMAC	ditallowdimethylammonium chloride ditallowdimethylammonium methylsulfate dialkyl (C = n) dimethylammonium chloride
imidazolium structure: CH_3, CH_2CH_2NHC-R with R-C, $\overset{\ominus}{O}SO_3CH_3$	IQAMS	imidazoliumquaternaryammonium methylsulfate (R = tallow alkyl)
$R\text{-}C\text{-}NHCH_2CH_2\text{-}\overset{\overset{CH_3}{\oplus\|}}{N}\text{-}CH_2CH_2NH\text{-}C\text{-}R$ with O, O, CH_2CH_2OH, $\ominus OSO_3CH_3$	EEQAMS	ethoxylatedethanaminiumquaternary ammoniummethylsulfate (R = tallow alkyl)

TABLE 2 Measured Physical/Chemical Properties of Cationic Surfactants

Compound	MW	Water solubility[a] (mg/L)	log K_{ow}[b]	BCF	K_d
A_{10}TMAB	235.5		−0.74		
A_{12}TMAC	263.5	5100			423[c]
					440–
					251,000[d]
A_{14}TMAB	291.5		−0.45		
A_{16}TMAC	319.5	440			
A_{16}TMAB	363.9				17,000;
					71,000[e]
A_{18}TMAC	347.5	130	1.50		66,000;
					226,000[e]
					18,000–
					49,000;
					2800–900C
					112–217[g]
DA_8DMAC	305.5	8100			
DA_{10}DMAC	361.5	700			
DA_{12}DMAC	417.5	77			
DA_{14}DMAC	473.5	12			
DA_{18}DMAC	585.5	2.7			3800;
					11,000;
					12,000[h]
					36–57[g]
DTDMAC	529.5–		2.69	256; 94;	
	585.5[i]			32; 13[j]	
IQAMS	577.0–	19[k]	2.15[k]		
	633.1[i]				
EEQAMS	639.1–	35[k]	2.48[k]		
	695.5[i]				
A_{12}PB	327.9			2010[l]	
A_{16}PB	383.9			21; 22; 13[m]	

[a]30°C; from Kunieda and Shinoda [146], except as noted.
[b]Woltering et al. [123], except as noted.
[c]Sediment from Little Miami River and/or Rapid Creek [147].
[d]Range of values for two sediments from Rapid Creek, ranging in sediment concentratio from 248,000 to 230 mg/L [6].
[e]Ohio River sediment and EPA$_{18}$ sediment, respectively [2].
[f]Ranges of five values for deactivated (mercuric chloride treated) activated sludge and seve values for raw domestic wastewater, respectively [1].
[g]Range of three values for oak leaf detritus from three pond sources [27].
[h]Ohio River, Rapid Creek, and EPA$_{18}$ sediment, respectively [2].
[i]C_{16}–C_{18} alkyl (tallow range).
[j]Inedible fish (bluegills) tissue, well water; inedible tissue, river water; whole body, wel water; whole body, river water [48].
[k]Temperature not stated [61].
[l]EPA$_{12}$ soil [5].
[m]Whole-body BCFs for clams, minnows, and tadpoles, respectively [134].

materials. Sorption is rapid in well-mixed test systems. Games et al. [1] determined equilibrium constants (K_d) for A_{18}TMAC sorption to activated sludge and influent wastewater solids (Table 2) and found that more than 98% of the A_{18}TMAC initially present at 20 mg/L was removed from solution within 90 min of its addition to a semicontinuous activated sludge (SCAS) system. Sorption was also rapid in flask experiments, equilibrium being reached within 30 min in both wastewater and activated sludge.

Cationics also sorb strongly to natural sediments and soils. Larson and Vashon [2] found that sorption was rapid, equilibrium being reached within a few hours. Interestingly, K_d values were higher for the monoalkyl quaternaries A_{16}TMAB and A_{18}TMAC than for DA_{18}DMAC, despite the much lower water solubility of the latter (Table 2). This suggests that sorption of cationics to sediments involves more than a simple surface area–dependent or solute-partitioning phenomenon, in which the chief variables are the hydrophobicity and organic carbon content of the solute and adsorbent, respectively.

Indeed, adsorption of organic cations such as surfactant quaternary ammonium compounds (QACs) to clay minerals, sediments, aquifer materials, and soil seems to occur mainly by an ion-exchange mechanism [3–6]. Adsorption depends primarily on the cation-exchange capacity (CEC) of the sorbent, the nature and concentration of the electrolyte, and the concentration and alkyl chain length of the organic cation [3,5]. As would be expected, solution pH has little effect on adsorption of QACs, which have a permanent positive charge [5,7].

Displacement of small cations such as H^+, Na^+, and Ca^{2+} from the surfaces of clay minerals by large organic cations may significantly alter the sorptive properties of the substrate. Smith and Bayer [8] observed that the presence of QACs with large alkyl groups enhanced sorption of the herbicide diuron by soil. Their interest was in controlling the movement of the herbicide through the soil profile. More recently, Boyd and co-workers [9–12] found that the substitution of QACs for Na^+ and Ca^{2+} in expanding lattice clay minerals effectively creates an organic medium at the mineral surface, and that this medium enhances sorption of neutral organics by a partitioning process.

These observations have been confirmed and extended in several laboratories [13–16], since there is a high potential for practical applications. This is so because the unexchanged surfaces of clay minerals used in landfill liners and slurry walls at waste disposal sites have relatively little capacity to retain neutral organics by sorption. For example, Lee et al. [14] recently demonstrated that treatment of a subsurface soil with hexadecyltrimethylammonium ion increased sorption coefficients for toluene, perchloroethene, ethylbenzene, and dichlorobenzene 200-fold. Alkyl chains of at least

12 carbons seem to be required for the QAC-exchanged material to serve as an effective partition medium [14,15]. This observation is consistent with that of Cowan and White [3], much earlier, who reported that the hydrophobicity of montmorillonite exchanged with *n*-primary alkylamines was most pronounced at chain lengths of 12 or greater.

Adsorption by ion exchange may also explain observations such as those of Weber and Coble [17] and Barbaro and Hunter [18], who reported little or no effect of kaolinite on biodegradation of QACs, in contrast to the inhibitory influence of montmorillonite. The latter is an expanding lattice clay and has a much higher CEC than kaolinite. Organic cations such as $A_{16}TMAB$ intercalate into the interlamellar space of montmorillonite, with the QAC forming two layers having parallel orientation of the alkyl chains [19]; as the degree of substitution approaches the CEC, there is increased bending and overlap of the alkyl chains [16]. Binding is essentially irreversible [14]. Thus organic cations apparently can be protected from microbial attack by adsorption on the inner surfaces of the expanding lattice. Further, it may be that the interlamellar phase is not available to the much bulkier dialkyl quaternaries (such as $DA_{18}DMAC$) for steric reasons. If so, this would explain the lower K_d values for dialkyl quaternaries reported by Larson and Vashon [2].

Numerous studies indicate that cationic surfactants are also rapidly and strongly sorbed by many other materials of environmental relevance. In addition to activated sludge, sediments, and clay, cationics sorb to minerals, including halides, sulfides, oxides, and sulfates [20–26], oak leaf detritus [27], proteins [28–33], and cell walls of microorganisms [34–38]. Sorption to mineral substrates and protein have been reviewed by Ginn [39] and Swisher [40], respectively. The substantial capacity of microorganisms to sorb QACs is well documented in the older literature. Specific rates of $A_{16}TMAB$ uptake at saturation in excess of 200, 300, and 400 mg/g dry weight have been reported for yeast [36], *Staphylococcus aureus* [34,35], and *Escherichia coli* [35,37], respectively. These figures are consistent with the data of Neufahrt et al. [41], who found levels of $DA_{18}DMAC$ on activated sludge in the range 250 to 470 mg/g dry matter, in an activated sludge plant dosed continuously with 10 mg/L $DA_{18}DMAC$ plus 40 mg/L linear alkylbenzene sulfonate (LAS).

A related property of cationic surfactants is that they form 1:1 complexes with anionic materials, especially anionic surfactants. The complexes are relatively hydrophobic but largely ionic in character [42]. Cationics should ordinarily exist in this form in surface waters as well as domestic sewage, since anionic surfactants are produced and released to the environment in much higher quantities [43], and similar removal efficiencies are expected for both. One result is that cationics may not be detectable by some analytical methods, possibly leading to a failure to demonstrate their presence in environmental samples or to confusion of complexation with degradation in fate studies. Analytical methods are discussed in Sec. II.

It has often been stated that these "electroneutral salts" precipitate near the equivalence point, but insolubility seems unlikely at levels of cationic surfactants in the microgram per liter range, which have been reported in several river monitoring studies [44–50]. In fact, the solubilities of 1:1 complexes of nonylxanthate (nonyldithiocarbonate) with $A_{16}TMAB$ and $A_{12}TMAB$ have been measured and found to be approximately 1 and 10 mg/L, respectively [42]. Uncertainty over so fundamental a property as solubility clearly demonstrates the need for a careful study of the properties of cationic–anionic surfactant complexes under environmental conditions.

II. ANALYTICAL METHODS FOR ENVIRONMENTAL APPLICATIONS

Numerous analytical methods have been developed for cationic surfactants. Many of the older methods are colorimetric, in which the cationic surfactants react with anionic dyes, the surfactant–dye complexes are extracted into an organic solvent, and the absorbance of the solution is measured spectrophotometrically [51]. However, these methods are generally unsuitable for monitoring levels of cationics in sewage or environmental samples or for laboratory studies in which anionic surfactants are also present, because the affinity of cationic surfactants for anionic surfactants is often greater than their affinity for the dyes. For this reason, the method of Waters and Kupfer [52] or modifications of it are preferred for environmental studies. In this procedure, samples are passed through an anion-exchange column to remove the anionic component from the cationic–anionic surfactant complex before colorimetric analysis for the cationic as the disulfine blue–cationic complex. The results are reported as disulfine blue–active substances (DBAS).

One drawback to this method is that it is not very specific. In particular, secondary and tertiary alkylamines, which are possible metabolites in QAC biodegradation [53], as well as other nitrogen-containing compounds that may be present in environmental samples, also react with disulfine blue [44]. To avoid this problem, cationics may be removed from their disulfine blue complexes by cation-exchange chromatography and analyzed by thin-layer chromatography (TLC). The combined disulfine blue–TLC procedure allows the determination of both DBAS and specific cationics [46,54].

Michelsen [55] also developed a method suitable for cationic surfactants in wastewater, based on TLC with quantitation by densitometry. Cationics are isolated by N_2-bubble stripping (sublation) and separated from anionic surfactants by ion exchange. After elution from the cation exchanger, the cationic material is concentrated and separated by TLC, and the plates are sprayed with Dragendorff reagent to detect the individual compounds.

All of these procedures are long and tedious. However, Wee and Kennedy [47] developed a method that appears to be superior to colorimetric methods in sensitivity, specificity, and ease of performance. This method uses high-performance liquid chromatography (HPLC) with conductometric detection in a nonaqueous medium, made possible because cationic surfactants are both soluble and ionized in organic solvents. Before HPLC analysis, cationics are extracted into a nonpolar medium after addition of LAS to the water sample, which enhances the extraction. Using this method, cationics can be determined at submicrogram quantities without derivatization. The detection limit for environmental samples is about 0.02 μg, or 2 μg/L in terms of the cationic concentration in the sample.

Wee [49] subsequently used the HPLC method to determine the concentration of DTDMAC in river water and sewage treatment plant influent and effluent, and compared the results to those obtained using a modified Waters–Kupfer DBAS procedure, and a DBAS–TLC procedure. These data are discussed in more detail in Sec. VI. The comparison showed that the DBAS procedure was less specific and sensitive than either DBAS–TLC or HPLC, and tended to overestimate DTDMAC levels in environmental samples. The DBAS–TLC and HPLC methods were comparable in specificity, but HPLC was more sensitive and convenient to perform than either DBAS or DBAS–TLC.

III. DISCHARGE TO THE ENVIRONMENT

The affinity of cationic surfactants for negatively charged surfaces makes them suitable for a wide variety of uses (Table 3). The primary market for QACs is as fabric softeners and antistats [56], but they are also potent germicides and deodorizers, and see widespread use in part because they are relatively nontoxic to humans, odor free, and stable in storage. Most industrial uses of QACs are related to textile manufacturing. The major uses of amine and amine oxide surfactants are as ore flotation aids, corrosion inhibitors, dispersing agents, and wetting agents for asphalt and tar emulsions. Amine oxides are also used in certain household products, such as dishwashing detergents, and in textile processing [57].

Most uses of QACs and some uses of the amines and amine oxides can be expected to lead to their release to wastewater treatment systems. If all uses of QACs with the exception of drilling muds result in their release to publicly owned treatment works (POTWs), it can be estimated that approximately 80 million pounds of QACs were sewered in the United

TABLE 3 Major Uses and Consumption of Cationic Surfactants in 1987

	Consumption ($\times 10^3$ metric tons)	
Use	United States	Western Europe
Softeners and detergents	79.8	72
Mineral processing and oil: flotation aids, drilling muds, etc.	30.4	16.5
Textiles/fibers: dye assistants, dye levelers, etc.	17.1	16.5
Road chemicals: asphalt emulsifiers	15.2	12
Biocides	11.4	10.5
Others	36.1	22.5
	190	150

Source: Ref. 148.

States in 1979 [43]. Since the total volume of water flowing through POTWs discharging to surface waters was 3.6×10^{13} L that year [58], the average concentration of QACs in domestic sewage should have been approximately 1 mg/L. Using the same approach, Gerike et al. [59] arrived at a figure of 1.4 mg/L for the approximate level of QAC fabric-softening agents (mainly DTDMAC) in German sewage. This difference in predicted QAC concentrations is expected, since Germany is the world leader in per capita consumption of fabric softeners [60].

IQAMS and EEQAMS (as well as their more highly ethoxylated derivatives, the polyethoxylated ethanaminium quaternary ammonium compounds) are also used mainly as fabric softeners. Procter & Gamble [61] estimated that consumer uses of IQAMS and EEQAMS should yield concentrations of these compounds in U.S. sewage of about 0.17 nd 0.28 mg/L, respectively. In a similar fashion, Woltering and Bishop [62] estimated approximate levels of C_{12-18} monoalkyl quaternaries in U.S., French, British, and German sewage as 0.11, 0.30, 0.42, and 0.50 mg/L, respectively.

Calculations for LAS indicate that the level of LAS in domestic sewage in the United States should have been about 9 mg/L in 1979; however, LAS levels are typically on the order of 3 or 4 mg/L. It is likely, therefore, that actual levels of QACs in sewage are also lower than the calculations above would suggest. Although somewhat limited, QAC monitoring data generally confirm this suspicion. These data are discussed in detail in Sec. VI.

IV. ENVIRONMENTAL FATE OF CATIONIC SURFACTANTS

Biodegradation is the ultimate fate of cationic surfactants released to wastewater treatment and the environment. The biodegradability of cationics has been studied extensively in the laboratory, especially in the last 10 years. Degradability ranging from 0 to 100% has been reported, often for the same or structurally related compounds. To some extent, this is a reflection of widely differing test systems and methods used to measure biodegradation. But in large measure it is also a result of failure to appreciate two key aspects of the environmental behavior of cationic surfactants. First, as mentioned previously, they are strongly sorbed by a wide variety of materials, and form complexes with anionic substances. This tendency has sometimes made it difficult to distinguish among sorption, complexation, and biodegradation as mechanisms to account for disappearance of parent compound, particularly in wastewater treatment systems. Second, acclimation may profoundly influence biodegradability, especially for cationics other than monoalkyl amines and quaternaries, since evidence shows that the other cationics are likely to be somewhat more resistant to degradation.

A. Fate in Wastewater Treatment

In acclimated activated sludge systems, the efficiency of removal of cationic surfactants at nontoxic levels should generally exceed 90%. Removal will normally correspond to biodegradation, but not necessarily to ultimate degradation. Degradation will occur mainly on sludge solids, since sorption is much faster than biodegradation. Biodegradation of cationics in anaerobic sludge digestion has not been investigated systematically. This is a potentially serious data gap in view of the expected partitioning of these compounds in wastewater treatment and the probable subsequent treatment of sludge by anaerobic digestion and/or landfilling. Removal data from key aerobic treatability studies are summarized in Table 4.

In several early studies [63–65], neither the analytical method nor the experimental design was adequate to distinguish removal due to biodegradation from removal by sorption, or analytical masking by complexation with anionic surfactants.

Barden and Isaac [66] were probably the first to provide good evidence that cationics could readily be biodegraded with acclimation. Concentrations of $A_{16}PB$ up to 25 mg/L were easily removed by model trickling filters acclimated to 30 mg/L of $A_{16}PB$. The detection of only traces of $A_{16}PB$ in the effluent at 25 and 30 mg/L was not the result of complexation with anionic substances in the sewage feed, since measurements of $A_{16}PB$ in

TABLE 4 Summary of Selected Studies on Fate of Cationic Surfactants in Wastewater Treatment

Compound	Concentration (mg/L)	Test system[a]	Removal (%)	Ultimate degradation[b] (%)	Ref.
A$_{16}$PB	25	TF, A	100		66
Dodecylpyridinium methylsulfate	8	CAS, A[c]		83	71
A$_{16}$TMAB	5–15	CAS, A	91–98		59
	10	CAS,A, C	98		59
	15	CAS, A		107	59
A$_{18}$TMAC	20	SCAS	>98 from soluble phase in 90 min; primary degradation $t_{1/2}$ = 2.5 h		1
	0.1–1.0	SCAS,[d] C		60–90[e] depending upon position of ^{14}C	1
ADMBAC	10–16	CAS, A, C	97; 94 primary degradation		69
A$_{12}$DMBAC	5	CAS, A, C	96		59
	15	CAS,A, C		83	70
	4	CAS, A[c]		101	71

Compound	Conc.	Test[a]	Biodegradation[b]		Ref.
A_{14}DMBAC	20	CAS, A	>70	Probably substantial	107
Dichlorobenzyldimethyldodecylammonium chloride	15.6	CAS, A[c]		57	71
	3.9	CAS, A[c]		98	71
DTDMAC	10–16	CAS, A, C	96–97; 62–82 primary degradation		69
	2.1	SBAS, C	>90 from soluble phase in 4 days; 70 primary degradation in 39 days	Probably substantial	73
DA_{10}DMAC	5	CAS, A, C	95		59
	15	CAS, A, C		0.3	59
	20	TF, A	100	94	67
DA_{18}DMAC	10	CAS, A, C	95	108	59
	20	CAS, A, C		99	59
	20	TF, A	100	84	67
	20	TF, A, C	100		67

[a]TF, trickling filter; A, acclimated; CAS, continuous-flow activated sludge (OECD Confirmatory Test); C, cationic/anionic surfactant complex; SCAS. semicontinuous activated sludge test; SBAS, semibatch activated sludge.

[b]Defined as percent DOC removal with respect to cationic surfactant, except as noted.

[c]Not stated whether cationic surfactant was tested by itself or as a cationic/anionic surfactant complex.

[d]Incubation period extended from 24 to 172 h; suspended solids level reduced to approximately 1 g/L.

[e]Percent of theoretical $^{14}CO_2$ evolution.

the feed always showed only slightly less than the added amount. Moreover, extraction of the slime on the filters failed to reveal significant quantities of $A_{16}PB$, ruling out sorption per se as the primary mechanism of removal. More recently, Gerike [67] examined the removal of $DA_{18}DMAC$ by model trickling filters. Filters were inoculated with effluent from a POTW, and cationic surfactant levels in the synthetic sewage feed were gradually raised from 5 mg/L to approximately 20 mg/L. In filters exposed to $DA_{10}DMAC$, $DA_{18}DMAC$, or $DA_{18}DMAC$ plus LAS, disappearance of the cationic surfactant reached 100% and mean DOC losses were 94, 99, and 84%, respectively. Subsequent extraction of the filters revealed only small amounts of cationics, suggesting again that removal was due mainly to biodegradation.

Most investigators have used model-activated sludge systems to study the removal of cationics in treatment. Gerike et al. [59] studied the fate of $A_{16}TMAB$, $DA_{18}DMAC$, $A_{12}DMBAC$, and $DA_{10}DMAC$ in the OECD Confirmatory Test. Without the simultaneous presence of LAS in the feed, removal of $A_{16}TMAB$ in excess of 90% was observed at 15 mg/L, but only if the system was acclimated by gradually raising the $A_{16}TMAB$ concentration to that level. In the presence of 20 mg/L of LAS, removal of all four cationics at 10 mg/L was high (>90%) and steady. With all four, removal apparently corresponded to biodegradation, since minimal levels of cationics were found in sludge from the aeration vessel. However, a tendency to sorb to sludge is suggested by the presence in the return sludge of $DA_{18}DMAC$ at up to 12% of the total throughput. Gerike [67] later calculated the maximum amounts of $A_{16}TMAB$, $A_{12}DMBAC$, and $DA_{10}DMAC$ that could have been removed on a continuous basis by sorption alone, using the mathematical model of Wierich and Gerike [68]. These calculations indicated that removal by sorption could account for only 8 to 29% of the total elimination of the three cationics listed above, suggesting that biodegradation was responsible for most of the observed loss.

May and Neufahrt [33] and Janicke and Hilge [69] obtained similar results, also using the OECD Confirmatory Test. May and Neufahrt [33] examined $DA_{18}DMAC$ degradation in mixtures consisting of $DA_{18}DMAC$ (2.8 mg/L) and secondary alkane sulfonate (8.0 mg/L), or $DA_{18}DMAC$ (1.8 mg/L) and dodecylbenzenesulfonate (5.8 mg/L), secondary alkane sulfonate (2.6 mg/L), and a nonionic surfactant (0.5 mg/L). Removal of $DA_{18}DMAC$ exceeded 90% in both systems. Janicke and Hilge [69] investigated the treatability of DTDMAC and $A_{8-18}DMBAC$. After acclimation to 10 to 16 mg/L of DTDMAC or $A_{8-18}DMBAC$ in the presence of an equivalent amount of alkylbenzenesulfonate (ABS), filtered effluents contained 3.5 to 4.5% of the influent DTDMAC and 3% of the influent $A_{8-18}DMBAC$. Net elimination—sorption plus biodegradation—was 96%

for DTDMAC and 97% for $A_{8-18}DMBAC$. As in the preceding studies, the authors found relatively low levels of cationics in the sludge and concluded that they were biodegraded.

Gerike et al. [59] also observed complete disappearance of $A_{16}TMAB$ and $DA_{18}DMAC$ as measured by loss of DOC. DOC removal was somewhat lower in the $A_{12}DMBAC$-amended test system (54%), but a subsequent study reported a value of 83% [70], and complete removal (101.2%) was reported for a recent study in which the level of cationic surfactant in the feed was reduced to 4 mg/L [71]. DOC removals for $A_{16}TMAB$ and $A_{12}DMBAC$ were also high in the Zahn–Wellens test [70]. Together with the failure to detect substantial amounts of cationics in sludge [59], these findings suggest that $A_{16}TMAB$, $DA_{18}DMAC$, and $A_{12}DMBAC$ underwent extensive ultimate degradation. For $DA_{10}DMAC$, primary degradation did not coincide with ultimate degradation, however. In contrast to its high removability as measured by loss of parent compound, DOC loss amounted to only 0.3%.

Use of ^{14}C-labeled cationics has resolved many of the uncertainties concerning their fate in biological treatment systems, particularly with respect to ultimate degradation and the role of sorption. Krzeminski et al. [72] conducted a detailed study of the environmental impact of Hyamine 3500, an alkyldimethylbenzylammonium chloride. That compound was degraded extensively at 10 mg/L in a semicontinuous activated sludge (SCAS) test, but only if the test system was exposed to increasing levels of the surfactant over a period of about 2 weeks. Proof of ultimate degradation was provided by the detection of 80% of the influent ^{14}C as $^{14}CO_2$, and 16% of the ^{14}C in the effluent.

Games et al. [1] combined SCAS studies with measurements of sorption in sewage and activated sludge to determine the fate of $A_{18}TMAC$ in wastewater treatment. Primary degradation was slower than sorption but still rapid, occurring with a half-life of 2.5 h in an acclimated system. Thus biodegradation was associated primarily with sludge solids. Ultimate degradation followed apparent first-order kinetics at initial $A_{18}TMAC$ concentrations of 0.1 and 1.0 mg/L, levels close to those normally expected in sewage. Similar results were obtained with a 1:1 complex of $A_{18}TMAC$ and LAS. Conversion of $A_{18}TMAC$ to $^{14}CO_2$ was extensive and established the ultimate biodegradability of both the methyl groups and the alkyl chain.

Sullivan [73] studied the biodegradation and sorption of ^{14}C-labeled DTDMAC in semibatch-activated sludge reactors. In these experiments DTDMAC was added only once at 1.2 mg/L as a complex with LAS, but synthetic sewage was fed daily to maintain organic loading rates characteristic of conventional and extended aeration-activated sludge treatment. Results were similar to those of Games et al. [1]. Degradation of DTDMAC was considerably slower in both systems than was A_{18}TMAC degradation in the latter study, however, suggesting that dialkyl quaternaries may be more recalcitrant than the monoalkyl derivatives in wastewater treatment.

The recent work of Ruiz Cruz [74] adds substantially to the list of cationic surfactants for which treatability data are available. Removal was high in both the Spanish Official Test (a SCAS method) and the OECD Confirmatory Test for nearly all test compounds, including an isoquinolinium and two imidazolium cationics, classes not tested previously. Of course, the high removal of these compounds is not a guarantee of ultimate biodegradability. Data for cationics not previously tested as either the chloride or bromide salt are summarized in Table 5.

B. Fate in Receiving Waters

Several generalizations can be made regarding the fate of cationic surfactants in receiving waters. First, monoalkyl quaternaries as a class should be most rapidly, and alkylpyridinium compounds least rapidly biodegraded, with alkyldimethylbenzyl and dialkyl quaternaries intermediate in degradability. Second, aquatic environments that receive effluent from

TABLE 5 Removal of Selected Cationic Surfactants in the Spanish Official and OECD Confirmatory Tests[a]

	Removal (%)	
Compound	Spanish Official Test	OECD Confirmatory Test
A_{12}TMAC	97	96
DA_{12}DMAC	96	96
A_{18}DMBAC	94	95
p-tert-Octylphenoxyethoxyethyldimethylbenzylammonium chloride	30	45
A_{12}PC	93	91
A_{14}PC	94	92
Dodecylisoquinolinium chloride	90	92
Didodecylimidazolium chloride	89	92
Dodecylhydroxyethylimidazolium chlorhydrate	94	97

[a]All cationics were present in the synthetic sewage feed at 5 mg/L, without anionic surfactant. Both the Spanish Official Test (a SCAS method) and the OECD Confirmatory Test (a continuous-flow method) require acclimation before determination of removal.
Source: Ref. 74.

sewage treatment plants should be acclimated to some cationics, such as DTDMAC, and biodegradation rates for these may be quite rapid. Finally, sorption of cationics to suspended solids could play an important role in their biodegradation, but despite recent progress, complete understanding has not been achieved.

Screening tests have been used in many studies of cationic biodegradability [2,70,74–84], but these reports must be interpreted with caution since river die-away studies have shown that screening tests may underestimate biodegradability [85]. The high toxicity of cationic surfactants to the test organisms accounts for this at least in part, since test chemical concentrations are typically in the milligram per liter range. Anionic surfactants seem to mitigate toxicity in screening tests just as they do in bench-scale treatability tests. This effect is illustrated in Fig. 1, which shows that $A_{18}TMAC$ alone inhibited endogenous metabolism at 10 mg/L (note the

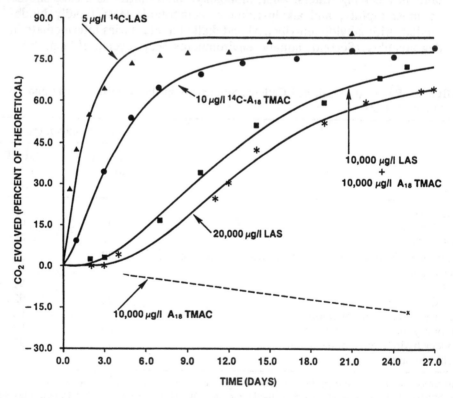

FIG. 1   Ultimate degradation of $A_{18}TMAC$ and LAS in screening tests and natural waters. (From Ref. 82.)

negative slope), but that toxicity was eliminated in the presence of an equal concentration of LAS. Nevertheless, taken together, the screening data do suggest the order of relative biodegradability given in the preceding paragraph. The biodegradability of amine oxides is not well documented, but both lauryl(dodecyl)dimethylamine oxide [76] and (1-methyldodecyl)-dimethylamine oxide [83] were rapidly biodegraded after acclimation in screening tests with activated sludge inocula. These findings are consistent with the large body of data on QACs.

The most useful information on the fate of cationic surfactants in receiving waters has come from die-away studies. Kaplin et al. [86] reported that $A_{18}TMAC$, $A_{17-20}TMAC$, $A_{18}DMBAC$, and an unidentified alkyldimethylbenzylammonium chloride disappeared from river water without added bottom sediment within 30 to 40 days. Several cationics disappeared from the water column more rapidly in microcosms containing sediment, but sorption probably accounted for some removal. Baleux and Caumette [87] studied the biodegradation of several cationics in river water and sewage. Among these were $A_{12}PC$, $A_{16}PB$, $A_{16}TMAC$, $DA_{18}DMAC$, dioctadecyl(ethoxy)$_{15}$methylammonium chloride, p-diisobutylphenoxyethoxy-ethyldimethylbenzylammonium chloride (Hyamine 1622), and an unspecified alkylimidazoline. $A_{16}TMAC$ and the alkylimidazoline disappeared completely from river water in 7 to 14 days, depending on the source of the water. Hyamine 1622 apparently was not degraded. The slow degradation of many of these compounds may have been the result of toxicity at the high concentrations employed (15 to 20 mg/L), but other tests failed to demonstrate a significant correlation between recalcitrance and toxicity.

Ruiz Cruz and Dobarganes Garcia [88] carried out the most extensive study to date of the influence of chemical structure on biodegradation of cationics in river water. Although removal by sorption cannot be ruled out, the kinetics of surfactant loss observed in this study and the low level of suspended solids in the river water (ca. 1.4 mg/L) suggest that biodegradation was the primary mechanism of removal. Of a series of quaternaries having C_{12} alkyl chains, $A_{12}TMAC$ and dodecylimidazolium chloride were the most rapidly degraded. Following these, in order of decreasing degradability, were $A_{12}DMBAC$, $DA_{12}DMAC$, $A_{12}PC$, and dodecylisoquinolinium chloride. Lag periods and half-disappearance times with unacclimated river water ranged from less than 1 to several days. These results suggest that monoalkyl quaternaries will be most rapidly degraded and alkylpyridinium quaternaries least rapidly degraded in receiving waters.

River die-away studies by Krzeminski et al. [72], Larson and Perry [89], and Larson and Vashon [2] also suggest that monoalkyl quaternaries should be degraded more rapidly in receiving waters than other QACs. Larson and Perry [89] used an electrolytic respirometer to follow ultimate degradation of A_{16}TMAC in Ohio River water. A_{16}TMAC was degraded with a half-life of 2.7 days after a 2-day lag, at A_{16}TMAC levels below the apparent toxicity threshold of 40 mg/L. Larson and Vashon [2] measured production of $^{14}CO_2$ from ^{14}C-labeled QACs, with initial concentrations ranging from 1 to 100 μg/L. Ultimate degradation of A_{16}TMAB and A_{18}TMAC in river water occurred with no detectable lag and half-lives of 2 to 3 days (Fig. 1). In contrast, DA_{18}DMAC was not degraded appreciably in 9 weeks in sediment-free water. Moreover, Krzeminski et al. [72] found that more than 3 weeks were required for complete ultimate degradation of ^{14}C-labeled Hyamine 3500 (an alkyldimethylbenzylammonium chloride), at initial concentrations from 0.01 to 1.0 mg/L.

Vives-Rego et al. [90] studied the biodegradation of anionic, nonionic, and cationic surfactants in seawater. Half-lives for loss of DBAS were in the range 4 to 9 days, even though the cationic tested, A_{16}TMAB, showed signs of toxicity at the high concentration employed (20 mg/L).

Biodegradation rates in the environment are also influenced by numerous factors other than chemical structure. Ruiz Cruz [91,92] investigated the influence of several experimental variables on disappearance of cationics from river water, at an initial concentration of 5 mg/L. The variables included biomass, acclimation, aeration, temperature, and organic and inorganic nutrients. Acclimation may be the most important of these variables. Degradation of A_{12}PC, A_{12}TMAC, A_{16}TMAC, A_{18}TMAC, A_{12}DMBAC, A_{14}DMBAC, A_{16}DMBAC, A_{14}PC, and A_{18}DMBAC proceeded more rapidly and without a lag in water samples previously exposed to these compounds. However, acclimation was lost if sufficient time passed before water was respiked with test compound.

Studies of the effects of acclimation on biodegradation of cationic surfactants have recently been extended to a laboratory model stream and field sites. Ventullo and Larson [93] observed effects of preexposure to A_{12}TMAC similar to those reported by Ruiz Cruz [91,92], at a lake ecosystem in southern Ohio. Upon initial exposure, ultimate degradation of A_{12}TMAC proceeded with a half-life of 90 h after a lag phase of 24 h, but degradation proceeded with no lag and a half-life of 20 h after the respike. Heterotrophic activity studies also showed that chronic exposure to A_{12}TMAC resulted in significant increases in maximum degradation rate (V_{max}) and number of A_{12}TMAC degraders relative to unexposed controls. Larson and Bishop [94] found that A_{12}TMAC was degraded with half-lives ranging from <2 to 15 h in the water column of a model stream dosed continuously with A_{12}TMAC for 15 months, whereas half-lives often exceeded 5000 h in the control. Responses to A_{12}TMAC exposure were similar in the epilithic and sediment compartments. The microbial community also developed resistance to "shock loads" of the surfactant, but acclimation was lost quickly when input of the surfactant was halted [95]. Acclimation of the epilithic microbial community was examined more closely at a second field site, the Little Miami River in Ohio. In situ exposure to $A_{12}TMAC$ produced an adapted community that degraded this compound with half-lives that averaged about 2 h, which was 14 to 50 times lower than in the unexposed control [94]. Subsequent studies [96] showed that both the number and specific activity of $A_{12}TMAC$ degraders in sediment increased during adaptation to the surfactant.

The tendency of cationics to sorb strongly to sludge, sediments, and other materials has been discussed. What has been less clear is the effect that sorption has on biodegradability. Investigators in several older studies [17,18,72] found that sorption inhibited biodegradation. But Larson and Vashon [2] reported the opposite effect of sediment on degradation of $DA_{18}DMAC$, and no significant inhibition of $A_{16}TMAB$ and $A_{18}TMAC$ degradation.

As discussed previously, differences in the sorbent (montmorillonite, an expanding-lattice clay, in the first two studies mentioned above) may be partly responsible for this apparent conflict. In any case, it seems clear that cationic surfactants are not necessarily unavailable for biodegradation in the presence of sediment. This question was addressed directly by Shimp and Young [97]. They found that biodegradation rates for $A_{12}TMAC$ were a function of the total amount of chemical present, sorbed and unsorbed, in sediment slurries, whereas in cores containing settled sediment with overlying water, sorbed material was not directly available for biodegradation. Shimp [98] extended this work to acclimation of sediment and water-column microbial communities to $A_{12}TMAC$ degradation. Higher levels of $A_{12}TMAC$ in the water column were required to elicit an adaptive response in settled sediments as compared to overlying water, but once acclimation occurred, the response was more prolonged than in the water column. Thus it appears that although sorption may reduce bioavailability, it can also increase the robustness of the adaptive response by buffering surfactant levels in interstitial water.

A potentially important observation made by Ruiz Cruz [92] relative to the influence of experimental variables concerns the effect of exposure of test vessels to sunlight. Disappearance of $A_{12}TMAC$ and $A_{12}DMBAC$—normally complete within several days—appeared to stop after loss of 50 to 60% of the amount initially present. This suggests that photodegradation may have resulted in the accumulation of toxic degradation products, or products not further degraded, but still detectable by the analytical method used. That cationics may be photooxidized was also suggested by Neufahrt (cited in Ref. 60), who observed a reduction in the carbon and hydrogen contents of $DA_{18}DMAC$ from 75 to 55.4% and 13.8 to 8.9%, respectively, with 16 to 72 h of exposure to UV light. In this case the resulting products were readily biodegradable. The potential for photodegradation of cationic surfactants should be studied in detail.

C. Fate in Soil and Groundwater

Data on the biodegradability of cationic surfactants in surface soils and subsurface environments are very limited. Half-lives for aerobic ultimate degradation of A_{18}TMAC were reported to be 1.4, 5.8, 5.3, and 3.2 to 8.7 days for compost, organic loam, sandy loam, and silt loam, respectively [94]. On the other hand, Larson and Vashon [2] found much slower degradation of A_{12}TMAC in aerobic sludge-amended soil ($t_{1/2}$ = 1 mon). Degradation is therefore likely to be fast enough in surface soil, relative to normal residence times, to prevent their accumulation. Only monoalkyl quaternaries have been tested, however.

Larson [99] and Federle and Pastwa [100] studied the biodegradation of QACs in subsurface samples. A_{12}TMAC was degraded in aerobic groundwater/subsurface soil slurries with a half-life of 1 week [99]. Federle and Pastwa [100] studied the aerobic biodegradation of A_{18}TMAC and DA_{18}DMAC as a function of depth in samples of subsurface sediment from below a laundromat discharge pond and a nearby control pond. Neither compound was mineralized to a significant extent in control sediment in 3 months, but extensive mineralization occurred in the laundromat profile. A_{18}TMAC was degraded more rapidly and extensively than was DA_{18}DMAC. These results provide another illustration of the effect of acclimation on biodegradation rates. But the lack of degradation in the control profile is disturbing and suggests that discharges of cationic surfactants could adversely affect pristine sites. In general, the lack of information on fate in the subsurface constitutes a serious data gap that must be addressed, since cationics may see greatly increased use in the future in landfill liners and slurry walls at waste disposal sites (see Sec. I).

It should be noted that photochemical instability, just discussed, could also affect the fate of cationics in surface soil. This would be particularly relevant for soil amended with wastewater sludge containing cationic residues. However, the potential for photodegradation must first be confirmed experimentally.

V. TOXICITY OF CATIONIC SURFACTANTS

A. Microbial Toxicity in Wastewater Treatment and Receiving Waters

An extensive literature, not reviewed here, amply documents the potent germicidal activity of many cationic surfactants [101]. This property probably accounts for the strong emphasis placed on the issue of toxicity in wastewater treatment in the older literature on environmental fate and effects of cationics. It can now be stated that under normal circumstances, cationic surfactants are unlikely to pose a significant risk of toxicity to microorganisms in wastewater treatment systems. The reasons are twofold. First, microbial populations in POTWs can be expected to be acclimated to low levels of commonly used cationics, as shown previously. This is significant because the high toxicity reported in the older literature is not observed in properly acclimated, high-biomass test systems. Second, the presence of anionic surfactants in sewage will greatly reduce cationic toxicity.

Nevertheless, sudden discharges of cationic surfactants resulting in temporarily high levels in treatment plants could upset plant function. This may be especially true for the nitrification process, which appears to be somewhat more sensitive than sewage purification in general. Slug doses are not likely to be a problem for cationics in consumer products, but other uses could lead to such discharges. Recent evidence also suggests that ecologically significant responses are possible in aquatic environments unacclimated to cationic surfactants, at levels below 1 mg/L.

An early study by Pitter [63] illustrates all of the major points outlined above in regard to wastewater treatment. Pitter studied treatability of $A_{16}PB$ and $A_{16}TMAB$ in a bench-scale activated-sludge system. The surfactant concentration was raised in steps from 1 to 20 mg/L. Without ABS, no effect of $A_{16}PB$ or $A_{16}TMAB$ was observed at 1 mg/L, but at 3 mg/L both QACs strongly inhibited nitrification. Even at 6 mg/L the ability of sludge microorganisms to reduce the biochemical oxygen demand (BOD) of synthetic sewage was reduced by only 6%, but nitrification was blocked completely. But when the QAC concentration was suddenly increased from 6 to 20 mg/L, sludge flowed from the system. More recently, Guhl and Gode [149] determined the EC_0 for inhibition of DOC removal in bench-scale activated-sludge units exposed to $A_{12}DMBAC$. The reported value of 5 mg/L was close to the value for inhibition of oxygen consumption by *Pseudomonas putida* (2 mg/L), as well as the tolerance limits reported by Pitter [63].

In contrast, Pitter [63] observed no inhibition of treatment processes by 6 mg/L of $A_{16}PB$ and $A_{16}TMAB$ in the presence of an equivalent amount of ABS. He therefore concluded that the quaternaries would not be toxic in wastewater treatment systems at concentrations up to 6 mg/L in the presence of equivalent or higher levels of anionic surfactants, which is the situation likely to prevail in domestic sewage treatment under normal circumstances. But sudden loading of the system with 20 mg/L of $A_{16}PB$ or $A_{16}TMAB$ and an equivalent amount of ABS again resulted in large amounts of sludge in the effluent, a drastic reduction in the oxidizing ability of the microbial population (BOD removal), and complete inhibition of nitrification.

Similar results have been obtained for these and other cationics—including $DA_{18}DMAC$, the major component of DTDMAC—by other investigators [59,66,102]. That nitrification in wastewater treatment plants may be especially sensitive to inhibition by cationic surfactants is also suggested by earlier reports [103,104], but interpretation of these studies is difficult because the authors provided little information on test methods. In contrast, Reynolds et al. [105] reported that a pure culture of nitrifying bacteria did not appear to be more sensitive to $A_{16}TMAB$ and $A_{12}DMBAC$ than many commonly used indicators of toxicity, such as respiration.

The EC_{50} values reported by Reynolds et al. [105] for inhibition of respiration by $A_{16}TMAB$ and $A_{12}DMBAC$ were generally in the range 10 to 40 mg/L. Other investigators have obtained similar values [66,106,107]. Larson and Schaeffer [108] used a novel method, inhibition of $[^{14}C]$glucose uptake, to determine the toxicity of chemicals to activated sludge. The EC_{50} for unacclimated sludge was 28 mg/L for $A_{14-18}TMAC$, in line with the respirometric data. Pitter and Horska [109] used still another method, inhibition of dehydrogenase activity in unacclimated sludge fed glucose or peptone, with similar results. The EC_{50} values were 9 and 19 mg/L for $A_{16}PB$ and $A_{16}TMAB$, respectively. Overall, the data suggest that cationic surfactants are unlikely to manifest significant toxicity in wastewater treatment at the levels normally expected (≤ 1 mg/L).

It also seems unlikely that cationics will inhibit anaerobic digestion under normal circumstances, but here the evidence is weaker. In studies by May and Neufahrt [33] and Janicke and Hilge [69], sludge from laboratory activated sludge systems in which treatability of QACs had been studied was incubated anaerobically, and gas production measured over time. May and Neufahrt [33] observed no difference between digestion of control sludge and sludge from systems acclimated to mixtures of $DA_{18}DMAC$ and anionic surfactants, and Janicke and Hilge [69] found no inhibitory effects of DTDMAC or $A_{8-18}DMBAC$. But since biodegradation during the activated sludge process resulted in the removal of approximately 90% of the cationic surfactant originally present in the influent in both studies, relatively little of the starting material was sorbed to the sludge and subsequently transferred to the anaerobic digestors. Thus the possibility exists that more recalcitrant cationics—or even normally biodegradable cationics, in the event of discharges of slug doses to the system—could reach inhibitory levels.

Studies using low-biomass test systems, most commonly involving measurement of the reduction in the biochemical oxygen demand (BOD) of natural or synthetic sewage, have generally indicated much higher toxicity under these conditions [44,63–65,75,106]. In most cases, however, the microbial populations were not acclimated by previous exposure to cationic surfactants. Moreover, the significance of such studies for wastewater treatment operations, where biomass levels are much higher, is doubtful. Similar statements can be made in reference to studies in which effects on bacterial growth were determined [87,88].

Data from test systems such as these may be more relevant to aquatic environments expected to receive cationic surfactants. In fact, in both of the studies just cited, river water was used as a source of microorganisms. For example, Ruiz Cruz and Dobarganes Garcia [88] incorporated QACs into agar plates and inoculated with river water. $A_{12}PC$ inhibited colony development at 1 mg/L, but all four compounds tested (the others were $A_{12}DMBAC$, $DA_{12}DMAC$, and $A_{16}TMAC$) reduced colony counts at levels below 10 mg/L. That cationics may elicit ecologically significant responses at concentrations below 1 mg/L was demonstrated first by Ventullo and Larson [93] and confirmed recently by Tubbing and Admiraal [110]. Ventullo and Larson [93] determined first observed effect concentrations (FOECs) for inhibition of heterotrophic activity in freshwater bacterial communities and reported FOECs of 0.1, 0.1, and 1.0 mg/L for TTMAC, $A_{12}TMAC$, and DTDMAC, respectively. Tubbing and Admiraal [110] observed decreases in the growth rate of bacterioplankton and photosynthesis rate of phytoplankton from the Rhine River at nominal DTDMAC concentrations of 0.03 to 0.1 mg/L. $A_{16}TMAB$ was also toxic to bacteria in sea water [90], but here this result is not so surprising, in view of the high (20 mg/L) surfactant concentration employed.

Toxicity probably accounts for the failure of some investigators to detect biodegradation of cationic surfactants, and for the frequent observation that biodegradability in low-biomass test systems is a function of initial concentration [38,77,78,86,89,91]. In fact, toxicity may account for the widespread failure to observe significant biodegradation of cationics in low-biomass biodegradability screening tests, where concentrations of test chemical are typically on the order of 20 mg/L [1,2,70,80]. Figure 1 is typical and shows that $A_{18}TMAC$ was not biodegraded and apparently was toxic at 10 mg/L in a biodegradability screening test.

B. Ecotoxicity

Many cationic surfactants are not only potent germicides, but also acutely toxic in the milligram per liter range and lower to aquatic organisms, including algae, fish, mollusks, barnacles, rotifers, starfish, shrimp, and others [44,48,72,111–123]. Cooper [124] has reviewed the older literature on environmental toxicity of cationics. Data from Taft [111] on the toxicity of alkyldimethylbenzylammonium chlorides to invertebrates are typical (Table 6).

Vallejo-Freire et al. [113] showed that a wide range of cationics were acutely toxic to snails (*Australorbis* sp.) at concentrations of 1 to 10 mg/ L. The dose-response curves were very steep, with 0 and 100% mortality occurring within a 10-fold change in concentration. However, Cooper [124] estimated that LC_{50} values for 2 of 15 compounds were approximately 0.3 mg/L and that no-effect levels could therefore be even lower. Moreover, if the dose-response curves for species listed in Table 6 are similar, no-effect levels for some species could be as low as 0.01 mg/L.

More recent data on fish and invertebrates are generally consistent with the older studies. Some of these results are given in Table 7. In addition, Woltering et al. [123] have published 48-h LC_{50} values for toxicity of $A_{12}TMAC$ to a wide variety of freshwater invertebrates. These data suggest that *Daphnia magna* may be more sensitive than many other invertebrates.

An aquatic safety assessment by Lewis and Wee [48] adds substantially to the database on aquatic toxicity of cationic surfactants, especially in regard to the effects of suspended solids on observed toxicity. Tests were conducted in which freshwater and marine species representing three trophic levels were exposed to dialkyl quaternaries, in surface waters containing suspended solids as well as high-quality waters free of solids. Acute toxicity data for fish and invertebrates are summarized in Table 8. The 96-

TABLE 6 Toxicity of Alkyldimethylbenzylammonium Chlorides to Invertebrates

Species (group)	Toxicity
Balanus (barnacle)	2.5 mg/L, no effect; 5 mg/L, 100% mortality "within a short time"
Euglena (protozoan)	20 mg/L, 100% mortality in 19 h
Rotifers	1 mg/L, 100% mortality in 19 h; 10 mg/L, 100% mortality in 2 h
Amphipoda (amphipods)	2.5 to 10 mg/L, 100% mortality "in a few hours"
Cypridae (ostracods)	50 mg/L, active after 5 h
Planaria (flatworm)	10 mg/L, 100% mortality in 1 h
Ascaris (roundworm)	20 mg/L, alive after 18 h
Enchytrae albidus (annelid)	1 mg/L, 100% mortality in ≤4 h
Asteroidea (starfish)	2.5 mg/L, 100% mortality in 24 h

Source: Ref. 111.

TABLE 7 Acute Toxicity of Cationic Surfactants to Fish and Invertebrates

Species	Compound	LC_{50} (mg/L)	Ref.
Daphnia magna (water flea)	ATMAC[a]	1.2–5.8	114
Idus melanotus (golden orfe)	ATMAC[a]	0.36–8.6	114
Planorbis corneus (water snail)	ATMAC[a]	0.73–23	114
Lepomis macrochirus (bluegill sunfish)	Hyamine 3500[b]	0.5	72
Carrasius auratus (goldfish)	Hyamine 3500[b]	2	72
Idus melanotus	A_{12}DMBAC	3.5,[c] 8[d]	115
Rasbora heteromorpha (harlequin fish); Salmo truta (brown trout); Idus melanotus; Carrasius auratus	Unidentified	Mean = 2.04–3.51	117
Lepomis macrochirus	Various	Mean = 1.65 (range = 0.33–5.90	124
Oncorhynchus mykiss (rainbow trout)	Various	Mean = 3.80 (range = 0.34–12.3)	124

[a]A_{10}TMAC, A_{12}TMAC, A_{14}TMAC, A_{16}TMAC, A_{18}TMAC, and $A_{20/22}$TMAC were tested. The ranges of LC_{50} given exclude the values for A_{10}TMAC.
[b]An alkyldimethylbenzylammonium chloride (i.e., ADMBAC).
[c]LC_0.
[d]LC_{100}.

h LC_{50} values for toxicity to bluegills (*Lepomis macrochirus*) in high-quality water varied from 0.64 to 3.0 mg/L, but toxicity was much reduced in Town River water. A similar effect was observed for *Daphnia magna*. Woltering et al. [123] later reported that data for monoalkyl quaternaries showed the same trends. Marine species appeared to be less sensitive to DTDMAC ([48]; Table 8), but mysid shrimp (*Mysidopsis bahia*) were as sensitive as *Daphnia*.

Although Marchetti [125] stated that the toxicity of cationic surfactants decreases with increasing water hardness, water hardness did not appear to be a factor in the Lewis and Wee [48] study. It is more likely that other factors—especially sorption to suspended solids and the tendency to form

TABLE 8 Acute Toxicity of DTDMAC to Freshwater and Marine Fish and Invertebrates

Species	Dilution water	Suspended solids (mg/L)	Endpoint	Effect level[a] (mg/L)
Freshwater				
Lepomis macrochirus	Well[b]	0	96-h LC$_{50}$	1.04
(bluegill)	Reconstituted	0	Same	0.62–3.00
	Reconstituted[c]	0	Same	1.23
	Town River	2–84	Same	10.1–>24
Daphnia magna	Well	0	48-h LC$_{50}$	1.06
(water flea)	Reconstituted	0	Same	0.19–0.48
	Reconstituted[b]	0	Same	0.16
	White River[b]	3–5	Same	3.1
Marine				
Crassostrea virginica larvae (oyster)	Estuarine water, salinity 16–26%		48-h EC$_{50}$	2.0
Penaeus duorarum (pink shrimp)	Same		96-h LC$_{50}$	36.0
Mysidopsis bahia (mysid shrimp)	Same		Same	0.22
Callinectes sapidus (blue crab)	Same		Same	>50
Cyprinodon variegatus (sheepshead minnow)	Same		Same	24.0

[a]Nominal level of active ingredient.
[b]Compound tested was DA$_{18}$DMAC, the major component of DTDMAC.
[c]Compound tested was DTDMAMS, not DTDMAC.
Source: Ref. 48.

complexes with anionic surfactants—were responsible for the reduction in acute toxicity in natural waters relative to high-quality water. Support for this conclusion was provided by experiments in which the effects of added river bottom silt and C$_{12}$ LAS on toxicity to bluegills were determined directly. For example, no mortality was observed at C$_{12}$ LAS/DTDMAMS molar ratios of 0.5:1 to 4:1 (absolute concentrations, in milligrams per liter, of 0.3:1.2 to 2.7:1.2). Knauf [114] obtained similar results with the golden orfe (*Idus melanotus*).

Chronic effects have been the focus of most recent ecotoxicity studies. The no observed effect concentration (NOEC) for *Daphnia* exposed to DA$_{18}$DMAC in river water was 0.38 mg/L [48], about one-tenth the 48-h LC$_{50}$ (Table 8); for A$_{12}$TMAC a NOEC of 0.065 mg/L was reported by

Pittinger et al. [126]. FOECs in 4- and 7-day exposures of *Ceriodaphnia dubia* to A_{12}TMAC ranged from 0.17 to 0.35 mg/L]127]. However, Woltering et al. [123] recently reported that acclimation or compensation may occur in *Daphnia* populations chronically exposed to cationic surfactants. Although there was some initial mortality in the mixing zone in a laboratory model stream, total numbers and biomass were not significantly different from control values after several generations, despite nominal surfactant levels that should have been associated with lethality.

In an earlier study, Hidu [128] determined FOECs for growth and development of clam (*Mercenaria mercenaria*) and oyster (*Crossostrea virginica*) larvae. These invertebrates seem to be even more sensitive than daphnids, with FOECs ranging from 0.01 to 0.25 mg/L for A_{12}PC and 0.10 to 1.0 mg/L for Hyamine 1622. The mean of the minimum concentrations causing a significant reduction in development of fertilized eggs and survival and growth of veliger larvae was 0.30 mg/L.

There were no significant differences in larval survival and adult emergence of midge (*Paratanytarsus parthenogenica*) exposed to sediments from Rapid Creek, relative to the controls [48]. DTDMAC levels in sediments and water were 2 to 67 μg/g and 8 to 92 μg/L, respectively. Similar results were reported for midge in the model stream study just mentioned [123], where emergence and population density were not affected despite levels of A_{12}TMAC in sediment that were 35 to 75 times above those associated with mortality in water. These data suggest that sorption to sediment reduces the bioavailability of cationic surfactants to benthic organisms. Lee [129] and Pittinger et al. [126] reached the same conclusion in studies of QAC toxicity to *Chironomus riparius*. Lee [129] found that exposure to DTDMAC in artificial sediments at concentrations up to 0.8% on solids had no effect on growth and development of the larval stages, the pupa, adult emergence, or the sex ratio. In the second study [126], NOECs of sorbed DA_{18}DMAC and A_{12}TMAC were 876 μg/g and 3084 μg/g (highest concentration tested; equivalent to 0.3% on solids), respectively, and were approximately 1000 and 10,000 times greater than the respective NOECs of the soluble surfactants.

Among vertebrates, only fathead minnows (*Pimephales promelas*) seem to have been tested for chronic toxicity of cationic surfactants. NOECs for exposure to DTDMAC were 0.23 mg/L with river water and 0.053 mg/L with filtered well water [48]; for A_{12}TMAC, a NOEC of 0.46 mg/L has been reported [126], presumably for river water. Table 9 summarizes data on chronic toxicity of cationic surfactants to fish and invertebrates. Lewis [130,131] has critically reviewed these data, as well as data on chronic toxicity of high-volume anionic and nonionic surfactants.

Cationic polyelectrolytes have been shown to elicit acute toxic effects in aquatic organisms by disrupting gill membranes, thus interfering with

TABLE 9 Chronic Toxicity of Cationic Surfactants to Fish and Invertebrates

Species	Compound	Effect concentration[a] (mg/L)	Test duration[b] (days)	Effect[b]	Ref.
Daphnia magna	A$_{12}$TMAC	N 0.065	NR	NR	126
	DA$_{18}$DMAC	N 0.38	21	Reproduction	48
		L 0.76			
Ceriodaphnia dubia	A$_{12}$TMAC	F 0.17–0.35	7	Reproduction	127
Mercenaria mercenaria	A$_{12}$PC	F 0.01–0.25	14	Larval growth and development	128
	Hyamine 1622[c]	F 0.25–1.0			
Crassostrea virginica	A$_{12}$PC	F 0.05–0.25	14	Larval growth and development	128
	Hyamine 1622[c]	F 0.10–0.5			
Paratanytarsus parthenogenica	DTDMAC	N >67(S)	20	Emergence	48
Chironomus riparius	A$_{12}$TMAC	N >0.092(OW)	24	Emergence	126
		N 0.23(NS)			
		L 0.62(NS)			
		N >3084(S)			
		N >2.3(IW)			
		N >0.90(OW)			
	DA$_{18}$DMAC	L 1.02(NS)			
		L 2708(S)			
		L 0.18(IW)			
		L 0.41(OW)			
Pimephales promelas	A$_{12}$TMAC	N 0.46	NR	NR	126
	DTDMAC	N 0.05(FWW)	28	Growth, hatching	48
		L 0.09 (FWW)			
		L 0.23(RW)			
		L 0.45(RW)			

[a]N, no observed effect concentration (NOEC); L, lowest observed effect concentration (LOEC); F, first observed effect concentration (FOEC); S, sediment (concentration in μg/g); OW, overlying water; NS, no sediment; IW, interstitial water; FWW, filtered well water; RW, river water.
[b]NR, not reported.
[c]Hyamine 1622, *p*-diisobutylphenoxyethoxyethyldimethylbenzylammonium chloride.

O_2 exchange [132]. Cationic surfactants presumably act by the same mechanism. However, the only published reports on uptake and distribution of cationics in aquatic species are those of Neufahrt et al. [133] and Knezovich et al. [134]. Neufahrt et al. [133] mixed fish food with dried activated sludge from a bench-scale wastewater treatment system dosed continuously with [^{14}C]DA$_{18}$DMAC, and followed the uptake and tissue distribution of radiolabel in carp (*Cyprinus carpio*). Radiolabel was concentrated in the intestinal tract and gallbladder, with some residual ^{14}C in the gills, probably due to filtration of contaminated dung particles or leftover feed. Knezovich et al. [134] studied the uptake and distribution of A$_{16}$PB in three species: clams (*Corbicula fluminea*), tadpoles (*Rana catesbeiana*), and fathead minnows. Organisms were exposed to a sublethal concentration (10 μg/L) of [^{14}C]A$_{16}$PB in the water for 24 h. Radiolabel was concentrated in the gills in all three species, with some residual ^{14}C in the intestine and skin of tadpoles. These results are consistent with the postulated mechanism of acute toxicity.

The data of Knezovich et al. [134] also suggest that A$_{16}$PB has a low bioaccumulation potential, since fat bodies and tissues of toxicological interest (e.g., liver and kidneys) contained only trace amounts of ^{14}C. In fact, whole-body bioconcentration factors (BCFs) for radiolabel from [^{14}C]A$_{16}$PB following a 24-h exposure were only 21, 22, and 13 for clams, minnows, and tadpoles, respectively. Although the exposure time was very short in this study, the results are similar to those of Krzeminski et al. [72] and Lewis and Wee [48]. The latter reported whole-body BCFs for DTDMAC in bluegills of 32 (well water) and 13 (river water) following a 49-day exposure period (Table 3). Such low BCFs are consistent with preferential binding of cationic surfactants by gill tissue.

The general mechanism by which cationic surfactants disrupt membrane structure and function is well known. Cationics bind preferentially to negatively charged membrane surfaces, and the hydrophobe intercalates into the lipid bilayer, with the alkyl groups oriented between the alkyl groups of the phospholipids. Thus, increasing concentrations of the surfactant progressively disrupt the bilayer and ultimately cause it to collapse [101,135]. A variety of effects on membrane-associated processes are possible at concentrations below those required to destroy the bilayer. For example, benzalkonium (dodecyldimethylbenzylammonium) bromide acts as an inhibitory uncoupler in mitochondria isolated from the fungus *Agaricus bisporus* [136] and in aged potato mitochondria [137], uncoupling at low concentrations and inhibiting electron transfer at higher concentrations. Similarly, in internodal cells of the alga *Nitellopsis obtusa*, amphiphilic derivatives of glycine esters having the general formula $(CH_3)_3N^+$ $CH_2COOC_nH_{2n+1}$ Cl$^-$ (n = 10 to 16) inhibit the electrogenic proton pump in the plasmalemma and enhance the passive efflux of Cl$^-$ from the cell [138]. The only study of sublethal effects in fish is that of Sutterlin et al. [139], who found that a variety of QACs blocked the sensory discharge evoked by amino acids in the olfactory epithelium of Atlantic salmon parr, at concentrations of 1.0 mg/L or lower.

J. P. Knezovich (unpublished data) examined the effect of sediment and montmorillonite on A$_{16}$PB bioavailability. The presence of a layer of bottom sediment reduced the amount of ^{14}C in gill tissue of clams and tadpoles relative to sediment-free controls by 36 and 29% respectively, in 7-day exposures. Montmorillonite (100 mg/L) had a much greater effect, reducing gill uptake by 96% in clams and 99% in tadpoles. However, the clay mineral had a much smaller effect on ^{14}C uptake by tadpole intestine (29% reduction), and settled sediment had no effect at all. The data suggest that sorption of cationics reduces bioavailability to gill tissue, the primary site at which acute toxicity is manifested, but not necessarily to the gastrointestinal tract. The effect of sediment on A$_{16}$PB availability to gill tissue thus is consistent with earlier observations of reduced toxicity in the presence of solids [48,123]. Binding of cationics to aquatic humic substances also reduces the rate of uptake and toxicity to fathead minnows [140].

Cationic surfactants are toxic to algae and possibly higher plants as well. Walker and Evans [116] found that A$_{16}$TMAB, A$_{12}$DMBAC, DA$_{10}$DMAC, and A$_{16}$PB inhibited growth of green algae (*Chlorella* sp.) and great duckweed (*Spirodela oligorhiza*) at 3 to 5 mg/L. The data of Lewis and Wee [48] on algal toxicity of DTDMAC are summarized in Table 10. In general, algae seem to be at least as sensitive as fish and invertebrates. An apparent effect of sorption to suspended solids again was observed, but even in water containing solids, algistatic concentrations were below 1 mg/L in some cases. Tubbing and Admiraal [110] studied the effect of DTDMAC on photosynthesis by phytoplankton in Rhine River water. In three of four water samples, EC$_{50}$ values for inhibition of photosynthesis were in the range 3 to 4 mg/L, but for one sample collected during a period of low algal abundance, an EC$_{10}$ value of 0.04 mg/L (EC$_{50}$ = 0.51 mg/L) was reported. Toxicity did not appear to be related to the amount of suspended solids in the water, but the difference in level of suspended solids between the two treatments was only about twofold.

Recent studies by Lewis and co-workers [120–122] have explored the relationship between laboratory and field data on toxicity of cationics to algae. Laboratory-derived EC$_{50}$ values for growth of selected species were compared to effect concentrations for measures of algal community structure [120] and function [121], derived from lake studies in which natural phytoplankton communities were exposed to cationics in situ. In general, laboratory-derived EC$_{50}$ values were many times lower than FOECs for

TABLE 10 Phytotoxicity of DTDMAC to Freshwater and Marine Algae

Alga	Dilution water	Suspended solids (mg/L)	Algistatic concentration[a] (mg/L)
Freshwater			
Selenastrum	Distilled	0	0.23
capricornutum	White River (autoclaved)	3–5	0.71
	Rapid Creek (autoclaved)	131	2.6
	Rapid Creek (filtered)[b]	131	>4.0
	Distilled[c]	0	>0.1, <0.5
Microcystis	Distilled	0	0.32
aeruginosa	White River (autoclaved)	3–5	0.21
	Distilled[c]	0	0.10
Navicula	Distilled	0	>1, ≤10
seminulum	Distilled	0	0.5
Marine			
Dunaliella tertiolecta	Artificial seawater	[d]	>0.5, ≤1.0

[a]Nominal level of active ingredient.
[b]Filtered through a 0.45-μm filter, then 131 mg/L of river sediment (dry weight) added to replace suspended solids removed by filtration.
[c]Compound tested was DTDMAMS, not DTDMAC.
[d]Not reported, but presumably zero.
Source: Ref. 48.

TABLE 11 Algal toxicity of Cationic Surfactants in Laboratory Tests and In Situ

Compound	Lake photosynthesis studies, 3-h EC_{50} (mg/L)		Laboratory growth studies, 96-h EC_{50}[a] (mg/L)		
	Mean	Range	Sc	Ma	Np
DTDMAC	6.4	0.4–31.9	0.06	0.05	0.07
A_{12}TMAC	2.2	0.4–6.1	0.19	0.12	0.20
A_{16}TMAB	0.6	0.1–2.6	0.09	0.03	
Saturated imidazolium compound	1.9	0.3–3.5	0.60	0.45	
Unsaturated imidazolium compound	13.5	11.0–18.4	0.30	0.21	

[a]Laboratory tests were conducted once with each test species. Sc, *Selenastrum capricornutum*; Ma, *Microcystis aeruginosa*; Np, *Navicula pelliculosa*.
Source: Ref. 121.

measures of community structure and function. Selected data are presented in Table 11. Lewis et al. [122] reported similar findings for river periphyton communities.

VI. RESIDUES IN THE ENVIRONMENT

The available data suggest that cationic surfactants are ubiquitous contaminants of sewage and surface waters, at least in populated areas. At the same time, levels are considerably lower than for anionic and nonionic surfactants, as would be expected. Monitoring has been conducted at several wastewater treatment facilities treating primarily domestic sewage, and the results generally confirm the expectation of high removal rates ($\geq 90\%$). Unfortunately, data on residues in aquatic sediments near sites of industrial or domestic effluent discharge are sparse.

Michelsen [55] analyzed random samples of Swiss sewage having various amounts of input from metallurgical processes and/or the textile industry. Levels of cationic substances, obtained by sublation–TLC, ranged from 0.04 to 0.45 mg/L. Huber [44] and Kupfer [45] described monitoring studies in Germany, and Wee [49] determined levels of DTDMAC in untreated sewage and final effluent from a plant in the United States by three analytical methods: DBAS, DBAS–TLC, and HPLC. These data are summarized in Table 12. Kupfer [45] found that levels of cationic surfactants in sewage varied with time of day and day of the week, the highest concentrations being recorded in the evening hours on the "main laundry days" of Monday and Tuesday.

Topping and Waters [46] conducted a detailed monitoring study of cationics in full-scale activated sludge plants in England and Germany. At a plant in Germany, overall removals of DBAS and $DA_{18}DMAC$ were 87.5 and 94.3%, respectively, from initial levels in raw sewage of approximately 1.6 mg/L. Most of the removal—85% for DBAS and 92.5% for $DA_{18}DMAC$—occurred in the biological treatment stage, not in the primary clarifier. At the Alderley Edge plant in England, overall removals were 92.5% for DBAS and 95.7% for $DA_{18}DMAC$. Other data discussed by the authors, from monitoring in Belgium, show that DBAS and $DA_{18}DMAC$ were removed to about the same extent in a trickling filter plant.

Matthijs and de Henau [141] carried out the most comprehensive study to date. Composite samples were collected daily for 8 consecutive days at the Ham plant in Taunton, England and a plant in Heidelsheim, Germany, and analyzed by HPLC for monoalkyl quaternaries. These QACs are used in lower quantities than fabric softening agents such as DTDMAC, so concentrations in sewage and surface water should be lower. The data are summarized in Table 13. Removal was lower in the trickling filter, in

TABLE 12 Cationic Surfactants in Treated and Untreated Sewage in Germany and the United States

Compound	Analytical method[a]	Mean, concentration [µg/L (range)]		Mean overall removal (%)	Comments	Ref.
		Influent sewage	Final effluent			
Total cationic surfactants	Michelsen-TLC	80–600 (50–1300)	10–100 (10–200)	83–88	Continuous sampling of plant near Frankfurt; functioning poorly	44
		200–540 (100–1,000)	10–20 (8–25)	94–96	Same; functioning well	44
		300–600	3–20	94–97	Continuous sampling of plant in Taunus	44
	DBAS and Michelsen-TLC	(800–1,300)			Continuous sampling of Niederrad plant, on Main River near Frankfurt; maximum values for laundry days	45
		(30–100)			Same; minimum values for all days	45
		540	20	96.3	Same; Laundry day	45
		230	12	94.8	Same; non-Laundry day	45
DTDMAC[b]	DBAS	720 (620–830)	65 (46–80)	91.0	Grab sampling at unidentified activated sludge plant	49
DTDMAC	DBAS-TLC	290 (250–330)	31 (22–48)	89.2	Same	49
DTDMAC	HPLC	240 (180–330)	40 (28–56)	83.4	Same	49

[a]See text for details.
[b]Total cationic substances reported as DTDMAC; however, note that DBAS alone cannot discriminate between DTDMAC and other cationics.

TABLE 13 Monoalkyl Quaternaries in Treated and Untreated Sewage in England (Ham Plant) and Germany (Heidelsheim Plant)[a]

| Alkyl group | Ham | | | | | Heidelsheim[b] | | |
| | Activated sludge | | | Trickling filter | | | | |
	Influent (μg/L)	Effluent (μg/L)	Mean Percent removal	Effluent (μg/L)	Mean Percent removal	Influent (μg/L)	Effluent (μg/L)	Mean percent removal
C_{12}	49	1.3	97.3	7.2	85.3	<1[c]	<0.5	—
C_{14}	33	1.1	96.7	6.3	80.9	<1	<0.5	—
C_{16}	25	1.0	96.0	5.7	77.2	26	1.1	95
C_{18}	26	1.1	95.8	6.6	74.6	50	2.2	95

[a]Daily composite samples at both locations (8 days at Ham, 7 days at Heidelsheim).
[b]Activated sludge plant.
[c]There were no C_{12} to C_{14} products on the German detergent market at the time of sampling.
Source: Ref. 141.

contrast to the results cited above: total removal of monoalkyl quaternaries at the Ham plant averaged 97% in activated sludge but only 80% in the trickling filter. A mass balance study was also performed at the Lüdinghausen plant in Germany, with results similar to those of Topping and Waters [46].

Levels of cationic surfactants in receiving waters are typically in the low microgram per liter range. Huber [44] reported concentrations of 5 to 20 µg/L (Michelsen–TLC analysis) for cationics in the Main River near Frankfurt, West Germany. Kupfer [45] later confirmed these figures. Schneider and Levsen [142] used field desorption in combination with collisionally activated decomposition in a tandem mass spectrometer to determine levels of $DA_{18}DMAC$ in water. The concentrations of $DA_{18}DMAC$ in sewage and surface water samples collected near Bonn, West Germany were 0.35 to 0.48 mg/L and 6 to 12 µg/L, respectively. The most recent European study is that of Matthijs and de Henau [141], who focused on C_{12} to C_{18} monoalkyl quaternaries. Eighty grab samples were collected from rivers in Germany, France, and England. German rivers had the highest levels of these QACs, with typical concentrations in the range 3 to 4 µg/L. Over 70% of the samples from France and England had levels below 1 µg/L.

Wee and Kennedy [47] collected random samples from several rivers in the United States and reported concentrations (HPLC analysis) of 5 to 30 µg/L for DTDMAC and <2 µg/L for $A_{16}PC$, $A_{12}TMAC$, and $A_{18}DMBAC$. Lewis and Wee [48] subsequently conducted a more extensive study, in which grab samples were collected at various distances downstream from wastewater treatment facilities. Mean DTDMAC levels (DBAS–HPLC method) were <2, 24, 17, and 33 µg/L for Millers River (MA), Otter River (MA), Blackstone River (MA), and Rapid Creek (SD), respectively. Wee [49] compared the results of analyses for DTDMAC in river water by three methods. Levels of cationic substances decreased with distance below a sewage treatment plant regardless of analytical method but were much higher with the relatively nonspecific DBAS method than with DBAS–TLC or HPLC. Apparent concentrations of DTDMAC in grab samples collected at distances from 4.4 to 55 miles downstream from the plant ranged from 191 to 100, 77 to 50, and 37 to 13 µg/L by DBAS, DBAS–TLC, and HPLC, respectively.

Relatively few data are available on cationic surfactants in aquatic sediments. Lewis and Wee [48] obtained sediment samples from Rapid Creek at distances from 0.8 to 88 km downstream from a sewage outfall. DTDMAC levels averaged 23 mg/kg over 18 samples (range = <3 to 67 mg/kg). More recently, Fernandez et al. [143] found that DTDMAC was a ubiquitous contaminant in coastal sediments collected near Barcelona, Spain (DTDMAC levels were not quantified), and Utsunomiya et al. [50]

determined levels of cationic substances in river water and sediment samples from Japan. In the latter study, samples were analyzed for total cationic substances by Orange II colorimetry in conjunction with ion-exchange chromatography. Levels of total cationic substances in influent sewage, river sediment, and overlying water were 0.10 to 0.15 mg/L (five samples), 6.2 to 69 μg/g (four samples), and 0.05 mg/L (one sample), respectively. The elevated levels reported for sediments relative to surface waters are of course expected, given the sorptive tendencies of these substances.

In another study, Federle and Pastwa [100] reported that DTDMAC was present (HPLC method) at 0.63 mg/L in surface water and 9.7 (0 to 0.6 m depth) and 7.4 (0.6 to 1.2 m depth) mg/kg dry weight in subsurface sediments, at a pond that had been receiving untreated wastewater from a laundromat since 1962. This sampling location is probably not representative of the majority of sites where effluent discharges containing cationics are expected, however. Cationic surfactants were also present in extracts of British drinking waters derived from river water and groundwater [144]. Field desorption mass spectrometry of HPLC fractions identified a series of dialkyl quaternaries ranging from DA_9DMAC to $DA_{19}DMAC$, with the dominant species being the dioctadecyldimethylammonium ion (principal component of DTDMAC).

VII. SUMMARY AND CONCLUSIONS

Consumption of cationic surfactants was estimated to be 190,000 and 150,000 metric tons in 1987 in the United States and Western Europe, respectively ([148]; Table 3). It is likely that most or at least a large portion of this was sewered. Thus cationics represent a major class of potential environmental contaminants. This was recognized by the U.S. Interagency Testing Committee in its 22nd report to EPA [145], which recommended that IQAMS and EEQAMS undergo testing for chemical fate and health and ecological effects, based on expected widespread release and exposure.

As a class, cationics sorb strongly and rapidly to suspended solids in well-mixed test systems and form 1:1 complexes with anionic surfactants. Most cationics should also be readily biodegraded in acclimated biological treatment systems, although ultimate degradation may not always be extensive. Overall, removal in wastewater treatment should generally exceed 90%. In receiving waters, mono- and dialkyl quaternaries should be biodegraded quite rapidly in the water column, with half-lives for ultimate degradation in the range of several days or less. However, the aquatic fate of alkyldimethylbenzyl and alkylpyridinium QACs, and other fabric-softening agents such as IQAMS and EEQAMS, is not as well characterized. All QACs should enter receiving waters largely as complexes with anionic surfactants, sorbed to sludge solids. Cationics that are not rapidly biodegraded are expected to partition to sediments following release to surface waters, but their fate there is not as well characterized as their fate in overlying water.

Whereas the germicidal activity of many cationics is well established, these chemicals will not pose a significant risk to wastewater treatment systems under normal conditions. But slug doses could upset plant function. Cationic surfactants are also toxic in the milligram per liter range and lower to aquatic organisms, including fish, invertebrates, and algae. Acute toxicity is reasonably well characterized for mono- and dialkyl quaternaries but not for other major classes. Data on chronic toxicity are also limited. Results from studies on tissue uptake and distribution are consistent with the anticipated mechanism of action (interference with gill function). Sorption to sediment apparently reduces acute toxicity and bioavailability to gill tissue, but bioavailability in the gut warrants further study.

In 1982, Huber [60] identified several areas in which definitive studies were needed. Among these were (1) biodegradability and behavior in anaerobic digestors and benthic sediments; (2) chronic ecotoxicity; (3) physical/chemical properties and behavior of "electroneutral" surfactant complexes; and (4) development of better analytical methods, with the aim of improving analysis of environmental samples. Through research efforts mainly in the last 10 years, much has been learned about the toxicity of cationic surfactants and their behavior in wastewater treatment and the environment. Yet definitive studies are still lacking in several areas. Of particular interest is the behavior of cationic surfactants in sediments. This topic has received some attention [97,98], but more work is needed, especially on anaerobic biodegradation.

REFERENCES

1. L. M. Games, J. E. King, and R. J. Larson, *Environ. Sci. Technol. 16*: 483 (1982).
2. R. J. Larson and R. D. Vashon, *Dev. Ind. Microbiol. 24*: 425 (1983).
3. C. T. Cowan and D. White, *Trans. Faraday Soc. 54*: 691 (1958).
4. J. P. Law, Jr., and G. W. Kunze, *Soil Sci. Soc. Am. Proc. 30*: 321 (1966).
5. B. J. Brownawell, H. Chen, J. M. Collier, and J. C. Westall, *Environ. Sci. Technol. 24*: 1234 (1990).
6. V. C. Hand, R. A. Rapaport, and R. H. Wendt, *Environ. Toxicol. Chem. 9*: 467 (1990).
7. T. Mangialardi and A. E. Paolini, *Water Air Soil Pollut. 53*: 139 (1990).
8. L. W. Smith and D. E. Bayer, *Soil Sci. 103*: 328 (1967).
9. M. M. Mortland, S. Shaobai, and S. A. Boyd, *Clays Clay Miner. 34*: 581 (1986).

10. S. A. Boyd, S. Shaobai, J.-F. Lee, and M. M. Mortland, *Clays Clay Miner.* 36: 125 (1988).
11. S. A. Boyd, J.-F. Lee, and M. M. Mortland, *Nature 333*: 345 (1988).
12. S. A. Boyd, M. M. Mortland, and C. T. Chiou, *Soil Sci. Soc. Am. J.* 52: 652 (1988).
13. D. C. Bouchard, R. M. Powell, and D. A. Clark, *J. Environ. Sci. Health A23*: 585 (1988).
14. J.-F. Lee, J. R. Crum, and S. A. Boyd, *Environ. Sci. Technol.* 23: 1365 (1989).
15. J. A. Smith, P. R. Jaffé, and C. T. Chiou, *Environ. Sci. Technol.* 24: 1167 (1990).
16. J. A. Smith and P. R. Jaffé, *Environ. Sci. Technol.* 25: 2054 (1991).
17. J. B. Weber and H. D. Coble, *J. Agric. Food Chem.* 16: 475 (1968).
18. R. D. Barbaro and J. V. Hunter, *Proc. Ind. Waste Conf.* 20: 189 (1965).
19. V. N. Moraru, S. A. Markova, and Ovcharenko, *Ukr. Khim. Zh.* 47: 1058 (1981).
20. D. W. Fürstenau, *J. Phys. Chem.* 60: 981 (1956).
21. D. W. Fürstenau and H. J. Modi, *J. Electrochem. Soc.* 106: 336 (1959).
22. H. Rupprecht, E. Ullmann, and K. Thoma, *Fortschrittsber. Koll. Polym.* 55: 45 (1971).
23. B. H. Bijsterbosch, *J. Colloid Interface Sci.* 47: 186 (1974).
24. L. D. Skrylev and E. A. Streltsova, *Zh. Prikl. Khim.* 52: 1493 (1979).
25. S. Takeda and S. Usui, *Coll. Surf.* 23: 15 (1987).
26. R. Schwarz, K. Heckmann, and J. Strnad, *Colloid Interface Sci.* 124: 50 (1988).
27. T. W. Federle and R. M. Ventullo, *Appl. Environ. Microbiol.* 56: 333 (1990).
28. F. W. Putnam, *Adv. Protein Chem.* 4: 79 (1948).
29. W. E. Knox, V. H. Auerbach, K. Zarudnaya, and M. Spirtes, *J. Bacteriol.* 58: 443 (1949).
30. I. M. Klotz, *Sci. Mon.* 70: 24 (1950).
31. C. E. Chaplin, *Can. J. Bot.* 29: 373 (1951).
32. H. Mueller and E. Krempl, *Fette Seifen Anstrichm.* 65: 532 (1963).
33. A. May and A. Neufahrt, *Tenside Deterg.* 13: 65 (1976).
34. K. McQuillen, *Biochim. Biophys. Acta 5*: 463 (1950).
35. M. R. J. Salton, *J. Gen. Microbiol.* 5: 391 (1951).
36. T. Fugita and S. Koga, *J. Gen. Appl. Microbiol.* 12: 229 (1966).
37. W. G. Salt and D. Wiseman, *J. Pharmacol. (Suppl.)* 20: 14S (1968).
38. J. A. Mackrell and J. R. L. Walker, *Int. Biodeterior. Bull.* 14: 77 (1978).
39. M. E. Ginn, in *Cationic Surfactants*, Vol. 4 (E. Jungermann, ed.), Marcel Dekker, New York, 1970, p. 341.
40. R. D. Swisher, *Surfactant Biodegradation*, Marcel Dekker, New York, 1970.
41. A. Neufahrt, K. Lötzsch, D. Pleschke, and G. Spaar, *Comun. J. Com. Esp. Deterg.* 14: 9 (1983).
42. R. V. Scowen and J. Leja, *Can. J. Chem.* 45: 2821 (1967).
43. SRI, *Chemical Economics Handbook*, SRI International, Menlo Park, Calif., 1981.

44. L. Huber, *Münch. Beitr. Abwasser Fisch. Fluss biol. 31*: 203 (1979).
45. W. Kupfer, *Tenside Deterg. 19*: 158 (1982).
46. B. W. Topping and J. Waters, *Tenside Deterg. 19*: 164 (1982).
47. V. T. Wee and J. M. Kennedy, *Anal. Chem. 54*: 1631 (1982).
48. M. A. Lewis and V. T. Wee, *Environ. Toxicol. Chem. 2*: 105 (1983).
49. V. T. Wee, *Water Res. 18*: 223 (1984).
50. A. Utsunomiya, S. Naitou, and I. Tomita, *Eisei Kagaku 35*: 152 (1989).
51. R. A. Llenado and R. A. Jamieson, *Anal. Chem. 53*: 174R (1981).
52. J. Waters and W. Kupfer, *Anal. Chim. Acta 85*: 241 (1976).
53. R. B. Cain, in *Treatment of Industrial Effluents* (A. G. Callely, C. F. Forster, and D. A. Stafford, eds.), Hodder & Stoughton, London, 1977, p. 283.
54. Q. W. Osburn, *J. Am. Oil Chem. Soc. 59*: 453 (1982).
55. E. R. Michelsen, *Tenside Deterg. 15*: 169 (1978).
56. R. Reck, in *Kirk-Othmer Encyclopedia of Chemical Technology* (M. Grayson, ed.), 3rd ed., Vol. 19, Wiley, New York, 1982, p. 521.
57. B. F. Greek, *Chem. Eng. News*, p. 21 (Jan. 25, 1988).
58. USEPA, *Needs Survey. Conveyance and Treatment of Municipal Wastewater. Summaries of Technical Data*, EPA-430/9-79-002, U.S. Environmental Protection Agency, Washington, D.C., 1979.
59. P. Gerike, W. K. Fischer, and W. Jasiak, *Water Res. 12*: 1117 (1978).
60. L. Huber, *Tenside Deterg. 19*: 178 (1982).
61. Procter & Gamble, unpublished information submitted to A. Stern (USEPA) by T. Mooney (The Procter & Gamble Co.), Sept. 12, 1984.
62. D. M. Woltering and W. E. Bishop, in *The Risk Assessment of Environmental and Human Health Hazards: A Textbook of Case Studies* (D. J. Paustenbach, ed.), Wiley, New York, 1989, p. 345.
63. P. Pitter, *Sb. Vys. Sk. Chem. Technol. Praze Technol. Vody 5*: 25 (1961).
64. A. Uhl and H. Sedlmayer, *Brauwelt 105*: 1529 (1965).
65. O. Pauli and G. Franke, *Gesundheitswes. Desinfekt. 63*: 150 (1971).
66. L. Barden and P. C. G. Isaac, *Proc. Inst. Civil Eng. 6*: 371 (1957).
67. P. Gerike, *Tenside Deterg. 19*: 162 (1982).
68. P. Wierich and P. Gerike, *Ecotoxicol. Environ. Saf. 5*: 161 (1981).
69. W. Janicke and G. Hilge, *Tenside Deterg. 16*: 117 (1979).
70. P. Gerike and W. K. Fischer, *Ecotoxicol. Environ. Saf. 3*: 159 (1979).
71. P. Gerike and P. Gode, *Chemosphere 21*: 799 (1990).
72. S. F. Krzeminski, J. J. Martin, and C. K. Brackett, *Household Personal Prod. Ind. 10*: 22 (1973).
73. D. E. Sullivan, *Water Res. 17*: 1145 (1983).
74. J. Ruiz Cruz, *Grasas Aceites 38*: 383 (1987).
75. W. D. Sheets and G. W. Malaney, *Sewage Ind. Wastes 28*: 10 (1956).
76. R. L. Huddleston and E. A. Setzkorn, *Soap Chem. Spec.*, p. 63 (Mar. 1965).
77. L. J. Gawel and R. D. Huddleston, *American Oil Chemists' Society National Meeting*, Los Angeles, Apr. 23–26, 1972.
78. D. Dean-Raymond and M. Alexander, *Appl. Environ. Microbiol. 33*: 1037 (1977).

79. F. Masuda, S. Machida, and M. Kanno, in *Proceedings of the 7th International Congress on Surface-Active Agents*, Moscow, 1978, p. 129.
80. R. J. Larson, *Appl. Environ. Microbiol. 38*: 1153 (1979).
81. K. Miura, K. Yamanaka, T. Sangai, K. Yoshimura, and N. Hayashi, *Yukagaku 28*: 351 (1979).
82. R. J. Larson and G. E. Wentler, *Soap Cosmet. Chem. Spec. 58*: 33 (1982).
83. V. Čupková, F. Devinsky, L. Sirotková, and I. Lacko, *Pharmazie 44*: 577 (1989).
84. C. G. van Ginkel and M. Kolvenbach, *Chemosphere 23*: 281 (1991).
85. R. J. Larson, *Residue Rev. 85*: 159 (1983).
86. V. T. Kaplin, L. S. Zernova, and A. S. Kosogova, *Gidrokhim. Mater. 44*: 196 (1968).
87. B. Baleux and P. Caumette, *Water Res. 11*: 833 (1977).
88. J. Ruiz Cruz and M. C. Dobarganes Garcia, *Grasas Aceites 30*: 67 (1979).
89. R. J. Larson and R. L. Perry, *Water Res. 15*: 697 (1981).
90. J. Vives-Rego, M. D. Vaque, J. Sanchez Leal, and J. Parra, *Tenside Deterg. 24*: 20 (1987).
91. J. Ruiz Cruz, *Grasas Aceites 30*: 293 (1979).
92. J. Ruiz Cruz, *Grasas Aceites 32*: 147 (1981).
93. R. M. Ventullo and R. J. Larson, *Appl. Environ. Microbiol. 51*: 356 (1986).
94. R. J. Larson and W. E. Bishop, *Soap Cosmet. Chem. Spec. 64*: 58 (1988).
95. R. J. Shimp, B. S. Schwab, and R. J. Larson, *Environ. Toxicol. Chem. 8*: 723 (1989).
96. R. J. Shimp and B. S. Schwab, *Environ. Toxicol. Chem. 10*: 159 (1991).
97. R. J. Shimp and R. L. Young, *Ecotoxicol. Environ. Saf. 15*: 31 (1988).
98. R. J. Shimp, *Environ. Toxicol. Chem. 8*: 201 (1989).
99. R. J. Larson, *Household Personal Prod. Ind. 21*: 55 (1984).
100. T. W. Federle and G. M. Pastwa, *Ground Water 26*: 761 (1988).
101. C. A. Lawrence, in *Cationic Surfactants* (E. Jungermann, ed.), Marcel Dekker, New York, 1970, p. 491.
102. N. Sayama, *Nippon Eiseigaku Zasshi 35*: 869 (1981).
103. G. Van Beneden, *Bull. CEBEDEAU 17*: 159 (1952).
104. R. Manganelli and E. S. Crosby, *Sewage Ind. Wastes 25*: 262 (1953).
105. L. Reynolds, J. Blok, A. de Morsier, P. Gerike, H. Wellens, and W. J. Bontinck, *Chemosphere 16*: 2259 (1987).
106. R. Manganelli, *Sewage Ind. Wastes 24*: 1057 (1952).
107. B. H. Fenger, M. Mandrup, G. Rohde, and J. C. Kjaer Sorensen, *Water Res. 7*: 1195 (1973).
108. R. J. Larson and S. L. Schaeffer, *Water Res. 16*: 675 (1982).
109. P. Pitter and M. Horska, *Sb. Vys. Sk. Chem. Technol. Praze Technol. Vody F14*: 19 (1968).
110. D. M. J. Tubbing and W. Admiraal, *Appl. Environ. Microbiol. 57*: 3616 (1991).
111. C. H. Taft, *Tex. Rep. Biol. Med. 4*: 27 (1946).
112. S. B. Pessoa, *Folia Clin. Biol. 18*: 137 (1952).

113. A. Vallejo-Freire, O. F. Ribeiro, and I. F. Ribeiro, *Science 119*: 470 (1954).
114. W. Knauf, *Tenside Deterg. 10*: 251 (1973).
115. W. K. Fischer and P. Gode, *Vom Wasser 48*: 247 (1977).
116. J. R. L. Walker and S. Evans, *Marine Pollut. Bull. 9*: 136 (1978).
117. B. Reiff, R. Lloyd, M. J. How, D. Brown, and J. S. Alabaster, *Water Res. 13*: 207 (1979).
118. T. U. Kappeler, *Tenside Deterg. 19*: 169 (1982).
119. J. Waters, *Tenside Deterg. 19*: 177 (1982).
120. M. A. Lewis, *Environ. Toxicol. Chem. 5*: 319 (1986).
121. M. A. Lewis and B. G. Hamm, *Water Res. 20*: 1575 (1986).
122. M. A. Lewis, M. Taylor, and R. Larson, in *Multispecies Toxicity Testing* (J. Cairns, ed.), ASTM STP 920, American Society for Testing and Materials, Philadelphia, 1986, p. 241.
123. D. M. Woltering, R. J. Larson, W. D. Hopping, R. A. Jamieson, and N. T. de Oude, *Tenside Deterg. 24*: 286 (1987).
124. J. C. Cooper, *Ecotoxicol. Environ. Saf. 16*: 65 (1988).
125. R. Marchetti, *Riv. Ital. Sostanze Grasse 41*: 553 (1964).
126. C. A. Pittinger, D. M. Woltering, and J. A. Masters, *Environ. Toxicol. Chem. 8*: 1023 (1989).
127. J. A. Masters, M. A. Lewis, D. H. Davidson, and R. D. Bruce, *Environ. Toxicol. Chem. 10*: 47 (1991).
128. H. Hidu, *J. Water Pollut. Control Fed. 37*: 262 (1965).
129. C. M. Lee, *Tenside Deterg. 23*: 196 (1986).
130. M. A. Lewis, *Ecotoxicol. Environ. Saf. 20*: 123 (1990).
131. M. A. Lewis, *Water Res. 25*: 101 (1991).
132. K. E. Biesinger and G. N. Stokes, *J. Water Pollut. Control Fed. 58*: 207 (1968).
133. A. Neufahrt, H. G. Eckert, H. M. Kellner, and K. Lötzsch, *CED Congress*, Madrid, Mar. 1978.
134. J. P. Knezovich, M. P. Lawton, and L. S. Inouye, *Bull. Environ. Contam. Toxicol. 42*: 87 (1989).
135. B. Isomaa, *Ecol. Bull. 36*: 26 (1984).
136. S. Steffann, J.-P. Joly, B. Loubinoux, and P. Dizengremel, *Pestic. Biochem. Physiol. 32*: 38 (1988).
137. J.-C. Pireaux and P. Dizengremel, *J. Plant Physiol. 136*: 349 (1990).
138. Z. Trela, T. Janas, S. Witek, and S. Przestalski, *Physiol. Plant. 78*: 57 (1990).
139. A. Sutterlin, N. Sutterlin, and S. Rand, *Technical Report 287*, Fisheries Research Board of Canada, St. Andrews, New Brunswick, Canada, 1971, pp. 1–8.
140. V. J. Versteeg, *Abstr. Pap. Am. Chem. Soc. 199*: 178 (1990).
141. E. Matthijs and H. de Henau, *Vom Wasser 69*: 73 (1987).
142. E. Schneider and K. Levsen, *Comm. Eur. Commun., EUR 10388 Org. Micropollut. Aquatic Environ.*, 1986, p. 14.
143. P. Fernandez, M. Valls, J. M. Bayona, and J. Albalgés, *Environ. Sci. Technol. 25*: 547 (1991).

144. B. Crathorne, M. Fielding, C. P. Steel, and C. D. Watts, *Environ. Sci. Technol. 18*: 797 (1984).
145. Interagency Testing Committee (ITC), *Fed. Regist. 53*: 18196 (1988).
146. H. Kunieda and K. Shinoda, *J. Phys. Chem. 82*: 170 (1978).
147. C. A. Pittinger, V. C. Hand, J. A. Masters, and L. F. Davidson, in *Aquatic Toxicology and Hazard Assessment* (W. J. Adams, G. A. Chapman, and W. G. Landis, eds.), Vol. 10, ASTM STP 971, American Society for Testing and Materials, Philadelphia, 1988, p. 138.
148. I. J. I. Roes and S. de Groot, in *Proceedings of the 2nd World Surfactants Congress, "Surfactants in Our World—Today and Tomorrow,"* Paris, May 24–27, 1988.
149. W. Guhl and P. Gode, *Vom Wasser 72*: 165 (1989).

III

Analytical Evaluation

JOHN CROSS Faculty of Sciences, University of Southern Queensland, Toowoomba, Queensland, Australia

The analytical evaluation section of this book is devoted almost entirely to a detailed examination of cationic surfactants themselves rather than to the somewhat complex mixtures that are often marketed under that heading. There are many parameters in the quality and process control of quats and of the materials from which they are synthesized that are measured routinely. These include such determinations as water content; total, primary, secondary, and tertiary amine content; acid or amine value (i.e., the amount of excess acid or amine that may be present); ash content; iodine value (i.e., unsaturation in fatty chains); nonvolatile matter; and pH.

There are standard test methods available for each of the parameters above, and these are constantly under review. Perhaps the most comprehensive collection is to be found in the *Annual Book of ASTM Standards*, published in the United States by the American Society for Testing and Materials, 1916 Race Street, Philadelphia, PA 19103. Many of them, however, are to be found in a companion volume to this book, *Analysis of Surfactants*, Thomas Schmitt, 1992 (Volume 40 of Marcel Dekker's Surfactant Science Series). Readers with an interest in surfactant analysis in general will find in it much useful complementary material to that presented in the more specialized volumes in the series.

5

Volumetric Analysis of Cationic Surfactants

JOHN CROSS Faculty of Sciences, University of Southern Queensland, Toowoomba, Queensland, Australia

I. INTRODUCTION

For routine determination of a wide range of materials, the traditional combination of burette and pipette has withstood the test of time and the relentless competition of a wide range of instrumental techniques: ionic surfactants provide no exception.

By way of summary, the volumetric analysis of cationic surfactants has largely evolved alongside the development of procedures for determination of the more common anionic counterparts, in which the well-known "antagonistic" relationship between these two types:

$$R_4N^+_{(aq)} + R'SO^-_{3(aq)} \longrightarrow R_4NR'SO_{3(s)} \quad \text{or} \quad R_4NR'SO_{3(org)}$$

has been used to full advantage. The reaction between these two species is followed by detection of either the disappearance of the last trace of titrand or the appearance of the first trace of excess titrant.

Many other ions have been shown to react quantitatively with quats, the most popular of these being the tetraphenylborate ion. Originally marketed as Kalignost, sodium tetraphenylborate (STPB) was introduced as a welcome alternative to perchloric acid for the gravimetric determination of potassium. Since ammonium and potassium ions are readily exchangeable in many crystalline structures, the reaction of STPB with ammonium and then other amines were soon investigated. Natural product chemists found it to be an ideal reagent for precipitating alkaloids from solution. Schall [1] was one of the first to suggest its application to the analysis of cationic surfactants.

Although these two titrants have dominated the scene for some decades, the *style* of titration has varied: a two-phase system involving acid-base indicators has given way to single-phase potentiometric techniques, although most standard methods continue to be based on the former.

One factor that may surprise even experienced analytical chemists meeting surfactant titrations for the first time is the extreme sensitivity of these methods. The majority of traditional volumetric analyses involve solutions with concentrations around $0.1 \ M$ for acid-base reactions: in some areas, such as water quality work, more dilute solutions are common [e.g., 0.01 M ethylenediaminetetraacetic acid (EDTA) for hardness and even 0.0125 N sodium thiosulfate for dissolved oxygen]. Solution concentrations for surfactant titrations, however, vary typically from $10^{-2} \ M$ down to a remarkable $10^{-6} \ M$, far below that found in any other common types of titration. Indeed, attempts to titrate $0.1 \ M$ surfactant solutions would be thwarted by foams and emulsions.

One essential requirement for the development of any analytical method is ready access to a supply of reliable primary standards. In surfactant analysis this has been a particular problem, due to the mixture of homologs that comprise the fatty chain when derived from natural sources. By way of example, a commercial sample of benzalkonium (alkyldimethylbenzyl-ammonium) chloride was shown to possess a $C_{12,14,16,18}$ alkyl distribution of 39.3, 46.45, 11.2, and 3.0%, respectively [2]. Particular credence, therefore, should be paid to work in which the purity of the materials was carefully controlled. Of particular current use are "specially purified" sodium lauryl sulfate (SLS) (1) in which the fatty alcohol has been carefully fractionated prior to sulfation to ensure that only the C_{12} component remains, and p(diisobutylphenoxyethoxyethyl)dimethylbenzylammonium chloride monohydrate (2) (mercifully more commonly known as benzethonium chloride or Hyamine 1622), which is entirely synthetic. Most practitioners regard both of these materials as primary standards.

$C_{12}H_{25}OSO_3^- \; Na^+$ C_8H_{17}⟨benzene ring⟩$OCH_2CH_2OCH_2CH_2\overset{\oplus}{\underset{CH_3}{\overset{CH_3}{N}}}CH_2$⟨benzene ring⟩ $Cl^- \cdot H_2O$

(1)   (2)

The various options available are examined in some detail in the ensuing pages. Tackling the task on a chronological basis, the cationic versus anionic systems using visual indicators are treated first. It is appropriate to begin with a few observations on the indicators.

II. THE SURFACTANT ION: INDICATOR REACTION

It has been known for many years that addition of surfactants to acid-base indicators of opposing ionic charge gives rise to color changes [3]. Mostly, these changes occur at surfactant concentrations orders of magnitude less than the critical micelle concentration. For example, Hiskey and Downey reported that the normal adsorption maxima of the acidic and basic forms of methyl orange (507 and 426 nm, respectively) were replaced by a new maximum (463 nm) that was stable over an extremely wide pH range (0 to 12) and for cationic surfactant concentrations as low as $3 \times 10^{-6} M$ [4].

These colored complexes can conveniently be regarded as stoichiometrically composed salts formed from the surfactant and indicator ions. For example, Mukerjee and Mysels isolated two compounds from the reaction between cetyltrimethylammonium (R_4N) bromide and bromophenol blue (BPB); elemental analysis of these corresponded very closely with that required for R_4NBPB and $(R_4N)_2BPB$, from acidic and alkaline solutions, respectively [5].

Such compounds are sparingly soluble in water, although they may be solubilized by becoming incorporated into the micelle of excess surfactant. They are, however, appreciably soluble in organic solvents such as di-chloroethane and chloroform. This fact forms the basis of colorimetric methods for the estimation of trace quantities of surfactants. Early versions (to 1967) of such reactions as they pertain to cationic surfactants were reviewed in Volume 4 of this series [6], and updated comments are presented elsewhere in this book.

Historically, anionic surfactants have always been of greater significance and manufactured in far greater quantities than have their cationic counterparts. It is not surprising, therefore, that emphasis on the well-known antagonistic reaction between the two types should have been directed toward the determination of anionic types.

The first reported titrimetric application of the cationic versus anionic reaction goes back more than 50 years. Hartley and Runnicles estimated cetylpyridinium bromide by titration with 0.001 M solutions of sodium cetanesulfonate or sodium cetyl sulfate until the royal blue color of the cationic–bromophenol blue complex was displaced to liberate the normal purple (basic) form of the indicator [3]. Some 10 years later, Epton reported a two-phase procedure for the determination of anionic surfactants containing eight or more carbon atoms in the alkyl chain [7]. Specifically, he used cetylpyridinium chloride as a titrant, chloroform as the second phase, and methylene blue as the indicator: the endpoint was taken empirically as the point at which the latter was equally distributed between the two phases. To promote good mass transfer between the chloroform and aqueous phases, vigorous shaking in a stoppered bottle or separatory funnel was required rather than the gentle continuous swirling associated with single-phase titrations. Almost simultaneously, Barr et al. published details of two procedures [8]: the first was similar to Epton's method except that the titration was continued until all of the blue color of the methylene blue–surfactant anion complex had been removed from the organic phase; and the second was a two-phase version of the Hartley and Runnicles method, in which the endpoint was taken as the first appearance in the organic phase of the blue color of the complex formed between excess cationic titrant and the indicator anion. The latter was preferred since it gave sharper endpoints and more accurate results.

An exhaustive and informative comparison of these three procedures was carried out by Lincoln and Chinnick using a variety of surfactants whose purity had been carefully controlled and checked [9,10]. The results, reproduced in Tables 1 to 3, show that the bromophenol blue method best approaches true stoichiometry provided that both alkyl chains contain 12 or more carbon atoms. More significantly, they show that choice of indi-

TABLE 1 Apparent Purities of Cationic Surfactants by the Barr, Oliver, and Stubbings Procedure (Bromophenol Blue Indicator)

Quaternary ammonium salt	Anionic surfactant			
	Sodium 2-ethylhexylsulfate	Sodium dodecylsulfate	Sodium tetradecane-2-sulfate	Sodium dioctylsulfosuccinate
Decyltrimethylammonium bromide	22.8	96.7	97.2	96.8
Dodecyltrimethylammonium bromide	22.1	99.5	99.5	99.6
Tetradecyltrimethylammonium bromide	23.2	101.2	100.2	101.2
Hexadecyltrimethylammonium bromide	23.8	100.2	99.8	100.4
Decylpyridinium bromide	26.0	97.0	96.4	96.4
Dodecylpyridinium chloride	25.6	100.1	99.7	100.8
Tetradecylpyridinium chloride	27.8	102.3	101.7	102.0
Hexadecylpyridinium chloride	26.6	102.1	101.5	101.8
(p-tert-Octylphenoxyethoxyethyl) dimethylbenzylammonium bromide		101.3	100.4	100.7
Didecyldimethylammonium bromide		99.7	99.2	100.9

Source: Abstracted from Refs. 9 and 10.

TABLE 2 Apparent Purities of Cationic Surfactants by the Epton Procedure (Methylene Blue Indicator)

	Anionic surfactant			
Quaternary ammonium salt	Sodium 2-ethylhexylsulfate	Sodium dodecylsulfate	Sodium tetradecane-2-sulfate	Sodium dioctylsulfosuccinate
Decyltrimethylammonium bromide	a	a	a	a
Dodecyltrimethylammonium bromide	93.7	90.4	86.9	94.6
Tetradecyltrimethylammonium bromide	98.4	94.8	91.6	100.5
Hexadecyltrimethylammonium bromide	101.2	97.4	93.8	105.2
Decylpyridinium bromide	a	a	a	a
Dodecylpyridinium chloride	100.1	96.2	93.1	100.1
Tetradecylpyridinium chloride	104.7	100.0	97.2	107.3
Hexadecylpyridinium chloride	108.7	106.4	102.6	114.3
(p-tert-Octylphenoxyethoxyethyl) dimethylbenzylammonium chloride		100.8	98.3	105.6
Didecyldimethylammonium bromide		97.8	95.7	105.7

aNo endpoint was observed even when two equivalents of quaternary ammonium salt had been added.

Source: Abstracted from Refs. 9 and 10.

TABLE 3 Apparent Purities of Cationic Surfactants by Hartley and Runnicles' Procedure (Dichlorofluorescein Indicator)[a]

	Anionic surfactant		
Quaternary ammonium salt	Sodium 2-ethylhexylsulfate	Sodium dodecylsulfate	Sodium dodecane sulfonate
Decyltrimethylammonium bromide	b	b	b
Dodecyltrimethylammonium bromide	b	46.5	48.5
Tetradecyltrimethylammonium bromide	b	97.4	97.7
Hexadecyltrimethylammonium bromide	b	98.1	98.3
Decylpyridinium bromide	b	b	b
Dodecylpyridinium chloride	b	96.2	96.8
Tetradecylpyridinium chloride	b	99.7	99.1
Hexadecylpyridinium chloride	b	99.8	99.1

[a]Dichlorofluorescein was substituted for bromophenol blue.
[b]No endpoint was observed even when two equivalents of quaternary ammonium salt had been added.
Source: Abstracted from Ref. 10.

cator is of the utmost importance and that the length of the alkyl chain is critical.

III. TITRATION WITH ANIONIC SURFACTANTS/ VISUAL INDICATORS

A. Methylene Blue

It was the methylene blue method of Epton that was to become the standard method for many years. Methylene blue (MB) (3) is one of the relatively few common indicators in which the large organic ion responsible for the color is cationic in nature. Its intense royal blue color fades upon reduction to a very pale yellow, enabling it to function as a redox indicator for the

(3)

volumetric determination of reducing sugars and dissolved oxygen, for example. It also shows a tendency to bond to negatively charged sites, making it a useful stain for microscopy. In the presence of anionic surfactants (An^-) it forms a 1:1 salt that can be extracted into an organic solvent:

$$An^-_{(aq)} + MB^+_{(aq)} \longrightarrow An^-MB^+_{(org)} \qquad (1)$$

This reaction forms the basis of the colorimetric estimation of what are called *methylene blue–active substances* (nominally anionic surfactants) in wastewaters and the like.

The basic method for titrating anionic surfactants using this indicator involves addition of MB and sulfuric acid (to prevent interference from soaps) to specified volumes of surfactant and chloroform. After shaking, the majority of the blue color is to be found in the organic layer as a result of reaction (1). As a standard solution of cationic surfactant (Cat^+) is added from the burette, the two antagonistic ions mutually react and transfer to the organic layer upon shaking.

$$A^-_{(aq)} + Cat^+_{(aq)} \longrightarrow An^-Cat^+_{(org)} \qquad (2)$$

At the endpoint, with the supply of free An^-MB^+ and the organic layer begins to decolorize as the liberated MB^+ transfers to the aqueous layer.

$$An^-MB^+_{(org)} + Cat^+_{(aq)} \longrightarrow An^-Cat^+_{(org)} + MB^+_{(aq)} \qquad (3)$$

The endpoint is taken (empirically) as the stage at which the intensity of the indicator appears to be the same in both layers.

It is important to realize that two-phase titrations such as this require transfer of material through phase boundaries, and hence vigorous agitation is necessary. Adaptation of this procedure to determination of cationic surfactants can be achieved either by reversing the foregoing process (standard anionic surfactant solution in the burette) or by a backtitration process (see below). As described by Milwidsky and Gabriel [11], these alternatives may be carried out as follows.

1. Solutions Required

(a) Dissolve 0.300 g of methylene blue monohydrate in water, make up to 100 mL, and mix thoroughly. Dilute 10 mL of this solution to about 300 mL; add 6.6 g of sulfuric acid and 50 g of sodium sulfate (anhydrous): shake well until dissolved, make up to 1 L, and mix thoroughly. The "working" solution is approximately 0.00009 M with respect to methylene blue.

(b) *Standard anionic surfactant*, 0.002 *M*. "Specially pure" sodium lauryl sulfate (SLS) contains at least 99% active matter, expressed as $C_{12}H_{25}SO_4Na$ (MW 288.4): samples assayed in this author's laboratory have averaged 99.7%. This may be regarded as sufficiently pure for many purposes, and a 0.001 M solution may be prepared by dissolving 1.154 g in water and making up to 2 L in a standard volumetric flask (SVF). The assay process is detailed in Sec. VIII.

(c) *Standard cationic surfactant*, 0.002 M. Hyamine 1622 (2) is the standard of choice. It is marketed as its monohydrate, and it is common practice to remove the water by drying it in an oven at 105°C prior to use, whereupon its molecular weight is 448. A 0.002 M solution may be prepared by weighing out 1.792 g of dried material, dissolving in water, and making up to 2 L in a standard volumetric flask.

2. Direct Method

Pipette an aliquot of solution containing approximately 0.04 mmol of cationic surfactant under test into a 100-mL stoppered measuring cylinder. Add 20 mL each of methylene blue solution and chloroform, insert the stopper, shake well, and allow to settle. Add a 2-mL portion of 0.002 M SLS and again stopper, shake, and allow to settle. Repeat the process, decreasing the increments of titrant as the decreasing stability of the chloroform–water emulsion indicates that the endpoint is near. Continue the titration until the indicator migrates from the upper layer to the stage where it is of equal intensity in both phases.

3. Indirect Method

Pipette an aliquot containing approximately 0.02 mmol of cationic surfactant under test into a 100-mL stoppered measuring cylinder. Pipette in 15 mL of 0.002 M SLS (excess) and add methylene blue and chloroform as in the direct method. Stopper, shake well, and allow to settle. Titrate the excess anionic surfactant with the 0.002 M Hyamine 1622 in a manner analogous to the direct method. In this case the indicator will migrate from the organic layer toward the aqueous layer.

Heinerth [12] points out that a certain amount of An^-MB^+ salt remains in the chloroform layer at the endpoint (whereas ideally, there should be none) and suggests that a correction factor be applied. For a satisfactory titration of this nature there are three basic requirements:

1. The 1:1 An^-Cat^+ salt should be quantitatively extracted into the organic phase, leaving no dissociated ions.
2. The salt An^-MB^+ must essentially be more soluble in the organic phase than in water.
3. The salt An^-MB^+ must exchange the indicator ion MB^+ rapidly for the titrant Cat^+, but not until all free An^- ions have reacted quantitatively.

   Heinerth suggests that in practice, although criteria 2 and 3 are achievable by selection of the appropriate indicator, varying degrees of departure from criterion 1 are observed. In particular, surfactants with short alkyl chains are more hydrophilic than their longer-chain homologs and are very soluble in water: the partition of the salts that they form with antagonist surfactant ions tends to favor the aqueous phase, and if sufficiently water favorable, may make the titration unfeasible. Li and Rosen [13] extended the range of chain lengths that anionic surfactants could possess and still be titrable by modifying the chloroform phase by addition of 1-nitropropane. Extensive treatments on the balance of equilibria involved in titrations of this type have been given by Han [14] and by Jensson et al. [15].
   Although popular for many years, the Epton method was plagued by a number of inherent disadvantages that were highlighted by the publication of the results of a collaborative study organized by the Society of Cosmetic Chemists in 1964 [16].

1. The endpoint is not easily seen since the shade of color in each of the two phases is slightly different. Consequently, careful standardization of lighting conditions is required and there is considerable risk of operator-to-operator variation.
2. Since the endpoint depends on partition between two phases, some degree of standardization in the final ratio of the volumes of those two phases is required (i.e., the titer will necessarily lie between two limits).
3. The exact technique of addition of titrant and of the style of agitation is critical and can cause deviations of up to 3%.
4. The visual endpoint is not necessarily identical with the equivalence point (stoichiometric endpoint) and the difference between these quantities varies with the length of the hydrophobic chain in the surfactant under test.
5. Alkylpyridinium salts give different endpoints to equimolar solutions of alkyltrimethylammonium salts. The length of the alkyl chain in the cationic titrant also affects the result.
6. Deviations obtained by various operators on the same standard solutions were as high as 5%.

   Practicing analysts also signified their dissatisfaction with methylene blue by suggesting a wide variety of alternative indicators. The nonexclusive list includes eosin and related substances [17–21]. (di)methyl yellow [22–24], thymolphthalein [25], neutral red [26], bromocresol green [27], calcium and magnesium complexes of Eriochrome Azurol B [28], bromocresol blue [29], pinacyanol bromide [30], azophloxine [31], Rose Bengal [32], and a mixture of disulfine blue VN and dimidium bromide [33,34].

B. Mixed Indicators

An extensive collaborative study by eight member laboratories of the Commission International d'Analysis of the Comité International des Dérivés Tensio-Actifs [34] found that the mixed-indicator system recommended by Herring [33] was superior in performance to all single indicators. The recommendations of the study group were tentatively adopted by the International Organization for Standardization (ISO) in 1971 and have been retained with minor modifications [35,36]. According to Cullum and Platt [37], this method, which has been popular for some 25 years, is likely to remain in evidence for some time.

The mixed-indicator functions along the following lines. In the determination of anionic surfactants, the cationic dimidium ion (4) reacts with the analyte to produce a pink, chloroform-soluble salt, while the anionic disulfine blue (5) remains in the aqueous phase. In the region of the end-

(4)   (5)

point, the cationic titrant (benzethonium chloride), having reacted with all of the free anionic surfactant, displaces the dimidium cation from the indicator–surfactant salt and the liberated pink indicator ion transfers to the aqueous phase. Almost simultaneously, a slight excess of titrant combines with the disulfine blue anion to produce a blue, chloroform-soluble salt. The endpoint is taken as the point at which the last trace of pink is seen to disappear from the chloroform, leaving a neutral gray color. This endpoint is infinitely more sharp than that given by methylene blue, but at the same time does not permit determination of surfactants with short chains as readily as the latter [12]. Li and Rosen recommended a modification to overcome that particular problem [13].

The precision required by the ISO for such a procedure is that any two determinations carried out in rapid succession by the same analyst using the same apparatus should not differ by more than 1.5% (relative) from the mean value. Results for the same sample run by two different laboratories should not differ by more than 3% (relative) from the mean value.

The American Society for Testing and Materials (ASTM) [38] adopted an almost identical procedure for determining the active matter in a sample of anionic surfactant but laid down different tolerances of variation. The standard deviation of results (each in duplicate) should not exceed 0.14% absolute at 40 degrees of freedom (df) when obtained by the same analyst on two different days: two such average results should be considered suspect (95% confidence level) if they differ by more than 0.40% absolute. Corresponding permissible variations between analysts in different laboratories are 0.26% and 0.87% respectively. The ISO recommend two versions of this technique for the titration of cationic surfactants [39,40]: both utilize the same standard solution of sodium lauryl sulfate and the same mixed-indicator solution.

1. Solutions Required

(a) *Sodium lauryl sulfate.* The specially pure grade of sodium lauryl sulfate as supplied by such recognized agencies as B.D.H. (British Drug Houses) contains only marginal quantities of alkyl chains other than C_{12} and may normally be used as a primary standard. For utmost accuracy it may be assayed via the increase in acidity following acid-catalyzed hydrolysis to lauryl alcohol and hydrogen sulfate ions [34,36]: see Sec. VIII. To prepare a 0.004 M solution, take 1.15 g, dissolve in 50 mL of water, and dilute to 1 L in a SVF. Mix thoroughly.

(b) *Mixed indicator.* Dissolve completely in separate 10-mL portions of hot ethanol 0.5 + 0.005 g of dimidium bromide (e.g., Burroughs Wellcome) and 0.25 + 0.005 g of disulfine blue VN 150 (e.g., ICI [Imperial Chemical Industries]). Transfer both solutions to the same 250-mL SVF, dilute to the mark with water, and mix thoroughly to provide a stock solution. The working solution is prepared by diluting 20 mL of the stock solution to 500 mL with water containing 20 mL 5 N H_2SO_4 (to suppress the activity of any soap present).

ISO 2871-1:1988 [39] is for what is described as high-molecular-mass cationic-active matter, especially quats such as distearyldimethylammonium salts, which contain two alkyl chains each with 10 or more carbon atoms, and salts of imidazoline or 3-methylimidazoline in which long-chain acylaminoethyl and alkyl groups are substituted into the 1- and 2-positions. A known weight (to the nearest milligram) of surfactant under test equivalent to 0.002 to 0.004 mol is dissolved in 20 mL of isopropanol, diluted with 50 mL of water, transferred to a 1-L SVF, and further diluted to the mark. After thorough mixing, this solution is transferred to the *burette* and used to titrate a mixture of 10 mL of standard 0.004 M sodium lauryl sulfate, 10 mL of water, 10 mL of mixed indicator, and 15 mL of chloroform in the manner already described until the last trace of pink disappears from the lower organic layer. The titer should lie between 10 and 20 mL.

ISO 2871-2:1990 [40] is preferred for low-molecular-mass (200 to 500) cationic material such as monoamines, amine oxides, simple quats, and alkylpyridinium salts which have one main alkyl chain of between 10 and 22 carbon atoms, but is also suitable for other cationic materials. A sample weight to the nearest 0.5 mg and containing 0.002 to 0.003 mol of cationic-active matter is dissolved in water, transferred to a 1-L SVF, diluted to the mark, and mixed thoroughly. A 25-mL aliquot of the solution is transferred to a stoppered measuring cylinder or bottle together with 25 mL of water, 10 mL of mixed indicator, and 15 mL of chloroform. Titrate with the 0.004 M sodium lauryl sulfate solution until the blue color is discharged from the lower layer and is replaced by a grayish pink.

Both of these methods carry the same limits of precision as that for the determination of anionic surfactants already described (i.e., within 1.5% of the average value for the same analyst and within 3% for interlaboratory comparison).

The ISO-adopted disulfine blue/dimidium bromide combination is not the only mixed-indicator system available, although it would be true to state that none other has been subjected to such vigorous interlaboratory testing. Wang and Panzardi [41] combined use of the anionic indicator methyl orange, which had been used by Uno et al. [42] and Cross [43], with the cationic indicator Azure A, which had been suggested by Steveninck and Riemersma [44] as an alternative to the closely related methylene blue. Excess titrand, buffered to pH 3.0, gives rise to a blue surfactant anion/Azure A complex in the organic phase, and this blue color migrates to the aqueous layer as the last trace of indicator is displaced from the complex by the titrant. At the same time, a slight excess of titrant is indicated by the appearance of a yellow methyl orange/quaternary ammonium salt in the organic layer.

Given that this technique can be reversed to make it applicable to the titration of cationic surfactants and is therefore relevant to this treatise, it has some features worthy of note.

1. The method is aimed at the estimation of linear alkylbenzenesulfonates (LAS) in wastewaters and covers concentrations in the range 10^{-5} to 10^{-4} M (i.e., about 10 ppm), thereby overlapping the concentration range normally covered by colorimetric techniques.
2. At these dilute levels of concentration, the rate of phase separation using Azure A is at least five times faster than with methylene blue.
3. The volume of benzalkonium chloride titrant (50 mg/L, approximately 1.3×10^{-4} M) used to titrate LAS aliquots in the range 80 to 500 μg

was not exactly directly proportional to the amount of anionic surfactant present. A calibration graph was required which reduced relative errors to less than 2%.
4.   Carbon tetrachloride could be exchanged for chloroform as the organic phase. A separate calibration curve was required and the relative error was somewhat higher.

C. Indicators Without Phase Transfer

1. (Di)methyl Yellow

Although the mixed-indicator method has been popular for many years, other indicators have been successful in trials. Prominent among these is dimethyl yellow (4-N,N-dimethylaminoazobenzene, also known as methyl yellow). Unlike other indicators discussed above, the yellow basic (azoid) form (6) does not possess any hydrophilic groups. Both it and the pink salt formed between a surfactant anion and the pink acidic (quinoid) form (7) are preferentially soluble in chloroform, and the endpoint is indicated by a pink–yellow transition in the organic phase only. The upper aqueous layer remains colorless throughout the titration.

(6)   (7)

Eppert and Liebscher [45] used 0.01 M Septonex (carbethoxylpenta-decyltrimethylammonium bromide) $C_2H_5OCOC_{15}H_{30}N(CH_3)_3{}^+Br^-$ to titrate a variety of alkyl- and alkylarylsulfonates at a pH of about 1.3 and found the endpoint clearer than that obtained with the Hyamine/mixed-indicator procedure. Markó-Monostory and Börzsönyi [46] used the same system to determine high-molecular-weight oil-soluble alkylbenzenesulfonic acids and their salts. Results showed no systematic deviation from those obtained using benzethonium chloride and the mixed indicator, but for dark-colored samples the authors found the pink-to-yellow color change of dimethyl yellow easier to discern than the disappearance of the pink dimidium compound.

Associated with optimizing the estimation of a variety of substituted ammonium compounds and basic drugs by titration with sodium lauryl sulfate using dimethyl yellow, Jansson et al. [15] have made one of the most complete studies of this style of titration yet to be reported. Whereas most methods of this class have evolved via an empirical approach, these authors measured the equilibrium constants for the formation/extraction of the quat–lauryl sulfate and indicator–lauryl sulfate salts and used them to calculate the most appropriate pH conditions such that the error between the equivalence point and the endpoint was minimized. Unfortunately, the analytes tested did not include a broad spectrum of typical cationic surfactants: the closest were decylamine and nonyltrimethylammonium, for which the optimum pH values were 1.99 and 2.17, respectively.

Their recommended procedure involves dissolving 10^{-5} to 10^{-4} mol of cationic salt in 10 mL of an appropriate buffer prepared from sodium dihydrogen phosphate and phosphoric acid to produce an ionic strength of 0.1. Add 20 mL of 2×10^{-5} M dimethyl yellow in chloroform and titrate the mixture using automatic stirring with a buffered 10^{-3} to 10^{-2} M standard solution of sodium lauryl sulfate in a photometric titrator. The first perceptible pink color is indicated by a rise in the absorbance at 545 nm to a value of 0.12. No compensation for indicator error is required unless the concentration of the sample is less than 10^{-3} M. The error (i.e., the difference between the equivalence point and endpoint) for visual determination can be about five times greater than for the photometric determination.

2. Tetrabromophenolphthalein

Another indicator in the same class as dimethyl yellow is the ethyl ester of tetrabromophenolphthalein (TBPP). It was used by Tsubouchi et al. [47] during titration of alkyl sulfates, alkylsulfonates, and alkylsulfosuccinates with Zephiran (tetradecyldimethylbenzylammonium chloride). In a typical estimation, a 10-mL aliquot containing 5 to 50 μmol of anionic surfactant would be mixed with 5 mL of a 0.3 M disodium hydrogen phosphate buffer, 2 drops of a 0.1% alcoholic solution of TBPP as its potassium salt, and 10 mL of chloroform. After initial shaking the yellow (undissociated) form of TBPP is to be found in the chloroform layer and the aqueous layer is colorless. As the titration proceeds with intermittent shaking, the chloroform layer turns green (endpoint) and, with a slight excess of titrant, to the sky blue color of the quaternary ammonium–indicator anion salt. The aqueous layer remains colorless.

Using Zephiran concentrations between 0.001 and 0.01 M, the method gave results identical to those of Wang and Panzardi's Azure A method [41] and some 3% less than those obtained with methylene blue. The pH could vary between 5.0 and 7.0 without affecting the titer values: a value of 6.0 was regarded as optimal and obtained by addition of sulfuric acid to the hydrogen phosphate solution. Variations in the phase volume ratios (10 to 25 mL of aqueous phase, 7 to 15 mL of chloroform) had no effect, nor did the presence of common cations and anions. A variety of water-immiscible solvents were tested (nitrobenzene, butanone, butyl acetate, benzene, carbon tetrachloride, dichloroethane, chlorobenzene, and *n*-hexane), but none was superior to chloroform.

IV. TITRATIONS WITH SODIUM TETRAPHENYLBORATE/VISUAL INDICATORS

The literature on the titration of surfactants using visual indicator(s) is so dominated by the relatively unfamiliar two-phase techniques that the reader may query what on earth had happened to the time-honored single-phase titration. Sodium tetraphenylborate (STPB) is the major alternative antagonist to surfactant anions such as lauryl sulfate, and provides a few such procedures, but most are still of the diphasic type. TPB$^-$ ions combine with long-chain quaternary ammonium ions much more strongly than do lauryl sulfate ions: even the low-molecular-weight tetramethylammonium TPB has a molar solubility in water of only 4×10^{-5} at room temperatures [48]. The precipitated TPB salts are usually contaminated by only a minimal amount of adsorbed materials, settle easily, and are readily filterable: a few procedures based on determination of the TPB content of such precipitates have been described [49], but this treatise will concentrate on the rapid, direct procedures.

   Sodium tetraphenylborate is an expensive reagent and the stability of its solutions has been the subject of some concern. In acid solutions it certainly is unstable, becoming cloudy and yielding an odor of benzene, but by adjusting the pH to between 9 and 10, long-term stability can be achieved [1,50–52]. It is not regarded as a primary standard by the majority of authors. Solutions are most conveniently standardized against a cationic surfactant of high purity and preferably of similar structure to the substances under test. Further consideration of the standardization of STPB are given in Sec. VIII.

A. Single-Phase Titrations

Uno et al. [42] described a simple titration using methyl orange as the indicator in a solution buffered at 3.0. At this pH the normal red (acid) form of the indicator changes to yellow in the presence of cationic surfactants. Ten- to 20-mL aliquots of 0.002 to 0.05 *M* solutions of benzethonium, benzalkonium, and alkyltrimethylammonium chlorides were titrated successfully with 0.002 to 0.05 *M* STPB solution, the sharp yellow-to-red color change of the endpoint being accompanied by coagulation of the precipitate. However, while attempting to expand the range of surfactants to which Uno's method was applicable, Cross [43] found that:

1. The endpoints for nonquaternary compounds such as dodecylamine, which would be expected to react quantitatively at pH 3.0, were very ill defined.
2. Dialkyldimethylammonium ions formed precipitates with TPB$^-$ which coagulated rapidly and adsorbed the indicator so strongly that the endpoint became indeterminate.
3. The presence of polyethoxylated nonionic surfactants prevented coagulation and resulted in a very ill-defined endpoint that was usually premature.

In another simple and rapid procedure specifically designed for the assay and quality control of quats during production, Metcalfe et al. [53] selected 2,7-dichlorofluorescein as the preferred indicator after having tested a wide range. Unlike most other indicators, which functioned well only between narrow pH limits (e.g., 2.5 to 3.3 for methyl orange), this material provided sharp color changes over a wide range of pH values.

In a typical analysis, 2 to 3 mmol of quat is weighed into a beaker and dissolved in 40 mL of water. If the sample is insoluble in water (e.g., dialkyldimethylammonium chloride), it is first dissolved in a minimal amount of isopropanol (which should not exceed 10% of the total volume at the endpoint) and diluted with water to produce a very fine dispersion. About 0.5 g of sucrose is added to prevent coagulation, together with 0.5 mL of a 0.2% ethanolic solution of dichlorofluorescein, and the resulting solution is titrated with standard (0.07 M) STPB until the indicator changes sharply from pink to yellow.

Results compared very favorably with those given by the familiar non-aqueous method using perchloric acid/acetic acid. Ethoxylated quats, whose indicator/ion salts are often too water soluble to allow good extraction into chloroform, performed well in this aqueous medium. Non-quaternary amine salts also titrated well provided that sufficient acid was added to overcome the effect of the alkaline stabilizer in the titrant. Over 20 years later a team from the same organization [54] stated that this procedure had been used for "many years to analyse common long-chain quats," but that some quats could not be thus determined (mainly because of fading endpoints and uncertain visual indicator changes) and gave preference to a potentiometric procedure (see Sec. VI.C).

Nevertheless, for many types of quats and for analyses in which the ultimate in accuracy is not necessarily required, these two methods, requiring only the most basic of laboratory glassware and no chlorinated solvents, have much to offer.

As an added advantage, the product of the titration is a readily filterable precipitate that can easily be removed, washed, and dried. A weighed portion of it can be dissolved in glacial acetic acid and titrated with standard perchloric acid to crystal violet. The latter exercise will permit calculation of the mean equivalent weight of the TPB salt and hence of the surfactant cation. Other quantitative treatments developed for precipitated TPB salts are discussed in Ref. [49]. Incidentally, this precipitation also provided a ready means of removing a surfactant cation from a solution for qualitative examination (e.g., infrared spectroscopy) [43,49]. Other types of single-phase titration are discussed in Sec. VI.

B. Diphasic Titrations

The first diphasic titration of cationic surfactants using STPB as a titrant was reported by Patel and Anderson [52]: they determined benzethonium and benzalkonium chlorides under weakly alkaline conditions to suppress any contribution from nonquaternary bases. They dissolved up to 300 mg of sample in 75 mL of water in a separatory funnel and added 0.4 mL of 0.05% bromophenol blue, 1 mL of 1 M sodium hydroxide and 10 mL of chloroform. This mixture was titrated with 0.02 M STPB with intermittent vigorous shaking until the last trace of the blue color of the quat BPB salt was removed from the lower layer: the aqueous phase turned from colorless to purple simultaneously with liberation of the basic BPB ion.

With this technique, several operators obtained consistent results that were within 0.1 and 0.5%, respectively, of those given by the ferricyanide method [55] (an established method involving addition of excess ferricyanide to a quat, filtering off the precipitate and measuring the excess ferricyanide in the filtrate by iodometric titration; it was adopted by the Association of Official Agricultural Chemists and by the British and U.S. Pharmacopoeias [49]).

The method of Patel and Anderson was recommended in the 1960 *U.S. Pharmacopoeia* [56] for the assay of benzethonium chloride and of cetyl-pyridinium chloride. In trials of the method for these and other cationic surfactants, Cross [43] obtained good recoveries for alkyltrimethylammonium salts but low, variable results for cetylpyridinium chloride. This proved to be due to deactivation, prior to decomposition, of the latter surfactant under the alkaline conditions. A study of the variation in titer of a selection of quaternary and nonquaternary materials across a wide range of pH values showed that they could be divided into three classes (Fig. 1). Group A consisted of tetraalkylammonium salts (including benzyl derivatives, and

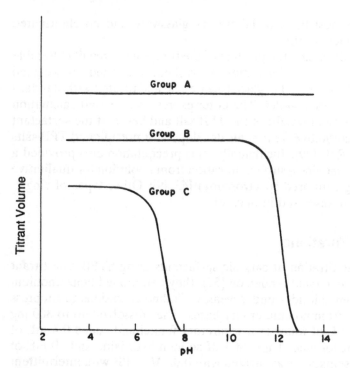

FIG. 1 Variation in titration with pH for three groups of surfactants. (From Ref. 43.)

compounds in which the quaternized nitrogen atom was contained in a nonaromatic ring system (e.g., alkylmorpholinium salts); group B contained those substances in which the nitrogen atom was part of an aromatic ring (alkylated pyridine, quinoline, isoquinoline, and biguanidine salts); and group C constituted nonquaternary amines which are only fully protonated at pH values below 6. By titrating aliquots of a sample at three selected pH values (3, 10, and 13), mixtures of surfactants from different groups could be analyzed. Bromophenol blue did not function well at pH values below 7, so the titration at pH 3 was carried out using methyl orange. This diphasic variation of Uno's method [42] (discussed above) overcame all of the problems associated with the single-phase titration.

Sakai et al. [57] preferred to use the hydrophobic ethyl ester of tetrabromophenolphthalein (TBPP) as the indicator. Numerous water-immiscible solvents were tested, of which chloroform and 1,2-dichloroethane proved to be the most useful and the latter was selected. In a typical titration, 10 mL of a 0.005 M quat would be mixed with 5 mL of a 0.3 M

phosphate buffer (pH 8), 2 to 3 drops of 0.1% TBPP, and 5 mL of di-chloroethane. The mixture was titrated with 0.005 M STPB with intermittent shaking until the blue color was discharged from the organic layer.

One of the interesting features of this investigation was the variation in the quality of the titration with the molecular weight of the quat under test. Tetramethylammonium salts did not give sharp endpoints at any concentration and tetraethylammonium salts not below 0.02 M: tetrabutylammonium, tetraphenylammonium, and tetraphenylarsonium could be determined satisfactorily in the range 0.002 to 0.02 M. Cationic surfactants such as benzethonium and cetyltrimethylammonium salts were recovered accurately using quantities as low as 10 mL of 5×10^{-4} M solution.

In a minor modification, Tsubouchi et al. [58], using a pH 8 buffer and only 0.5 to 1.5 mL of dichloroethane to 20 mL of aqueous layer, were able to use concentrations of titrant and titrands as low as 2×10^{-5} M. For a typical quat this corresponds to about 7 mg/L, a concentration at which colorimetric analysis might well be expected to be the dominant technique.

C. Cationic and Anionic Surfactants in Admixture

By this stage readers will scarcely need to be reminded that cationic and anionic surfactant ions react together to produce water-insoluble salts. Although such compounds might conveniently settle out in a titration flask, in wastewaters they may be present as a fine turbidity or perhaps solubilized by the surfactant in excess. Tsubouchi and his colleagues have developed methods by which both the cationic and anionic species in such samples may be determined volumetrically. The processes have been specifically applied to waste liquors in a hospital where quats have been used liberally for disinfecting and anionic surfactants for cleaning, but should be generally applicable.

Two different approaches are taken, depending on whether the cationic or anionic surfactant is in excess. Both require two separate titrations.

1. Anionic Surfactant in Excess
This procedure makes full use of the fact that quats bond much more strongly to TPB$^-$ ions than they do to surfactant anions [59], and hence

$$\text{Quat-An} + \text{TPB}^- \longrightarrow \text{Quat-TPB} + \text{An}^-$$

The indicator used is Victoria Blue B, a hydrophobic cationic dyestuff that stays in the organic layer throughout the titration. Typically, to suit the concentration range of the titrant, the surfactant should be in the region 1 to 12×10^{-5} M.

1. To estimate the excess anionic surfactant, 10 mL of sample is mixed with 5 mL of 0.3 M dihydrogen phosphate/borax buffer (pH 9.0), 1

to 2 drops of 0.01% ethanolic Victoria Blue B, and 3 mL of 1,2-dichloroethane in an Erlenmeyer flask. The mixture is titrated with a standard cationic surfactant (e.g., 5×10^{-5} or 1×10^{-4} M benzethonium chloride) with vigorous shaking after each addition until the organic layer changes from blue (indicator–surfactant anion salt) to red (free indicator).

2. To estimate the cationic surfactant, 10 mL of the same solution is mixed with 5 mL of 6 M sodium hydroxide, 1 to 2 drops of indicator and 2 mL of dichloroethane. The mixture is titrated with standard STPB solution (typically, 5×10^{-5} or 1×10^{-4} M) with vigorous shaking until the color changes from red to blue.

In titration 2 the TPB$^-$ anion displaces the surfactant anion from the cation–anion salt. Under the alkaline conditions, the indicator does not react with the surfactant anion and remains in the red (free) form until excess TPB$^-$ is present, whereupon it forms a blue ion pair. The total anionic surfactant present is obtained from the sum of titrations 1 and 2.

2. Cationic Surfactant in Excess

As in the first approach, discussed above, this procedure [60] depends on the TPB$^-$ ion displacing the surfactant anions, which are already combined as an ion pair with the quat ions. Without explanation, the indicator tetrabromophenolphthalein ethyl ester (TBPP) is used instead of Victoria Blue B.

1. To estimate the total cationic surfactant, 5 mL of sample (10^{-5} to 10^{-4} M) is mixed with 5 mL of buffer (pH 12.5), 1 drop of 0.03% TBPP in alcohol, and 3 mL of dichloroethane in a small separatory funnel. The mixture is titrated with 5×10^{-5} M STPB with intermittent shaking until the organic phase changes from blue to colorless.

2. To estimate the excess cationic surfactant, 5 mL of the same sample is mixed with 5 mL of buffer (pH 6.0), indicator, and dichloroethane as above. It is titrated with 5×10^{-5} M sodium dodecyl sulfate until the indicator changes from blue to pale yellow.

The concentration of combined anionic surfactant in the sample is given by the difference between titrations 1 and 2. Excellent recoveries on artificial mixtures were obtained in both types of sample, 1 and 2. Typically, the standard deviation for a 5-mL titration is only 0.05 mL. A 10-mL burette is needed for that level of precision.

In both samples 1 and 2, part of the procedure is carried out at a very high pH, at which the quat–indicator salt is not decomposed by surfactant anions; consequently, this restricts the class of cationic surfactant that will respond to this method to the tetraalkylammonium class, group A (see Sec. IV.B).

V. ALTERNATIVE ANTAGONISTS

This chapter has been dominated unashamedly by the use of anionic surfactants and of sodium tetraphenylborate as titrants. Prior to the advent
of these, most methods were based on the precipitation of quaternary
ammonium ions with a selection of inorganic anions. Such processes are
seldom used currently because they are lengthy and involve more manip-
ulative stages than current methods.

In general, they involve addition of a known, excessive amount of precipitating agent followed by removal of the quat-anion salt by filtration
before either (1) estimating the excess reagent remaining behind in the
filtrate and subsequently calculating how much had been precipitated, or
(2) analyzing the precipitate to determine how much of the precipitating
anion it contained.

Precipitating anions in this category include dichromate [61–63], te-
traiodocadmiate [64,65], ferricyanide [55], ferrocyanide [66], and triiodide
[67]. The ferricyanide procedure, in particular, enjoyed a period of popularity and was adopted as a standard method by the Association of Official Agricultural Chemists and the British and U.S. Pharmacopoeias. It
is still occasionally cited in the literature as the standard method against
which the results of a proposed new method have been checked (e.g.,
[68]). Other precipitating inorganic anions have been Reineckate,
$[Cr(NH_3)_2(SCN)_4]^-$ [69], and phosphotungstate [70,71].

These and related processes were reviewed in 1970 [6] and have largely
remained buried in the annals of analytical chemistry until being revitalized
by the work of Selig (see Sec. VI.B), which indicates that some of them
may yet have a future in the analytical chemistry of cationic surfactants.

A much newer reagent whose application to the analysis of quats appears
yet to remain untested is sodium tetrakis (4-fluorophenyl)borate. It has
been applied successfully to the two-phase titration of ethoxylated nonionic
surfactants in the presence of potassium ions, which combine to give ill-
defined coordination complex cations which have much in common with
quat cations [72–75]. The original method uses Victoria Blue B as the
indicator and dichloroethane as the organic phase [72]. O'Connell [73]
believed that it should become the method of choice for quality control of
nonionics, but commented on the very high price of the reagent [U.S. $395
for 5 g (1986)] and substituted potassium tetrakis (4-chlorophenyl)borate
(U.S. $26 for 5 g): he noted that quats interfered quantitatively with the
estimation but found that benzethonium chloride was seemingly 30% more
concentrated than indicated by other methods despite giving sharp endpoints. Further investigation is clearly warranted.

VI. INSTRUMENTAL DETECTION OF ENDPOINTS

The titrimetric methods described thus far have all used visual indicators and in general have involved the use of a water-immiscible organic solvent, usually chloroform, as a second phase.

Over the years, this type of titration has been subjected by many authors to a number of criticisms [76].

1. Long-term exposure to chloroform is now recognized as a health hazard. Even though in most cases 1,2-dichloroethane could be substituted with little effect, the general trend is to move away from the use of chlorinated solvents.

2. In comparison to a simple, one-phase titration, the process is relatively lengthy. Frequent stopping for periods of vigorous shaking is the only way to ensure adequate transfer of the various species between the two phases: less-than-vigorous agitation usually results in the addition of more titrant than should really be required (i.e., a positive error in the titer).

   There is no equivalent to adding dropwise with continual swirling as the endpoint is approached.

3. The endpoint is not always clearly defined and consequently is rather subjective. This is particularly true of the classic Epton procedure, in which the transfer of methylene blue between the two phases occurs gradually over the addition of a few milliliters of titrant. Detection of the endpoint, which is deemed empirically to be the point at which the intensity is equal in both phases, is complicated by the fact that the two shades of blue are slightly different. The quality and positioning of the lighting in the laboratory also plays a role.

4. Attempts to use concentrations over about $0.02\ M$, such as might be desirable in quality control/product evaluation laboratories, usually result in emulsification problems between the two phases. The presence of nonionic surfactants in admixture may have a similar effect.

5. The processes do not appear to lend themselves readily to automation.

There have been a number of proposals that address some or all of these problems.

A. Self-Contained Apparatus

The most recent ISO method (1989) for the titration of anionic surfactants with benzethonium chloride [36] offers a choice between the traditional stoppered bottle/measuring cylinder and a sealed system in which the necessary agitation is provided by a mechanical stirrer. The titration vessel, the dimension and design of which are fully specified in the standard, contains inlet ports for the addition of indicator solution, chloroform, and titrant and a drain for removal of the mixture to waste. With this apparatus, agitation can be standardized and exposure to chloroform all but eliminated. The endpoint, however, is still judged visibly.

B. Photometric Titrations

Photometric determination of the endpoint avoids the subjective nature of the operator's personal perception of the color change. Furthermore, it cannot only monitor change in color, but also in turbidity.

The work of Jansson et al. [15] has already been cited in Sec. II.C. They used an EEL titrator during the determination of various quaternary ammonium salts with sodium lauryl sulfate, selecting a number 605 filter with a nominal transmission at 545 nm which closely corresponded with the absorbance maximum of the pink methyl yellow–lauryl sulfate ion pair. The detection of the first perceptible red color in the organic phase against the yellow background color of the free indicator was signaled by a rise in absorbance to 0.12. The precision of the endpoint location by this technique was about five times better than by visual detection.

Mohammed and Cantwell developed an apparatus that permitted continuous monitoring of either of two phases in a vigorously agitated mixture by passing it through a spectrophotometer flow cell [77] and used it to titrate several cationic drugs by following the concentration of the titrant (picrate ion) in the aqueous phase [78]. They extended their work to include the titration of anionic and cationic surfactants using methylene blue and picric acid, respectively, as titrants [68]. For these analyses it was the chloroform layer that was monitored, at 658 nm for methylene blue and 400 nm for picric acid. The concentration of each of these species in the organic layer increases linearly as the titration proceeds due to the buildup in concentration of the titrant–surfactant anion salt. At and beyond the endpoint, the absorbance remains steady since additional titrant remains as the free ion in the water. The endpoint is located by the intersection of the two straight lines on a graph of absorbance versus volume of titrant.

Assays of commercial samples of benzethonium chloride and cetyltrimethylammonium bromide diluted down to the 10^{-4} M level averaged at $92.1 \pm 0.4\%$ and $95.6 \pm 0.1\%$, respectively, by this technique compared to $92.0 \pm 0.1\%$ and $96.0 \pm 0.4\%$, respectively, by the ferricyanide precipitation method (Sec. V). This technique again removes the subjectivity of visual interpretation of the color changes at the endpoint, but unfortunately, requires the use of a large volume of chloroform (70 mL) for each estimation.

The methods described above still make use of chloroform, and have merely substituted an instrumental method for location of the endpoint without necessarily reducing the complexity of the analysis or the time taken for an analysis. The salt formed between the antagonistic surfactant cation and anion is relatively hydrophobic. In the absence of a solvent such as chloroform in which to dissolve, it forms a finely divided precipitate in the aqueous layer which is presumably stabilized in suspension or solubilized by the excess analyte material. In other words, it causes turbidity. Following some initial studies on a two-phase system by Seguran [79], Hendry and Hockings [80] investigated the variation in turbidity in the single aqueous phase during the course of titration of some anionic surfactants with the $4 \times 10^{-3} M$ benzethonium chloride. In the absence of high concentrations of salts, the turbidity increased slowly during the initial stages of the titration, then increased rapidly before leveling off to a steady reading: the endpoint was taken as the point of inflection between the end of the sudden rise and the steady, post-endpoint reading.

An interesting point about this titration results from the limitation of the ability of the spectrophotometer used to measure high degrees of turbidity. To keep the absorbance reading below the maximum of 1.99, the working range of anionic surfactant to which the technique could be conveniently applied was 3×10^{-6} to $4 \times 10^{-5} M$. This corresponds roughly to 1 to 10 ppm, a range normally reserved for colorimetric techniques.

The presence of nonionic surfactants caused complications: the magnitude of the sharp rise in turbidity at the endpoint decreased with increasing concentration of ethoxylated alkylphenol, disappearing altogether when the latter was present in tenfold molar excess. This was tentatively assigned to solubilization of the precipitated salt by the nonionic surfactant but is in direct contrast to the observations of Volkmann [81]. Using $4 \times 10^{-3} M$ solutions and following the progress of the titration by means of transmittance rather than absorbance, the latter author found that the addition of 5 drops of Tween 80 caused the transmittance to rise steadily during the course of the titration and then to increase sharply at the endpoint.

Hendry and Hockings claimed that the repeatability of their technique was equal to that obtained by an experienced operator with the ISO method [36], but that the result could be obtained in less than half the time. The discrepancy between the results of the two methods, however, was ±5%.

In adapting the technique to the determination of cationic surfactants, Hendry and Read [82] investigated the titration of five different cationic materials in the concentration range 2×10^{-6} to $8 \times 10^{-5} M$ with 4×10^{-3} sodium lauryl sulfate: this extremely low range was once again dictated by the limitation of the spectrophometer employed.

As before, the repeatability was comparable to that of the standard, two-phase method, but serious discrepancies between the results of the two methods were observed. The results for alkylbenzyldimethylammonium chloride, benzethonium chloride, and laurylpyridinium chloride deviated by -6, -4, and -2%, respectively. Of the other two analyte materials, N-hydroxyethyloleylimidazoline gave no response to the standard method, and diethylheptadecylimidazolinium ethyl sulfate gave a result 22% higher. Both of these compounds were shown by HPTLC to contain substantial amounts of nonionic material: the authors suggested that although this would not be enough to upset the turbidimetric technique, it could have invalidated the results of the standard method. Clearly, further study is required before this very sensitive technique can be recommended without reserve [83].

C. Potentiometric Titrations

There are many levels to which the study of electrodes can be carried. The in-depth researcher may be concerned with such factors as the mechanism that causes an electrode to respond to a particular analyte, with the degree of selectivity that it shows and the concentration range over which it exhibits a Nernstian response. Somewhere toward the other end of the scale comes the analytical chemist, who is concerned only with a sharp change in potential as the equivalence point is reached.

The construction of surfactant-sensitive electrodes and their applications, including potentiometric titrations, are dealt with in a definitive manner in Chapter 6. The information presented therein should be given due regard when evaluating the remarks made in the next few pages.

The use of potentiometry to locate the endpoint of a titration is receiving considerable attention. In the words of Cullum and Platt [84], "over the next few years any major advance in the titrimetry of surfactants seems more likely to lie in the field of potentiometry and potentiometric titration."

The traditional potentiometric titration assembly is a cell consisting of the analyte solution, into which are placed two probes. One of these is a reference electrode of constant potential, such as the classic calomel electrode. The other is an indicator electrode, the potential of which will vary according to either the diminishing concentration of titrand or the increasing concentration of titrant. For the potentiometric titration of cationic surfactants, SLS and STPB maintain their popularity as titrants, though not exclusively. Suitable indicator electrodes therefore respond to changes in concentrations of LS^-, TPB^-, and quaternary ammonium ions. Experience has shown that a wide variety of electrodes other than those purpose-designed for these three types of ion show a useful response [85–106]. For example, both potassium-selective [85] and perchlorate-sensitive [86] electrodes have been used for the titration of organic cations with STPB. Some electrodes show responses that are super-Nernstian (i.e., slopes of potential versus concentration are greater than 59 mV per decade change in concentration, e.g., a mercury-coated platinum electrode used by Pinzauti et al. [91] for TPB$^-$). The mechanism by which this electrode operates is not fully understood and the relationship between the concentrations of ion and the potential is empirical: this lack, however, does not stop the electrode from being a useful indicator.

As with any other method, reliable valuations of the potentiometric method ultimately rely on the use of standard solutions of surfactants, the concentration of which can be relied upon with confidence. Unfortunately, many authors have been content to judge their results on the doubtful assumption that the activity of the quat under test was exactly that claimed by the manufacturer.

1. Endpoint Location

One of the major problems seems to be location of the endpoint. Cullum and Platt [76] comment that a potentiometric titration "is easily performed using a modern autotitrator that uses a piston burette and is capable of drawing titration curves. The latter is important because autotitrators will nearly always find a spurious end point when there is no real one, and inspection of the curve is the only way to be sure that the instrument is telling the truth."

Most of the potentiometric methods presented in the scientific literature are accompanied by one or more figures depicting a "typical" curve of cell potential versus titrant volume. In many cases the endpoint needs to be located from a steep (but not vertical), seemingly linear change in electromotive force (EMF) over a region that represents 10% or more of the titer value. Figure 2, representing the titration of 10^{-3} M prifinium bromide with 10^{-2} M STPB (piston burette) using a commercial perchlorate ion-selective electrode, is a typical example [86]. Isolating one point at which the slope reaches a maximum would seem to be a daunting task, even for a microprocessor.

Figure 3 was derived using a 10^{-2} M solution of three different quats (cetylpyridinium chloride, cetyltrimethylammonium chloride, and Hyamine 1622) as titrants and a fluoroborate ion-sensitive electrode as the indicator. The three curves have been displaced laterally for clarity. Again the extensive linear steep section of the curves around the endpoint region is evident. It is also clear that the EMF range over which the potential

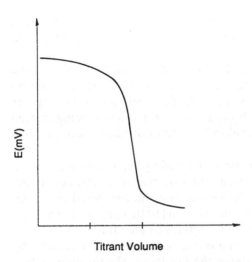

FIG. 2   Potentiometric titration of 10^{-3} M prifinium bromide with 10^{-2} M STPB. (From Ref. 86.)

break occurs is different in all three cases. This has two consequences: (1) it is inappropriate to preset an EMF to which all quats would be titrated; and (2) the fact that Hyamine 1622 shows the smallest sharp jump casts doubt on its generally accepted role as the most appropriate titrant for estimation of anionic surfactants. This study also provides an example of

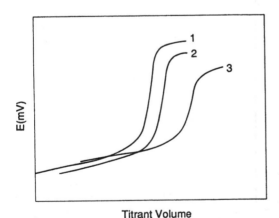

FIG. 3   Potentiometric titration of 10^{-3} M sodium dodecyl sulfate with 10^{-2} M solutions of three different quats: (1) cetylpyridinium chloride, (2) cetyltrimethyl-ammonium chloride, and (3) Hyamine 1622. (From Ref. 92.)

the use of 10^{-2} M solutions of surfactants (where solubilities permit) in a piston burette: foam problems would be considerable in a manually operated system.

Shoukry et al. [100] introduced phosphotungstic acid as an alternative titrant to STPB and SLS: being a triprotic acid, it forms a 1:3 ion association salt with singly charged quat ions (Fig. 4). The endpoint again appears difficult to locate, but this time with good reason: the solution being titrated was 5×10^{-5} M cetylpyridinium chloride, a concentration corresponding to a mere 20 mg/L.

Shoukry saw the need for a means of locating the endpoints of such titrations without having to rely on automated electronic devices to calculate the first- or second-order derivative data, and resorted to an imaginative variation in the electrode combination [103]. The traditional combination of an indicator electrode and a reference electrode was replaced by two indicator electrodes, one being sensitive to the titrand ion and the other to the titrant ion. Figure 5 shows the results for the titration of 10^{-3} M cetylpyridinium chloride with 10^{-2} STPB: the electrodes were both of the membrane type, one containing cetylpyridinium TPB and the other

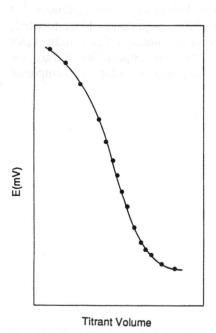

FIG. 4 Potentiometric titration of 5×10^{-5} M cetylpyridinium chloride with 5×10^{-5} M phosphotungtic acid. (From Ref. 100.)

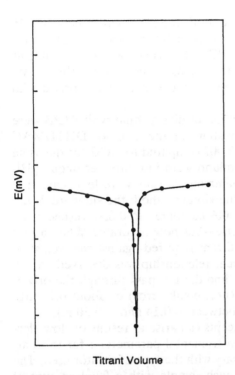

E(mV)

Titrant Volume

FIG. 5 Potentiometric titration of 10^{-3} cetylpyridinium chloride with 10^{-2} M STPB using two indicator electrodes. (From Ref. 103.)

(cetylpyridinium)$_3$ phosphotungstate. This innovative approach could well find application for analyses outside the field of surfactants.

2. Precision and Accuracy

One study that has paid particular attention to the precision and repeatability of the results is that of Wang et al. [54,99]. These researchers made the observation that the low solubility of the dialkyl quats [e.g., di(hydrogenated tallow)dimethylammonium chloride, DHTDMAC] in water made the use of quat-sensitive membrane electrodes impractical. They used a platinum ring electrode as the indicator electrode for TPB$^-$ ions. A second platinum electrode, in the form of a wire, functioned as a reference: this was located in the titrant stream via a plastic connection between the burette tap and tip. To complete the necessary circuit, the tip was necessarily sited below the surface of the titrand solution. An advantage of this concept is that the reference electrode never comes into contact with the precipitated quat-TPB salt and therefore cannot be contaminated.

The quat, equivalent to up to 15 mL of 0.05 M STPB, was dispersed in 15 mL of glacial acetic acid and diluted to about 100 mL with boiling water. By keeping the temperature above 70°C and stirring vigorously, problems caused by adhesion of the precipitate to the sensor electrode and the burette tip were avoided. After each titration these two items were rinsed with acetone followed by water.

Potential changes during the titration of dialkyl quat with STPB were particularly sharp: the standard deviation for the assay of DHTDMAC with an activity of 1.42 mEq/g was 0.007 compared to 0.032 for the same material determined by the dichlorofluorescein indicator technique [53]. Potential changes for alkyltrimethylammonium salts were less sharp, but still gave more precise results than the visual method: the standard deviation for a 2.81-mEq/g sample was 0.009 compared to 0.030, respectively.

The results were, therefore, precise—but how accurate? When a plot was made of the volume of STPB titrant required against the weight of DHTDMAC taken, an excellent linear relationship was observed (correlation coefficient 0.099999), but the line did not pass through the origin: rather, it gave a positive intercept (i.e., blank error) of about 0.13 mL. Other quats gave rise to intercepts between −0.14 and +0.30 mL.

In potentiometric titrations, intercepts can arise as results of slow electrode response or wrongly chosen instrumental parameters: for this particular titration they were found to vary with the type of titrator used. The problem was overcome by spiking each sample with a fixed amount of Hyamine 1622 and subtracting from the total titration value the titer yielded by the spike alone. Hyamine, being precipitated subsequent to the dialkyl quats, had the effect of normalizing the shape of the titration curve in the region of the endpoint to that of Hyamine 1622 itself: hence the titration error was made consistent for all types of analyte and then eliminated in the same treatment.

3. Coordination Complex Ions as Titrants?

No treatise on the potentiometry of surfactants would be complete without reference to the work of Selig. Over the last 15 or so years, he has been systematically examining the use of cationic surfactants (mainly cetylpyridinium chloride and cetytrimethylammonium bromide) as titrants for a wide variety of anions, most of which are complexes of metal cations and anionic ligands, such as $TiCl_4^-$, $Fe(CN)_6^{3-}$, $SnCl_6^{2-}$, and $PtCl_6^{2-}$. Many of these titrations show substantial potential breaks at the endpoint (e.g., quat versus $TiCl_4^-$, 200 mV) but none as substantial as quat versus TPB^-, 450 mV. Reviews of applications of these reactions are given in Refs. [109] and [110]. Reversing the titrant/titrand roles leaves the potential for salts such as potassium ferricyanide to act as a suitable titrant for quats. Sensors tested varied from expensive commercial solid-state and membrane electrodes, through homemade coated graphite rods, down to common lead pencils. The latter, which benefit from activation by soaking in potassium permanganate for a few minutes, can be turned into electrodes by removing the eraser from the end and filling the cavity with solder to make good contact [111]—devices that can be used not only for the analysis but also to record the result.

4. Discrepancies Between Electrodes

To compound the difficulties in selecting the endpoint of a potentiometric titration comes the observation that individual electrodes of the same type may respond to each other somewhat differently. In recent years ASTM has accredited potentiometric titration procedures for the determination of both anionic surfactants [107] and cationic fabric softeners [108]. In the former, a commercial nitrate ion-selective electrode is used during titration with Hyamine 1622. The results for using three different electrodes for the same test anionic surfactant are presented in Table 4, together with the result given by the standard two-phase mixed-indicator titration.

Two different procedures are given for the analysis of quats. Diamidoamine-based quats are dissolved in a little isopropanol, diluted with water, buffered at pH 6.0, and titrated with standard SLS: dialkyldimethylammonium salts are similarly dissolved but are then treated with a known (excess) quantity of SLS and the excess is measured by back titration with Hyamine 1622. Both techniques are claimed to have a precision of 1.24% relative at 20 df for the same analyst operating on different days: the overall precision for an eight-member multilaboratory analysis was 2.38% relative at 9 df.

TABLE 4 Apparent Purity (%) of a Sodium Alkylarylsulfonate Sample as Determined by Titration with Hyamine 1622[a]

Nitrate electrode 1	Nitrate electrode 2	Nitrate electrode 3	Two-phase mixed-indicator titration
94.15	96.26	94.91	95.12
95.31	95.67	95.50	95.12
94.73	96.26	94.91	

[a]Standard deviation for single operator 0.66% relative at 28 dF.
Source: Ref. 107.

5.   Some Critical Factors

Other points to emerge from the studies referenced above include the following:

1.   Wherever a comparative study has been made using different titrants, STPB always produced steeper and larger potential breaks than SLS, which in turn was superior to others, such as picrate and Reineckate (e.g., Ref. 93). This is consistent with the ability of TPB to displace LS from quat-LS salts (Sec. IV.C), which has been suggested as resulting from the TPB$^-$ ion being more electronegative than the LS$^-$ ion, due to the negative charge being localized on the central boron atom in TPB$^-$, whereas it is delocalized through three oxygen atoms in LS$^-$ [106].

2.   The stirring rate is crucial to this type of precipitation reaction. It should be as high as possible, without the formation of air bubbles [95].

3.   The success of the titration depends on the degree of attention paid to the electrodes. Some (membrane-type) require one or two "conditioning" runs prior to quantitative use (e.g., Ref. 92): others, such as mercury-coated platinum [91], require regeneration between runs. Most benefit from a rinse in acetone, alcohol, and/or water followed by a wipe with a tissue (depending on the material of construction) between runs (e.g., Ref. 95). It is an advantage to avoid connections via ceramic or asbestos plugs since these are easily blocked by deposited reaction product resulting in a sluggish response [107].

4.   In addition to requiring no chloroform, being applicable in the colored and turbid solutions and amenable to automation, the technique can give good results at low concentrations (e.g., to 10^{-5} M).

VII.   NONAQUEOUS TITRATIONS

The principles involved in the titration of weak (in aqueous media) organic bases by titration with perchloric acid in nonaqueous solvents, usually glacial acetic acid, are well established and will not be discussed here.

Since it is vital that water be all but eliminated from such titrations, the technique has no application for aqueous solutions. It has, however, been applied in quality control laboratories as part of the evaluation process of solid materials. The ASTM standard method [112] for determination of the average molecular weight of fatty quaternary ammonium chloride, D-2080-64, was reapproved as recently as 1987 and at the time this chapter was written, was still current.

The basis of the process is to dissolve a known weight of quat halide in glacial acetic acid and to react it with a slight excess of mercuric acetate to yield the quat acetate, which is a strong base in that solvent, and the

unreactive mercuric chloride. The mixture is then titrated potentiometri-
cally with standard perchloric acid in glacial acetic acid.

The technique works successfully only for halide salts, not for sulfates
[53] or methylsulfates [113], and in the absence of appreciable quantities
of inorganic salts [53]. By using a high percentage of dioxane in the solvent,
Pifer et al. [114,115] were able to obtain better breaks than with glacial
acetic acid alone, enabling sulfate and methylsulfate salts to be thus titrated.

Before the average molecular weight can be calculated, several other
quality control results (percent nonvolatile matter, percent ash, acid value,
amine value) need to be integrated into the equation.

An alternative procedure to determine the average molecular weight of
a quat, which in many samples may be presented in aqueous solution, is
to precipitate the quat by means of a suitable anion and then test the solid
obtained. The tetraphenylborate salts have much to offer in this regard.
They may be precipitated, filtered, and dried with ease and may be dissolved directly from a sintered glass filter or filter pad with a few milliliters
of acetone. In the presence of acetic acid, the TPB ion breaks down to
the weak phenylboric acid and the base may then simply be titrated with
perchloric acid/acetic acid to crystal violet [116,117].

Amine oxides receive scant attention in this book; indeed, the literature
specifically relating to the determination of this class of surfactants is sparse.
A brief review current to 1984 has been given by Wang and Metcalfe [118].
These authors describe a potentiometric method for the nonaqueous titration of amine oxides and any unreacted amines therein. The process is
based on the observation by Bezinger et al. [119] that the potentiometric
titration of amine oxides in certain solvents shows two potential breaks,
indicating that the reaction takes place in two stages. The first, which
requires only 1 mol of perchloric acid for 2 mol of oxide, corresponds to
the formation of an anomalous salt:

$$2R_3N \rightarrow O + HA \rightarrow [R_3N\text{---}O\text{---}H\text{---}O\text{---}NR_3]^+A^-$$

Wang and Metcalfe confirmed these findings provided that the test molecule contained an N-methyl group (e.g., alkyldimethylamine oxide) and
observed that any free amine present contributed to the second stage of
the titration. Samples were dissolved in either butanone or acetonitrile and
titrated with 0.1 M perchloric acid in acetonitrile using a glass electrode
as the indicator electrode and a platinum electrode in the titrant stream
(see Sec. VI.C) as a reference.

Many years earlier, Wimer [120] had concluded that amine oxides were
generally insufficiently basic in glacial acetic acid to permit titration and
had used acetic anhydride as the solvent and perchloric acid in dioxane as
the titrant.

VIII. SELECTED REAGENTS: PRACTICAL DETAILS

Hyamine 1622, sodium lauryl sulfate, and sodium tetraphenylborate are cited so frequently throughout the literature that a few lines about each of them in one section seems appropriate.

A. Hyamine 1622

t-Octylphenoxyethoxyethyldimethylbenzylammonium chloride monohydrate, better known as Hyamine 1622, is a totally synthetic quaternary ammonium salt and suffers from none of the uncertainties in mean molecular weight caused by a mixed alkyl chain length. It is regarded by many authors as a primary standard in its monohydrate form, but since it often contains some anhydrous salt, the more fastidious prefer to remove the water by drying at 105C to constant weight. Powdering of the caked material at least twice during the drying process is recommended to ensure complete dehydration [99]: its molecular weight may then by confidently taken as 448.

B. Sodium Lauryl Sulfate

Specially pure sodium lauryl sulfate is made by sulfation of dodecanol which has been carefully fractionally distilled. The contribution of alkyl chains other than C_{12} is negligible. The manufacturers claim that it contains not less than 99.0% active material (MW 288.4). This is a cautious claim and assays are usually closer to 99.8% [34]. It is, however, somewhat hygroscopic and will absorb a few percent of moisture from the air if not stored properly. For many purposes, therefore, it may satisfactorily be regarded as a primary standard.

For additional accuracy it is necessary to assay the SLS by measurement of the increase in acidity following acid-catalyzed hydrolysis [34].

$$C_{12}H_{25}SO_4{}^- + H_2O \xrightarrow{H+} C_{12}H_{25}OH + SO_4{}^{2-} + H^+$$

Weigh accurately about 5 g of solid into a 250-mL round-bottomed flask. Add exactly (volumetric pipette) 25 mL of 0.05 M sulfuric acid and reflux the solution under a water condenser. The initial heating must be gentle, as boiling is accompanied by substantial foaming, but as the hydrolysis proceeds, this effect will lessen and heating can be more vigorous. Lauryl alcohol will be seen clouding the condensed distillate. Continue refluxing for 90 min after foaming ceases. Cool, and wash down the condenser with a little ethanol followed by water. Add phenolphthalein and titrate with standard (ca. 1 M) sodium hydroxide.

Run a blank titration of 25 mL of sulfuric acid against the sodium hydroxide:

$$\% \text{ purity } = \frac{28.84\left(\dfrac{\text{sample}}{\text{titration}} - \dfrac{\text{blank}}{\text{titration}}\right) M_{\text{NaOH}}}{\text{sample weight}}$$

C. Sodium Tetraphenylborate

In contrast to the two previous materials, STPB cannot be regarded as a primary standard. Solutions are best made up in water containing a small amount of alkaline material such as sodium hydroxide or sodium carbonate (not a potassium salt) to keep the pH above 9, whereupon they are stable for several weeks at least. Solutions frequently exhibit a slight turbidity, but this does not affect their efficiency. If necessary, clarification can be achieved by stirring with chromatographic grade alumina and filtering after standing overnight. Storage in a dark glass bottle is recommended [60]. A detailed study on the stability of STPB solutions was made by Cooper [50].

Standardization is usually achieved by titration against a pure quat, typically Hyamine 1622, under the same experimental conditions as those under which the analyte is to be determined. For some purposes it may be considered desirable to standardize the STPB by a technique entirely independent of surfactants. Both Cross [43] and Tsubouchi and Mallory [60] have used a gravimetric determination of the potassium salt for this purpose. The latter authors added 1 mL of 1 M acetic acid and 25 mL of 0.1 M potassium hydrogen phthalate to 50 mL of 0.02 M STPB solution. After standing for 2 h the precipitate was filtered through a tared sintered glass crucible, washed with saturated KTPB solution, dried at 105°C to constant weight and weighed. The result is both accurate and precise. The precipitate is easily removed from the crucible by dissolution in acetone.

For an independent volumetric standardization, both silver and thallium salts may be used as titrants. There have been some doubts cast on the validity of silver as a quantitative titrant for STPB [121, 122], but Holten and Stein [123] found that a sensor comprising tetramethylammonium TPB and tritolyl phosphate in PVC coated onto a graphite rod was a satisfactory indicator electrode for the straightforward argentiometric titration. By such a process a nominal 0.01 M solution was found to be 0.00995 ± 0.00001 M.

Vytras [124] mistrusted silver nitrate as a titrant and preferred to use thallium nitrate or thallium sulfate. Both salts are available in a pure form but can easily be recrystallized and dehydrated if necessary. Satisfactory titration curves with steep potential breaks were observed using an electrode consisting of valinomycin as a neutral ion carrier in a PVC matrix containing dipentyl phthalate as a plasticizer.

IX. CONCLUDING REMARKS

By way of summary, this author can find no tidier comment than that provided, once again, by Cullum and Platt [84]: "It is sometimes suggested that this type of analytical chemistry (titrimety) will soon be completely superseded by the many very powerful high-tech instruments now available. But high tech can be prohibitively expensive, especially for the smaller laboratory, and in any case, the great merit of high-tech methods is not that they can do what classical techniques can do, only better, but that they can do with ease many things that classical techniques can only do with great difficulty, or not at all."

REFERENCES

1. E. D. Schall, *Anal. Chem. 29*: 1044 (1957).
2. E. C. Jennings and H. Mitchner, *J. Pharm. Sci. 56*: 1595 (1967).
3. G. S. Hartley and D. F. Runnicles, *Proc. Roy. Soc. (London) Ser. A 168*: 420 (1938).
4. C. F. Hiskey and T. A. Downey, *J. Phys. Chem. 58*: 835 (1954).
5. P. Mukerjee and K. Mysels, *J. Am. Chem. Soc. 77*: 2937 (1956).
6. J. T. Cross, in *Cationic Surfactants*, (E. Jungermann, ed.), Marcel Dekker, New York, 1970, p. 440.
7. S. R. Epton, *Trans. Faraday Soc. 44*: 226 (1948).
8. T. Barr, J. Oliver, and W. V. Stubbings, *J. Soc. Chem. Ind. 67*: 45 (1948).
9. P. A. Lincoln and C. C. T. Chinnick, *Lab Pract. 3*: 364 (1954).
10. P. A. Lincoln and C. C. T. Chinnick, *Proceedings of the First International Congress on Surface Activity*, Paris, 1954, Vol. 1, Sect. 2, p. 209.
11. B. M. Milwidsky and D. M. Gabriel, *Detergent Analysis*, George Godwin, London, 1982, p. 120.
12. E. Heinerth, in *Anionic Surfactants: Chemical Analysis* (J. Cross, ed.), Marcel Dekker, New York, 1977, p. 223.
13. Z. Li and M. J. Rosen, *Anal. Chem. 53*: 1516 (1981).
14. K. W. Han, *Tenside Deterg. 3*: 36 (1966).
15. S. O. Jansson, R. Modin, and G. Schill, *Talanta 21*: 905 (1974).
16. Scientific Committee, *J. Soc. Cosmet. Chem. 15*: 33 (1964).
17. J. B. Wilson, *J. Assoc. Off. Agric. Chem. 37*: 374 (1954).
18. W. J. Harper, P. R. Elliker, and W. K. Mosely, *Soap Sanit. Chem. 24*: 159 (1948).
19. M. Dolezil and J. Bulandr, *Chem. Listy 51*: 255 (1957).
20. H. Iwasenko, *J. Assoc. Off. Agric. Chem. 36*: 1165 (1953).
21. H. Iwasenko, *J. Assoc. Off. Agric. Chem. 37*: 534 (1954).

22. F. Pellerin, J. A. Gautier, and D. Demay, *Ann. Pharm. Franc. 22*: 495 (1964).
23. F. D. Carkhuff and W. F. Boyd, *J. Am. Pharm. Assoc. Sci. Ed. 43*: 240 (1954).
24. J. R. Gwilt, *J. Appl. Chem. 5*: 471 (1955).
25. W. Kupfer, *Proceedings of the 3rd International Congress on Surface Activity*, Cologne, 1960, Vol. III, p. 62.
26. T. Uno and Y. Miyajima, *Chem. Pharm. Bull. (Tokyo) 10*: 467 (1962).
27. H. Y. Lew, *J. Am. Oil Chem. Soc. 41*: 297 (1964).
28. J. A. Hart and E. W. Lee, *Tappi 34*: 77 (1951).
29. P. W. O. Wija, *Chem. Weekblad 45*: 477 (1949).
30. M. R. J. Salton and A. E. Alexander, *Research (London) 2*: 247 (1949).
31. G. R. Edwards and M. E. Ginn, *Sewage Ind. Wastes 26*: 945 (1954).
32. B. Borrmeister and R. Schaffner, *Dtsch. Textiltech 17*: 303 (1967).
33. D. E. Herring, *Lab. Pract. 11*: 113 (1962).
34. V. W. Reid, G. F. Longman, and E. Heinerth, *Tenside 4*: 292 (1967).
35. *Determination of Anionic-Active Matter: Direct Two-Phase Titration Procedure*, Draft ISO Recommendation 2271, International Organization for Standardization, Geneva, 1971.
36. *Determination of Anionic-Active Matter by Manual or Mechanical Direct Two-Phase Titration Procedure*, ISO 2271:1989, International Organization for Standardization, Geneva, 1989.
37. D. C. Cullum and P. Platt, in *Recent Developments in the Analysis of Surfactants* (M. R. Porter, ed.), Elsevier, London, 1991, p. 12.
38. *Standard Test Method for Synthetic Anionic-Active Ingredient by Cationic Titration Procedure*, ASTM D 3049-89, 1991 Annual Book of ASTM Standards, American Society for Testing and Materials, Philadelphia, 1991.
39. *Determination of Cationic-Active Matter Content*, Pt. 1, *High Molecular Mass Cationic-Active Matter*, ISO 2871-1:1988, International Organization for Standardization, Geneva, 1988.
40. *Determination of Cationic-Active Matter Content*, Pt. 2, *Cationic-Active Matter of Low Molecular Mass (200–500)*, ISO 2871-2:1990, International Organization for Standardization, Geneva, 1990; also adopted as British Standard 3762, Section 3.4, 1991.
41. L. K. Wang and P. J. Panzardi, *Anal. Chem. 47*: 1472 (1975).
42. T. Uno, K. Miyajima, and H. Tsukatani, *Yakugaki Zasshi 80*: 153 (1960); *Chem. Abstr. 54*: 11824e (1960).
43. J. T. Cross, *Analyst 90*: 315 (1965).
44. J. V. Steveninck and J. C. Riemersma, *Anal. Chem. 38*: 1250 (1966).
45. G. Eppert and G. Liebscher, *Z. Chem. 18*: 188 (1978).
46. B. Markó-Monostory and S. Börzsönyi, *Tenside Deterg. 22*: 265 (1985).
47. M. Tsubouchi, N. Yamasaki, and K. Matsuoka, *J. Am. Oil Chem. Soc. 56*: 921 (1979).
48. L. C. Howick and R. T. Pflaum, *Anal. Chim. Acta 19*: 342 (1958).
49. J. T. Cross, in *Cationic Surfactants* (E. Jungermann, ed.), Marcel Dekker, New York, 1970, p. 463.

50. S. S. Cooper, *Anal. Chem. 29*: 446 (1957).
51. G. H. Gloss and B. Alson, *Chem. Anal. 43*: 3 (1954).
52. D. M. Patel and R. A. Anderson, *Drug Std. 26*: 189 (1958).
53. L. D. Metcalfe, R. J. Martin, and A. A. Schmitz, *J. Am. Oil Chem. Soc. 43*: 355 (1966).
54. C. N. Wang, L. D. Metcalfe, J. J. Donkerbroek, and A. H. M. Cosijn, *J. Am. Oil Chem. Soc. 66*: 183 (1966).
55. J. B. Wilson, *J. Assoc. Off. Agric. Chem. 29*: 310 (1960).
56. *U.S. Pharmacopoeia*, 16th ed. U.S. Government Printing Office, Washington, D.C., 1960.
57. T. Sakai, M. Tsubouchi, M. Nakagawa, and M. Tanaka, *Anal. Chim. Acta 93*: 357 (1977).
58. M. Tsubouchi, H. Mitsushio, and N. Yamasaki, *Anal. Chem. 53*: 1957 (1981).
59. M. Tsubouchi and Y. Yamamoto, *Anal. Chem. 55*: 583 (1983).
60. M. Tsubouchi and J. H. Mallory, *Analyst 108*: 636 (1983).
61. E. Flotow, *Pharm. Zentrahalle 83*: 181 (1942).
62. A. Spada, D. Coppini and M. Montorosi, *Farmaco (Pavia) Ed. Sci. 12*: 582 (1957).
63. I. Renard, *J. Pharm. Belg. 7*: 403 (1952).
64. B. Budesinsky and E. Vanickova, *Chem. Listy 50*: 1241 (1956).
65. T. Uno and K. Miyajima, *Chem. Pharm. Bull. (Tokyo) 9*: 236 (1961).
66. A. Lottermoser and R. Steudel, *Kolloid Z. 83*: 37 (1938).
67. O. B. Hager, E. M. Young, T. L. Flanagan, and H. B. Waler, *Anal. Chem. 14*: 885 (1947).
68. H. Y. Mohammed and F. F. Cantwell, *Anal. Chem. 52*: 553 (1980).
69. J. B. Wilson, *J. Assoc. Off. Agric. Chem. 35*: 455 (1952).
70. P. A. Lincoln and C. C. T. Chinnick, *Analyst 81*: 100 (1956).
71. P. A. Lincoln and C. C. T. Chinnick, *Proceedings of the First International Congress on Surface Activity*, Paris, 1954, Vol. 1, Sec. 2, p. 209.
72. M. Tsubouchi, N. Yamasaki, and K. Yanagisawa, *Anal. Chem. 57*: 783 (1985).
73. A. W. O'Connell, *Anal. Chem. 58*: 669 (1986).
74. R. D. Hei and M. N. Janisch, *Tenside Deterg. 26*: 288 (1989).
75. J. Cross, in *Nonionic Surfactants: Chemical Analysis* (J. Cross, ed.), Marcel Dekker, New York, 1987, p. 59.
76. Ref. 37, p. 19.
77. F. F. Cantwell and H. Y. Mohammed, *Anal. Chem. 51*: 218 (1979).
78. H. Y. Mohammed and F. F. Cantwell, *Anal. Chem. 51*: 1006, (1979).
79. P. Seguran, *Tenside Deterg. 22*: 67 (1985).
80. J. B. M. Hendry and A. J. Hockings, *Analyst 111*: 1431 (1986).
81. D. Volkmann, *GIT Fachz. Lab. 28*: 278 (1984).
82. J. B. M. Hendry and H. Read, *Analyst 113*: 1249 (1988).
83. Ref. 37, p. 13.
84. Ref. 37, p. 15.
85. K. Vytras, *Collect. Czech. Chem. Commun. 42*: 3168 (1977).
86. E. Benoit, P. Leroy, and A. Nicolas, *Ann. Pharm. Franc. 43*: 177 (1985).

87. C. Gavach and P. Seta, *Anal. Chim. Acta 50*: 407 (1970).
88. A. G. Fogg, A. S. Pathan, and D. Thorburn Burns, *Anal. Chim. Acta 69*: 238 (1974).
89. A. S. Pathan, *Proc. Soc. Anal. Chem. 11*: 143 (1974).
90. S. Pinzauti and E. La Porta, *Analyst 102*: 938 (1977).
91. S. Pinzauti, E. La Porta, G. Papeschi, and R. Biffoli, *J. Pharm. Belg. 35*: 281 (1980).
92. W. Selig, *Fresenius Z. Anal. Chem. 300*: 183 (1980).
93. K. Vytras, M. Dajkova, and V. Mach, *Anal. Chim. Acta 127*: 165 (1981).
94. T. P. Hadjiioannou and P. C. Gritzapis, *Anal. Chim. Acta 126*: 51 (1981).
95. T. K. Christopoulos, E. P. Diamandis, and T. P. Hadjiioannou, *Anal. Chim. Acta 143*: 143 (1982).
96. K. Vytras, J. Kalous, and J. Symersky, *Anal. Chim. Acta 177*: 219 (1985).
97. E. P. Diamandis and T. P. Hakjiioannou, *Mikrochim. Acta II*: 27 (1980).
98. C. J. Dowle, B. G. Cooksey, J. M. Ottaway, and W. C. Campbell, *Analyst 112*: 1299 (1987).
99. J. J. Donkerbroek and C. N. Wang, presented at *CESIO 2nd World Surfactants Congress*, Paris, 1988.
100. A. D. Shoukry, S. S. Badaway, and R. A. Farghali, *Anal. Chem. 60*: 2399 (1980).
101. C. J. Dowle, B. G. Cooksey, and W. C. Campbell, *Anal. Proc. 25*: 78 (1988).
102. S. Pinzauti, G. Papeschi, E. La Porta, P. Mura, and P. Gratteri, *J. Pharm. Biomed. Anal. 6*: 957 (1988).
103. A. D. Shoukry, *Analyst 113*: 1305 (1988).
104. M. Morak, *Tenside Deterg. 26*: 215 (1989).
105. U. Denter, H. J. Buschmann, and E. Schollmeyer, *Tenside Deterg. 28*: 333 (1991).
106. H. H. Y. Oei, I. Mai, and D. C. Toro, *J. Soc. Cosmet. Chem. 43*: 309 (1991).
107. *Active Matter in Anionic Surfactants by Potentiometric Titration*, ASTM D 4251-89, American Society for Testing and Materials, Philadelphia, 1991.
108. *Synthetic Quaternary Ammonium Salts in Fabric Softeners by Potentiometric Titration*, ASTM D 5079-90, American Society for Testing and Materials, Philadelphia, 1991.
109. W. Selig, *Frezenius Z. Anal. Chem. 312*: 419 (1982).
110. W. Selig, *Microchem. J. 36*: 42 (1987).
111. W. Selig, *J. Chem. Ed. 61*: 80 (1984).
112. *Standard Method for Determination of the Average Molecular Weight of Fatty Quaternary Ammonium Chlorides*, ASTM D 2080-64, 1991 Annual Book of ASTM Standards, American Society for Testing and Materials, Philadelphia, 1991.
113. M. E. Puthoff and J. H. Benedict, *Anal. Chem. 36*: 2205 (1964).
114. C. W. Pifer and E. G. Wollish, *Anal. Chem. 24*: 300 (1952).
115. C. W. Pifer, E. G. Wollish, and M. Schmall, *J. Am. Pharm. Assoc. Sci. Ed. 42*: 509 (1953).
116. J. A. Gautier, J. Renault, and F. Pellerin, *Ann. Pharm. Franc. 13*: 725 (1955).

117. F. Pellerin, *Ann. Pharm. Franc. 14*: 193 (1956).
118. C. N. Wang and L. D. Metcalfe, *J. Am. Oil Chem. Soc. 62*: 558 (1985).
119. N. N. Bezinger, G. D. Galpern, N. G. Ivanova, and G. A. Sameshkina, *Z. Anal. Khim. 23*: 1538 (1968).
120. D. C. Wimer, *Anal. Chem. 34*: 873 (1962).
121. F. E. Crane, *Anal. Chim. Acta 16*: 370 (1957).
122. W. J. Kirsten, A. Berggren, and K. Nilsson, *Anal. Chem. 31*: 376 (1959).
123. C. L. M. Holten and H. N. Stein, *Analyst 115*: 1211 (1990).
124. K. Vytras, *Am. Lab.*, p. 93 (Feb. 1979).

6

Potentiometry of Cationic Surfactants

GWILYM JAMES MOODY and J.D.R. THOMAS, School of
Chemistry and Applied Chemistry, University of Wales College of
Cardiff, Cardiff, UK

I. INTRODUCTION

The determination of surfactants in a variety of samples is of increasing importance. The classic method for anionic and cationic surfactants is the two-phase titration method [1]. The modifications currently in use are still subject to the question as to whether the process is time-consuming, laborious, influenced by accuracy, and laboratory personnel are exposed to hazardous solvents. Thus alternative, safer methods have been sought to overcome such problems, in particular, potentiometry based on ion-selective electrodes (ISEs). This chapter is concerned especially with the preparation of sensor material, the formation and evaluation of a variety of alphatic and aliphatic cationic surfactant ISEs, and their application for the potentiometric analysis of cationic surfactants, as well as critical micelle concentration studies [2–9].

6

Potentiometry of Cationic Surfactants

GWILYM JAMES MOODY and J. D. R. THOMAS   School of
Chemistry and Applied Chemistry, University of Wales College of
Cardiff, Cardiff, Wales

I. INTRODUCTION

The determination of surfactants in a variety of samples is commercially
important. The classic method for anionic and cationic surfactants is the
two-phase Epton titration [1]. The modifications currently in use are still
subject to many of its disadvantages: the process is time consuming; colored
samples influence its accuracy; and laboratory personnel are exposed to
hazardous solvents. Thus alternative, safer methods have been sought to
overcome such problems: in particular, potentiometry based on ion-selective electrodes (ISEs). This chapter is concerned especially with the preparation of sensor materials, the fabrication and evaluation of a variety of
aliphatic and aromatic cationic surfactant ISEs, and their application for
the potentiometric analysis of cationic surfactants, as well as critical micelle
concentration studies [2–56].

II. EXPERIMENTAL ASPECTS

A. Preparation of Cationic Sensor Materials

Almost invariably the electrode sensor material comprises an ion association complex formed by precipitation of a cationic surfactant with an anionic surfactant:

$$R_4N^+ + SO_3R'^- \longrightarrow R_4NSO_3R'$$

In effect, the respective salts are chosen according to the need for selectivity toward cationic or anionic surfactants [40]. Thus tetrabutylammonium dodecyl sulfate has been utilized for cationic rather than anionic surfactant sensing [39]. It is readily produced by heating a mixture of $10^{-3}\,M$ solutions of tetrabutylammonium hydroxide and sodium dodecyl sulfate on a steam bath for about 30 min. The resulting salt is vacuum filtered, washed with the minimum volume of distilled water and oven dried at 80°C [39]. Such a liquid ion-exchanger sensor needs to be matched with a compatible solvent mediator for a profitable response to the surfactant of interest. In this case, tritoyl phosphate (TPP) proved suitable.

Another useful sensor material is hexadecyltrimethylammonium dodecyl sulfate, formed by the reaction of hexadecyltrimethylammonium bromide with sodium dodecyl sulfate. The sodium bromide formed is readily removed by dialysis with a Visking membrane for 48 h [7]. After air drying at 100°C, the sensor material is suitable for incorporation into a silicone rubber matrix.

B. Fabrication of Cationic Ion-Selective Electrodes

Several distinct models have been developed for the determination of cationic surfactants.

1. Orion 92-20 Model

The original calcium ISE marketed by Orion is a plastic unit designed to accommodate a calcium liquid ion-exchanger sensor [57]. However, in its own right the 92-20 body structure provides a very simple means of accommodating virtually any liquid sensor material for potentiometric purposes. Thus Birch and Clarke [5] utilized the following cell for critical micelle concentration studies:

Ag/AgCl	tetradecyltrimethyl- ammonium bromide	tetradecyltrimethyl- ammonium dodecyl sulfate/ 1,2-dichlorobenzene in Orion 92-20 body	test solution	SCE

$$\longleftarrow \text{surfactant ISE} \longrightarrow$$

Christopoulos and co-workers [10] have also used the same Orion unit for potentiometric titrations wherein tetrapentylammonium tetraphenylborate dissolved in the solvent mediator 4-nitro-m-xylene was supported on a Millipore Teflon membrane. The reference solution was a mixture of sodium tetraphenylborate (10^{-2} M) with sodium chloride (10^{-1} M). Conditioning was achieved by soaking the electrode in sodium tetraphenylborate (10^{-2} M) for 1 h prior to use, its operative lifetime being about 3 to 4 months [10].

2. Poly(Vinyl Chloride) Matrix Models

The original Orion 92-20 ISE requires a considerable load (0.4 cm^3) of often-expensive sensor cocktail. The system can be greatly simplified [58] by casting thin, low-cost, flexible master membranes from poly(vinyl chloride) (PVC) plus the sensor cocktail dissolved in tetrahydrofuran (THF). Up to 10 PVC ISEs can then be fabricated from just one such PVC sensor cocktail membrane (Fig. 1a) as previously described for nonionic surfactant ISEs [59]. Even more sensor electrodes can be produced (Fig. 1b) from a comparative batch of PVC–cocktail in THF by dip coating onto metal wires [47,53] or graphite rods [38,39]. Hollow graphite tubes internally coated with PVC sensor also provide excellent electrodes for flow injection analysis [40].

The fabrication of a PVC hexadecylpyridinium (HDPy) cationic ISE as described by Shoukry and co-workers [45] is typical of the art. The ion association sensor PT(HDPy)$_3$, made by precipitating phosphostungtic acid (PTA) with hexadecylpyridinium bromide (HDPyBr), was employed in conjunction with dioctylphthalate (DOP) as the solvent mediator and high-molecular-mass Aldrich PVC. Mixtures of four different compositions made up in THF were poured into a petri dish (9.5 cm diameter) and the THF allowed to evaporate. The sensor membranes (about 0.15 mm thick and diameter about 12 mm) were then cut from the larger master membrane and glued to one end of a hollow PVC tube. The construction of each type of ISE was finally completed by filling with a mixed internal reference solution of 10^{-1} M NaCl plus 10^{-3} M HDPyBr (Table 1) and conditioning effected by presoaking in 10^{-3} M HDPyBr for 2 h [45].

Membrane II provided the best electrode in terms of Nernstian slope and linear range (6.3×10^{-6} M to 3.1×10^{-3} M). Its performance was examined further as a function of soaking time in HDPyBr or PTA after 5 min and 2, 3, 6, and 24 h, respectively. The electrode became unreliable after exposure to PTA, whereas soaking in HDPyBr resulted in improved Nernstian behavior. However, 50 days of continuous use in HDPyBr resulted in impaired performance (e.g., the calibration slope fell to 45 mV/decade and the linear range shrank from 10^{-5} M to 6.3×10^{-4} M). Dry storage in a refrigerator is recommended when not in use [45].

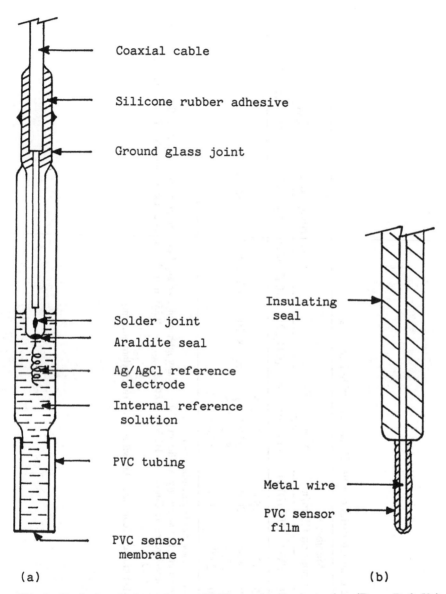

(a)   (b)

FIG. 1 Typical cationic surfactant PVC membrane electrodes. (From Ref. 59.)

TABLE 1 Some Characteristics of PVC Hexadecylpyridinium ISEs

Membrane	Membrane composition (% m/m)			Calibration slope (mV/decade)[a]	Selectivity coefficient, $k_{HDPy,B}^{pot}$ [b]				
	DOP	PVC	PT(HDPy)$_3$		$(CH_3)_4N^+$	$(C_4H_9)_4N^+$	$(C_2H_5)_3C_6H_5CH_2N^+$	HDMA^{+}[c]	NH_4^+
I	42.5	42.5	15.0	54.3 (1.1)	5.7×10^{-3}	9.3×10^{-3}	3.7×10^{-3}	0.35	1.2×10^{-3}
II	40.0	40.0	20.0	59.0 (1.1)					
III	37.5	37.5	25.0	52.9 (0.75)					
IV	35.0	35.0	30.0	53.6 (0.67)					

[a]Relative standard deviations (%) in parentheses ($n = 5$).
[b]Separate solution method. The selectivity coefficient relates to any interference arising from the presence of a cation B: the smaller its value, the better. Thus more interference is caused by HDMA$^+$ ($k_{HDPy,HDMA}^{pot} = 0.35$) than by an equivalent quantity of, say, tetramethylammonium ($k_{HDPy,(CH_3)_4N}^{pot} = 5.7 \times 10^{-3}$).
[c]Hexadecyltrimethylammonium (commonly known as cetryltrimethylammonium).
Source: Ref. 45.

An analogous PVC hexadecyltrimethylammonium bromide (HDABr) ISE system has also been evaluated (Table 2) and conditioning was realized by soaking in 10^{-1} M HDABr [46]. Membrane VI provided the best ISE in terms of Nernstian slope and linear range: namely, 3.2×10^{-6} M to 8.3×10^{-4} M. As with the hexadecylpyridinium ISE, interference from the short-chain cationic surfactants as well as ammonium was negligible (Tables 1 and 2).

The development of durable ISEs based on wires (only 1 to 2 mm in diameter) coated with a thin PVC film of a calcium cocktail in 1971 [60] followed that of the earlier PVC calcium macro model [58]. Since then the range of coated wire electrodes (CWEs) has been extended to cationic surfactant models [51–53] and where PVC could be replaced by epoxy resin or poly(methyl methacrylate) [53]. Thus a CWE for protonated amines was realized by dipping an exposed length of insulated copper wire into a cocktail comprising the amine, PVC, dinonylnaphthalenesulfonic acid, and dioctyl phthalate dissolved in THF. After conditioning in 10^{-3} M or 10^{-4} M analyte for several days, the CWEs were ready to use and it is claimed that such CWEs are equal to, and in some cases better than, the conventional PVC macro models employing an Ag/AgCl internal reference electrode. Thus the Nernstian slopes for a CWE and a PVC macro ISE based on tributylammonium [TBA$^+$] sensor were 58.59 ± 0.50 versus 59.53 ± 0.67 mV/decade for individual electrodes, while their detection limits $(4 \times 10^{-6}$ $M)$ and linear ranges $(10^{-5}$ M to 10^{-2} $M)$ were identical [52]. Again, the selectivity of the CWE model established that $k^{\text{pot}}_{\text{TBA,B}}$ values were closely related to the carbon atom content of different amines [e.g., decyltrimethyl ammonium $(0.54) >$ tetrapropylammonium $(0.02) >$ tetraethylammonium $(4.4 \times 10^{-4}) >$ tetramethylammonium (1.2×10^{-4})], as observed by other workers [32,45,46]. Interference from common inorganic cations (M$^+$) was low $(k^{\text{pot}}_{\text{TBA,M}^+} < 10^{-4})$ [52].

Cocktails may also be deposited on silver, gold, aluminum, or platinum [47] and graphite [38,39]. The sensor layers are typically 20 to 1000 μm thick [47]. Dowle and co-workers [38,39] developed ISEs wherein the tip (about 30 mm) of a graphite rod, otherwise largely sheathed in a larger glass tube (100 mm long), is dip coated with a cocktail of PVC–tritolylphosphate solvent–tetrabutylammonium dodecyl sulfate (0.1% m/m) in THF (Fig. 2). It is imperative to employ THF containing a stabilizer (e.g., 0.025% tri-t-butyl-p-cresol) (C. J. Dowle, private communication). Conditioning was realized by their use in duplicate titrations of sodium dodecyl sulfate $(4 \times 10^{-3}$ $M)$ with diisobutylphenoxyethoxyethyldimethylbenzylammonium chloride $(4 \times 10^{-3}$ $M)$ and commonly known as Hyamine 1622. The near Nernstian response ranged from about 10^{-6} M to 10^{-3} M (i.e., below the critical micelle concentration (CMC) of Hyamine 1622).

TABLE 2 Some Characteristics of PVC Hexadecyltrimethylammonium ISEs

Membrane	Membrane composition (% m/m)			Calibration slope (mV/decade)[a]	Selectivity coefficient, $k^{pot}_{HDA,B}$ [b]				
	DOP	PVC	PT(HDA)$_3$		$(CH_3)_4N^+$	$(C_4H_9)_4N^+$	$(C_2H_5)_3C_6H_5CH_2N^+$	$HDPy^+$	NH_4^+
V	45.0	45.0	10.0	53.3 (1.1)	2.7×10^{-3}	1.6×10^{-3}	9.2×10^{-3}	1.47	4.4×10^{-3}
VI	42.5	42.5	15.0	56.5 (0.94)					
VII	40.0	40.0	20.0	49.9 (0.87)					
VIII	37.5	47.5	25.0	37.0 (1.2)					

[a]Relative standard deviations (%) in parentheses ($n = 5$).
[b]Separate solution method.
Source: Ref. 46.

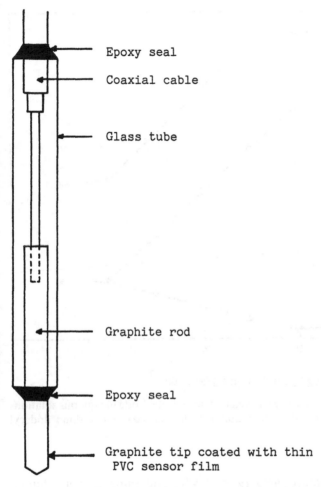

FIG. 2 Graphite PVC-coated cationic surfactant ion-selective electrode. (From Ref. 39.)

As expected, little interference arose from tetrabutyl ammonium ($k_{H,TBA}^{pot}$ $\approx 10^{-2}$; Fig. 3) owing to the smaller number of carbon atoms. Some anionic response was also evident [39].

The particular value of these electrodes lies in a sustained performance in mixed solvent and highly acid media (Fig. 4). This is particularly pertinent in quantifying the cationic and anionic fractions eluted from an ion-exchange column that resolves commercial packages into their anionic, cationic, and nonionic components. In this case, elution had been effected

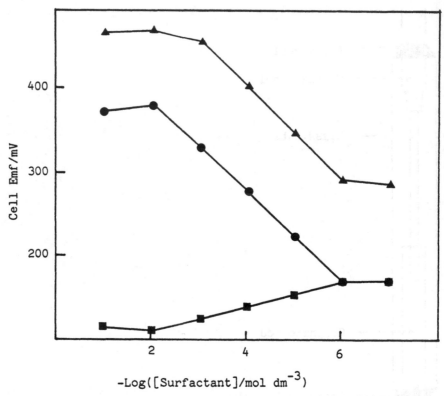

FIG. 3 Calibration of a cationic surfactant-selective electrode in aqueous solutions of Hyamine 1622 (▲); tetrabutylammonium hydroxide (●), and sodium dodecyl sulfate (■). (From Ref. 39.)

with ethanol–hydrochloric acid (9 + 1 v/v) and samples were diluted tenfold before analysis [38,39].

3. Nylon Matrix Models

In analogous fashion, nylon–Hyamine 1622–responsive ISEs can easily be made from the ion-pair complex of Hyamine–pentadecylbenzenesulfonate and nonylphenol mediator mixed with nylon and dissolved in methylene dichloride and methanol (1:1 v/v). The electrode body consisted of a 10-cm³ disposable pipette with its conical tip cut back to give an internal diameter of 1 to 2 mm. This was dipped into the nylon cocktail and after drying for 3 to 4 min, the process was repeated four more times [43]. The electrode was finally assembled with a Ag/AgCl internal reference electrode immersed in potassium chloride (10^{-1} M). Conditioning was then

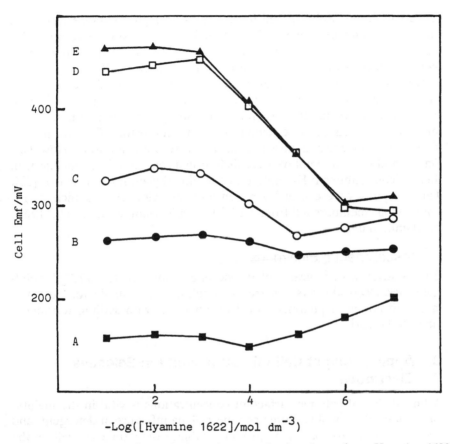

FIG. 4 Response of the cationic surfactant-selective electrode to Hyamine 1622 in ethanol and water. [Ethanol]: A (100), B (80), C (60), D (20), and E (0% v/v). (From Ref. 39.)

effected by overnight immersion in 10^{-4} M sulfonate. Calibration slopes were near-Nernstian and lifetimes 2 to 3 weeks [43].

4. Silicone Rubber Models

The use of flexible, tough silicone rubber matrices to disperse electrode-sensitive materials was first exploited by Pungor and Hollos-Rokosinyi [61]. The same material also proved suitable for fabricating cationic surfactant ISEs [7]. The master membrane was conveniently prepared by curing a mixture of hexadecyltrimethylammonium dodecyl sulfate and silicone elastomer for 12 to 24 h. A sensor disk (diameter 1.3 cm) was cut from the cured membrane and attached to one end of a high-resistance lead glass tube with a silicone sealant. The internal reference system comprised a Ag–AgCl reference electrode with 10^{-1} M KCl and not the cationic surfactant (which stripped AgCl from the reference wire) [7].

Disks comprising 40% m/m of the surfactant provided the best electrodes, with a Nernstian slope of 56 mV/decade in the range 10^{-5} M to 10^{-3} M hexadecyltrimethyl bromide or hexadecyltrimethylpyridinium bromide. Two aspects of these silicone ISEs are interesting; first, no solvent mediator is required and the expected anionic response is absent. However, Davies and Olejnik employed 1,2-dichlorobenzene as solvent mediator in their silicone rubber ISEs based on decyltrimethylammonium sensor [32]. Their responses were near-Nernstian and the selectivity coefficients for homologous members fell from 4.13 for tetradecylammonium to 0.419 for decylammonium.

5. Miscellaneous Electrodes

Both a silver wire indicator electrode and a homemade Ag/Ag$_2$S solid-state disk electrode versus mercury(I) sulfate are suitable for potentiometric titrations of pharmaceutical formulations with sodium tetraphenylborate [18,24].

C. Applications of Cationic Surfactant Ion-Selective Electrodes

A knowledge of cationic surfactant concentrations is vital in the maintenance of product quality during commercial manufacture of detergents and the associated raw materials' product stability over time as well as the formulation of new surfactants [48]. The use of cationic surfactant ISEs in the potentiometric titration mode offers a very simple and reliable means of achieving these goals. Sodium tetraphenylborate is preferred as a titrant to the older picric acid titrant since more compounds (e.g., tetramethyl- and tetraethylammonium ions) are not precipitated with picric acid [26]. Analysis time is typically 2 min per sample compared with about 30 min for two-phase titrations [48]. Minor, yet important applications relate to critical micelle concentration studies [5,32,42,49].

1. Potentiometric Titrations

Gavach and Seta first used liquid membrane ISEs based on alkyltrimethylammonium picrates or tetraphenylborates in nitrobenzene for monitoring titrations of long-chain alkyltrimethylammonium salts with sodium tetraphenylborate [2]. Their pioneering technique enabled the measurement of dodecyltrimethylammonium over the range 10^{-3} M to 2×10^{-6} M at $\pm$

0.1% and ± 0.5%, respectively [2]. Grans plots have also been utilized for determining tetramethylammonium in the range 20 to 200 μmol [9].

The Orion 92-20 electrode loaded with a cocktail of 10^{-2} M tetrapentylammonium/tetraphenylborate plus 4-nitro-m-xylene has been used for basic titration studies with 10^{-2} M sodium tetraphenylborate (Table 3) [10]. The necessary preconditioning for the first titration of the day was realized by one titration of a relatively concentrated sample, while two to three were needed for titrations of dilute samples which gave small potential breaks. Electrodes exhibiting sluggish responses after the semiautomatic titrations caused by adhering precipitate formed in the titration could be restored with a thorough water rinse. Pharmaceutical samples could also be handled, for example, cetylpyridinium chloride in lozenges (RSD = 1%, n = 6), again with about 470-mV endpoint breaks [10].

The same 92-20 series model with an alternative cocktail of either tetraphenylphosphonium 3,5-dinitrosalicylate in p-nitrocumene or dioctadecyldimethylammonium 3,5-dinitrosalicylate in decan-1-ol has been employed to titrate hexadecyltrimethylammonium, hexadecylpyridinium, and didoceyldimethylammonium with sodium 3,5-dinitrosalicylate [33].

The PVC ISE counterparts have also been employed for the potentiometric titration of hexadecylpyridinium [45] and hexadecyltrimethylammonium [46] in pharmaceutical preparations (Table 4). PVC proved to be a better matrix than poly(vinylbutyral) for various ISEs with 2-nitrophenyl-2-ethylhexyl ether as mediator, but interestingly, no sensor was introduced at the casting stage. Here the ion-pair material formed in the titrations saturated the solvent in the thin PVC film after two or three titrations and so completed the sensor cocktail. Should the titrimetric sample be altered,

TABLE 3 Potentiometric Titration of Aqueous Cationic Surfactant Standards with Sodium Tetraphenylborate

Cationic compound	Sample (μmol)	Relative standard deviation (%)	Potential endpoint break (mV)
Tetrabutylammonium	7–25	0.6 (n = 3)	170
Tetradecyltrimethylammonium	5–20	1.3 (n = 4)	380
Hexadecylpyridinium	10–20	1.3 (n = 3)	470
Trimethylphenylammonium	10–20	0.4 (n = 3)	80
Tetraphenylphosphonium	5–20	0.5 (n = 3)	400
Tetraphenylarsonium	5–20	0.3 (n = 4)	400

Source: Ref. 10.

TABLE 4 Potentiometric Determinations of Cationic Compounds in Pharmaceutical Preparations

Sample	Standard addition		Potentiometric titration	
	Taken (mg)	Recovery (%)	Taken (mg)	Recovery (%)
Pure hexadecylpyridinium	0.2–20.0	98.0 (1.1)[a]	0.2–2.0	100.1 (0.82)[a]
Cetapharm	2.0–6.0	101.0 (2.8)	3.5–2.0	99.9 (0.2)
Acnex 99 powder	2.0–10.0	102.8 (2.7)	3.5–15.0	98.6 (0.48)
Feathers cleaner	1.5–10.0	97.9 (1.7)	3.5–25.0	99.8 (0.18)
Pure hexadecyltrimethylammonium	2.0–20.0	98.0 (1.1)[b]	0.2–20.0	101.1 (0.85)[b]
The same in disinfectant	5.0–15.0	102.8 (1.2)	5.0–30.0	97.0 (0.48)

[a] $n = 5$ for first set.
[b] $n = 6$ for second set.
Source: Refs. 45 and 46.

TABLE 5 Cationic Surfactant Content of Fabric Conditioners

| | Surfactant (%) | |
| | Two-phase titration | Potentiometric[a] titration |
Fabric conditioner		
1	7.10	7.15
2	5.40	5.40
3	5.80	5.80
Raw material	64.7	64.8

[a]Coefficient of variation = 2%.
Source: Ref. 2.

only one or two titrations are required to set up the necessary sensor saturation of the PVC film [16].

Pure samples of cationic surfactants were easily titrated with sodium tetraphenylborate, but technical samples needed slow titration around the endpoint. Unlike 2-chloroethyltrimethylammonium, hydrophilic components (e.g., 2-hydroxy-3-choloropropyltrimethylammonium) could not be determined, owing to the poor-quality titration curves [16]. Cationics in fabric conditioners may be titrated [2] directly with sodium dodecyl sulfate or indirectly after adding a known excess of sodium dodecyl sulfate and backtitrating with standard cationic surfactant (Table 5).

PVC-coated graphite electrodes are suitable sensors for triplicate ion-pair titrations of industrial cationic surfactants and anionic surfactants after a preliminary automatic ion-exchange separation (Tables 6 and 7). The reproducible results compare favorably with the two-phase titration tech-

TABLE 6 Analysis of Industrial Cationic Surfactants by Potentiometric and Two-Phase Titration Techniques

| | Endpoint (arbitrary volumes, $n = 3$) | | | |
| | Aqueous solution | | 10% Eluant solution | |
Sample	Potentiometric	Two-phase	Potentiometric	Two-phase
Alkyl benzene quaternary salt	19.1	19.3	22.0	22.3
Amine oxide	18.2	18.8	15.9	15.7
Alkyl quaternary salt	20.2	20.0	16.5	16.4

Source: Ref. 39.

TABLE 7 Analysis of Real Industrial Samples by Potentiometric and Two-Phase Titration Techniques

	Endpoint for total cationics (arbitrary volumes, $n = 3$)	
Sample	Potentiometric	Two-phase
Washing-up liquid (1) ⎫ Washing-up liquid (2) ⎭	None present	None present
Washing powder (1)	4.2	4.2
Washing powder (2)	0.52	0.70
Washing powder (3)	1.39	1.40

Source: Ref. 39.

nique; endpoint breaks were observed from 80 mV up to about 200 mV; inorganic salts did not interfere, but fatty amines caused a ±10% error [38,39].

PVC-coated wires incorporating dimidium bromide plus disulfine blue (1:1 molar ratio), or either component alone, and with tritolyl phosphate mediator have also been employed for the analysis of cationic surfactants in raw material quaternary ammonium salts.

The results using all three types of CWE agreed well with those obtained using the two-phase method and for a given model precision was ±2% ($n = 3$). Numerous titrations could be undertaken without noticeable electrode wear or degradation and none of the CWEs showed any loss in performance after any storage for 9 months. However, CWEs fabricated without any ion-exchange material showed no response during the titration of Hyamine 1622 with sodium lauryl sulfate [48].

2. Critical Micelle Concentration Studies

The critical micelle concentrations (CMCs) for cationic surfactants are very readily evaluated from the sharp breaks in the electromotive force (EMF) versus [cationic surfactant] calibrations as reported previously for nonionic surfactants [59]. A selection is shown in Table 8. Palepu and co-workers [49] have used three types (A to C) of cells with cationic surfactant sensor to evaluate the effective degree of micelle dissociation (α) of tetradecyl-pyridinium (Table 9). The PVC used to fabricate the TDPy ISEs was specially conditioned to neutralize the inherent negatively charged groups in the matrix. In cell A, the five solutions, each with a fixed but different sodium bromide level (e.g., 10^{-4} M) was spiked with TDPy dissolved in 10^{-4} sodium bromide. The CMC was then taken as the sharp break point

TABLE 8 Assessment of CMCs for Cationic Surfactants with Cationic Surfactant ISEs

Surfactant	CMC (M)	Ref.
$[C_{16}H_{33}N^+(CH_3)_3]Br$	1.3×10^{-3}	42
	8.6×10^{-4}	
	8.2×10^{-4a}	32
$\left[C_{14}H_{29}N^+(CH_3)_2CH_2\langle\bigcirc\rangle\right]Cl$	2.1×10^{-3}	42
	2.6×10^{-3b}	5
$\left[C_{12}H_{25}N^+(CH_3)_2CH_2\langle\bigcirc\rangle\right]Br$	4.8×10^{-3}	42
$[(C_8H_{17})_3N^+CH_3]Cl$	1.9×10^{-3}	42
$\left[C_{14}H_{29}N\langle\bigcirc\rangle\right]Br$	2.85×10^{-3}	49
	2.82×10^{-3c}	

[a]Value obtained from conductivity measurements.
[b]Value for bromide salt, which agreed with the value measured with a bromide ISE.
[c]Literature value.

in the two linear sections of the EMF versus log[TDPy] calibration plot. The EMF data relate to the monomer surfactant concentrations in the micellar region, which decrease with total [TDPy].

Cell A:

TDPyBr	test solution	sodium ISE
ISE	+ fixed [NaBr]	(reference)

TABLE 9 Effective Degree of Micelle Dissociation (α) of Tetradecylpyridinium

[NaBr] (M)	CMC (M)	Method for measuring degree of micelle dissociation			
		1	2	3	4
1.0×10^{-4}	2.85×10^{-3}	0.21		0.21	0.28
1.0×10^{-3}	2.55×10^{-3}	0.11		0.24	0.22
5.0×10^{-3}	1.60×10^{-3}	0.19	0.22	0.24	0.20
1.0×10^{-2}	1.10×10^{-3}	0.20		0.26	0.24
2.0×10^{-2}	7.50×10^{-4}				

Source: Ref. 49.

Similarly, EMF measurements with cell B gave the free concentrations of the bromide counterions in the micellar regions:

Cell B:

| TDPyBr ISE | test solution + fixed [NaBr] | Bromide ISE (reference) |

Palepu and co-workers [49] derived the following expression for each surfactant concentration in the micellar region:

$$\frac{E_1}{S_1} = \text{constant} - (1 - \alpha)\frac{E_2}{S_2}$$

where E_1 and E_2 refer to the EMFs measured in cell C for TDPy$^+$ and Br$^-$, respectively, while S_1 and S_2 are the Nernstian slopes of the E_1 versus [TDPy$^+$] and E_2 versus [Br$^-$] calibrations as measured below the CMCs:

Cell C:

| [TDPyBr] ISE or bromide ISE | test solution + fixed [NaBr] | KNO$_3$ or NH$_4$NO$_3$ | Ag\|AgCl\|KCl(saturated) |

$\longleftarrow$————reference electrode————$\longrightarrow$

The α values were then finally calculated from the slopes of the plots of E_1/S_1 versus E_1S_2. Three other techniques were designed to evaluate the α values in Table 9 from the other cell measurements.

III. CONCLUSIONS

Potentiometric titrations of cationic surfactant samples with surfactant ion-selective electrodes provide a very satisfactory alternative to the more traditional methods. Sodium tetraphenylborate is a particularly useful titrant. Substantial advantages accrue in terms of simplicity and speed, as well as freedom from the health hazards associated with dichlorobenzene and chloroform used in Epton 2-phase titrations. Moreover, important information on the critical micelle concentrations and effective degree of micelle dissociation (α) can be obtained from electrochemical cells fitted with the appropriate cationic surfactant ISEs.

REFERENCES

1. S. R. Epton, *Trans. Faraday Soc. 44*: 226 (1948).

2. C. Gavach and P. Seta, *Anal. Chim. Acta 50*: 407 (1970).
3. B. J. Birch and R. N. Cockroft, *Ion-Selective Electrode Rev. 3*: 1 (1981).
4. K. Vytras, *Ion-Selective Electrode Rev. 7*: 77 (1985).
5. B. J. Birch and D. E. Clarke, *Anal. Chim. Acta 67*: 387 (1973).
6. B. J. Birch and D. E. Clarke, *Anal. Chim. Acta 61*: 159 (1972).
7. A. G. Fogg, A. S. Pathan, and D. T. Burns, *Anal. Chim. Acta 69*: 238 (1974).
8. K. Vytras, *Collect. Czech. Chem. Commun. 42*: 3168 (1971).
9. E. P. Diamandis and T. P. Hadjiiaonnu, *Anal. Lett. 13*(B15): 1317 (1980).
10. T. K. Christopoulos, E. P. Diamondis, and T. P. Hadjiioannou, *Anal. Chim. Acta 143*: 143 (1982).
11. C. J. Coetzee and A. J. Basson, *Anal. Chim. Acta 126*: 217 (1981).
12. W. Selig, *Mikrochim. Acta*, 133 (1980/II).
13. B. S. Smolyakov, Yu A. Dyadin, and L. S. Aladko, *Izv. Sib. Otd. Akad. Nauk SSSR Ser. Khim. Nauk*, (6)66 (1980).
14. M. Kataoka, M. Kudoh, and T. Kambara, *Denki Kagaku 46*: 548 (1980).
15. K. Vytras, M. Dajkova, and M. Remes, *Ceck. Farm. 30*: 61 (1981).
16. K. Vytras, M. Dajkova, and V. Mach, *Anal. Chim. Acta 127*: 165 (1981).
17. W. Selig, *Anal. Lett. 15*(A3): 309 (1982).
18. S. Pinzauti and E. La Porta, *Analyst 102*: 938 (1977).
19. S. Pinzauti and E. La Porta, *J. Pharm. Pharmacol. 31*: 123 (1979).
20. S. Pinzauti, E. La Porta, M. Casini, G. Papeschi, and R. Biffolis, *J. Pharm. Belg. 35*: 281 (1980).
21. S. Pintauti, E. La Porta, and M. Casini, *Boll. Chim. Farm. 120*: 43 (1981).
22. S. Pinzauti and E. La Porta, *Pharm. Acta Helv. 56*: 155 (1981).
23. S. Pinzauti, E. La Porta, and G. Papeschi, *Pharm. Acta Helv. 56*: 337 (1981).
24. S. Pinzauti, G. Papeshi, and E. La Porta, *J. Pharm. Biomed. Anal. 1*: 47 (1983).
25. S. Pinzauti and E. La Porta, *J. Pharm. Pharmacol. 31*: 573 (1979).
26. I. A. Gur'ev and I. A. Sbitneva, *Zh. Anal. Khim. 37*: 141 (1982).
27. I. A. Gur'ev, A. A. Kalugin, and E. A. Gushina, *Zavod. Lab. 45*: 309 (1979).
28. I. A. Borisova and I. A. Gurev, *Zavod. Lab. 45*: 309 (1979).
29. I. A. Gur'ev and T. S. Vyatchanina, *Zh. Anal. Khim. 34*: 976 (1979).
30. I. A. Gur'ev and Z. M. Gur'eva, *Zh. Anal. Khim. 38*: 1289 (1983).
31. I. A. Gur'ev and M. I. Drofa, *Zh. Anal. Khim. 38*: 1659 (1983).
32. S. S. Davies and O. Olejnik, *Anal. Chim. Acta 132*: 51 (1981).
33. T. Hadjiioannon and P. C. Gritzapis, *Anal. Chim. Acta 126*: 51 (1981).
34. M. Kataoka, S. Ueda, and T. Kambara, *Nippon Kagaku Kaishi*, 1442 (1982).
35. H. Fruhner, K. H. Kerche, and M. Moesglich, *German (East) Patent DD204,265* (Oct. 22, 1986).
36. T. Maeda and I. Satake, *Bull. Chem. Soc. Jpn. 61*: 1933 (1988).
37. L. Oniciu, D. A. Leng, I. A. Silberg, and D. F. Aughel, *Analusius 14*: 456 (1986).
38. C. J. Dowle, B. G. Cooksey, and W. C. Campbell, *Anal. Proc. 25*: 78 (1988).
39. C. J. Dowle, B. G. Cooksey, J. M. Ottaway, and W. C. Campbell, *Analyst 112*: 1299 (1984).

40. W. C. Campbell and C. J. Dowle, *British patent* 2,207,250, to ICI plc (Jan. 25, 1989).
41. V. N. Ivanov, Yu S. Pravashin, D. P. Stogmushka, and Ya. A. Shuster, *Zavod. Lab. 52*: 12 (1986).
42. X. Qian, J. Hu, P. Gan, C. Gong, and Z. Hu, *Anal. Chem. 16*: 873 (1988).
43. S. H. Hoke, A. G. Collins, and C. A. Reynolds, *Anal. Chem. 51*: 859 (1979).
44. A. F. Shoukry, S. S. Badawy, and Y. M. Issa, *Anal. Chem. 59*: 1078 (1987).
45. A. F. Shoukry, S. S. Badawy, and R. A. Farghali, *Anal. Chem. 60*: 2399 (1988).
46. S. S. Badawy, A. F. Shoukry, and R. A. Farghali, *Microchem. J. 40*: 181 (1989).
47. R. A. Garrison and M. A. Phillippi, *U.S. patent* 4,810,331, to Clorox Company (Mar. 7, 1989).
48. M. A. Phillippi, *U.S. patent* 4,948,473, to Clorox Company (Aug. 14, 1990).
49. R. Palepu, D. G. Hall, and E. Wyn-Jones, *J. Chem. Soc. Faraday Trans. 86*: 1535 (1990).
50. T. Maradone, T. Imoto, and N. Ishibashi, *Bunseki Kagaku 40*: 1 (1991).
51. C. R. Martin and H. Freiser, *Anal. Chem. 52*: 562 (1980).
52. L. Cunningham and H. Freiser, *Anal. Chim. Acta 132*: 43 (1981).
53. L. Cunningham and H. Freiser, *Anal. Chim. Acta 180*: 271 (1986).
54. X. Chen, W. Dong, P. Li, Y. Guo, and S. Wang, *Xuaxue Shijie 31*: 178 (1990).
55. H. H. Y. Oei, I. Mai, and D. C. Toro, *J. Soc. Cosmet. Chem. 42*: 309 (1991).
56. H. Gharibi, R. Palepu, D. M. Bloor, D. G. Hall, and E. Wyn Jones, *Langmuir 8*: 782 (1992).
57. J. W. Ross, *Science 156*: 137 (1966).
58. G. J. Moody, R. B. Oke, and J. D. R. Thomas, *Analyst 95*: 910 (1970).
59. G. J. Moody and J. D. R. Thomas, in *Nonionic Surfactants: Chemical Analysis* (J. Cross, ed.), Marcel Dekker, New York, 1986, pp. 117–136.
60. R. W. Cattrall and H. Freiser, *Anal. Chem. 43*: 1905 (1971).
61. E. Pungor and E. Hollos-Rokosinyi, *Anal. Chim. Acad. Sci. Hung. 27*: 630 (1961).

7

Tensammetric Determination of Cationic Surfactants

M. BOS   Chemical Technology, University of Twente, Enschede, The Netherlands

I. INTRODUCTION

Tensammetry is the general name for a number of electrochemical techniques that deal with the study of adsorption phenomena at the metal electrode–solution interface. The name was originally proposed by Breyer and Hacobian [1,2] and is a composite of *tens* (surface tension) and *ammetry*, as in voltammetry. As most surfactants exhibit adsorption behavior at a metal–electrolyte interface, these techniques can be used in the analysis of those compounds.

Tensammetry has a number of advantages over other techniques in the determination of tensides of which the main ones are its speed and its wide applicability. Factors that have seriously hindered its widespread use are that it is judged to be complex and to require sophisticated equipment and highly skilled operators. Compared to chromatography or related separation techniques, the selectivity of tensammetry is poor. Sometimes this is clearly a disadvantage, but in other cases, as in environmental analysis, this opens the possibility to define characteristic parameters that characterize a sample in terms of the total concentration of a group of related pollutants or surface-active compounds. When needed, the lack of specificity of tensammetry can be used to advantage as a detection method in chromatography or in titrations.

Sensitivity is an important issue in the evaluation of an analytical method: a low detection limit is mandatory to regard a method suitable for trace analysis. Originally, tensammetry with the dropping mercury electrode (DME) suffered from poor detection limits, but with the advent of computer-controlled hanging mercury drop electrodes (HMDEs), adsorptive preconcentration of surfactants came into use, enabling detections limits as low as 0.1 ppm.

II. THEORY

A. The Electrical Double Layer

The adsorption phenomena on which tensammetry is based occur at the electrode–electrolyte interface. This phase boundary is characterized by the presence of an electrical double layer, first described by Helmholtz, who attributed it to electrochemical processes associated with a potential difference between a solid electrode and the solution. Mobile ions with a charge opposed to the surface charge of the solid electrode are attracted by electrostatic forces to the surface of the electrode and form a second charged layer at a distance from the electrode that is determined by the radius of the solvated ions (Fig. 1). Such a double layer acts as a parallel-plate condenser with the electrode surface and a single solvated ion layer as its plates and a dielectricum consisting of solvent molecules. Between the two plates the potential, E, changes linearly with the distance from the

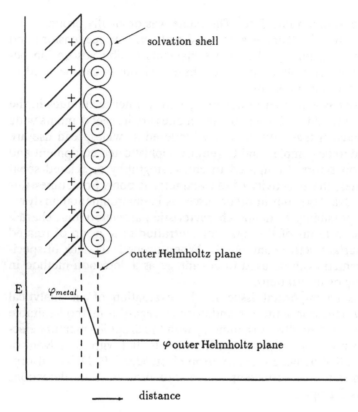

solvation shell

outer Helmholtz plane

E

φ_{metal}

φ outer Helmholtz plane

distance

FIG. 1 Helmholtz model of electrical double layer.

electrode surface. The distance between the two plates is determined primarily by the solvation of the ions, and therefore it changes with the concentration of the electrolyte, because the ions are more solvated at low concentrations. This simple model describes the situation with sufficient accuracy only for high concentrations of electrolytes of which neither the cations nor the anions are specifically adsorbed.

Generally, the situation is more complicated. Due to thermal motion the ions that constitute the outer part of the double layer are not fixed in one plane, but tend to disperse in the direction of the solution. The Helmholtz concept of the electrical double layer was later modified by Gouy [3] and Chapman [4] to accommodate the decrease of the density of a charge opposite to that on the electrode in the diffuse part of the double layer. A combination of the Helmholtz and Gouy–Chapman models was formulated by Stern [5] which describes the electrical double layer in the vicinity of the electrode as a fixed part in which the potential changes linearly with distance and a diffuse part in which the potential changes exponentially to reach the potential of the bulk of the electrolyte (Fig. 2).

In this model the fixed part of the double layer is most important for high ionic concentrations where the electrical field starting from the electrode surface is almost completely compensated by the oppositely charged ions in the fixed layer. For low electrolyte concentrations this field extents further into the solution due to the low density of charged particles. Its influence becomes zero if a volume element of the electrolyte contains the same amount of charge from cations as well as from anions.

Grahame [6] noticed a different behavior of cations and anions in the electrical double layer. In general, anions are only weakly solvated and therefore can interact with the electrode surface in a prominent way. Cations, however, have a smaller radius and are solvated more strongly than

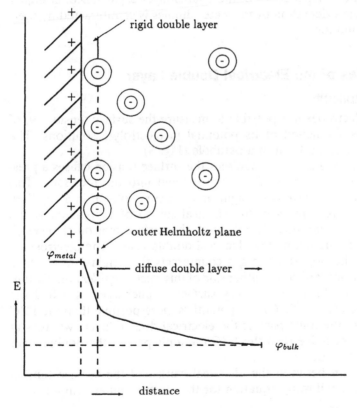

FIG. 2 Stern model for electrical double-layer electrocapillary curve.

anions for the same charge and have more difficulty to reach the electrode surface. According to Grahame, the anions are more easily adsorbed and can reach the electrode more closely, forming the inner Helmholtz layer, whereas the solvated cations can only be bound by electrostatic forces and form the outer Helmholtz plane.

A comparison of electrolytes of the same concentration but with different cations of the same charge shows that the double-layer capacitance varies only slightly with the identity of the cationic species and its radius. This is due to the orientation of the water dipoles in the direction of the electrical field between the electrode surface and the ions in the double layer. This orientation increases with field strength and lowers the relative dielectric constant. The net effect on the double-layer capacitance is that the smaller distance of the smaller ions to the electrode surface (in the parallel-plate condenser model a smaller plate distance) is more or less compensated by the lower dielectric constant. This effect also explains the strong temperature dependence of the double-layer capacitance: at higher temperatures the orientation of the water dipoles is counteracted by thermally induced motion.

B. Properties of the Electrical Double Layer

1. Electrocapillarity

For mercury electrodes it is possible to measure the surface tension, γ, of the mercury as a function of its potential in electrolyte solutions. The resulting curve has the form of a parabola (Fig. 3).

On the uncharged mercury meniscus the surface tension causes a pressure perpendicular to the surface and directed into the meniscus. This pressure can be measured as a height of a mercury column in a capillary that equals this pressure as in the classical set up of Lippmann. If the mercury electrolyte interface is polarized, the surface tension changes as a result of the formation of the electrical double layer. The electrostatic force between the opposite charges counteracts the original force, thus reducing the surface tension on either side of this *electrocapillary maximum*. The potential at which the mercury surface is uncharged is called the *potential of zero charge* (PZC). At potentials more positive than the PZC anions will form the ionic part of the electrical double layer, whereas at potentials more negative than the PZC the ionic part of the double layer will be formed by cations.

The equilibrium between the electrical work and the surface tension work leads to the following equation for the electrocapillary curve:

$$d\gamma + \sigma \, dE = 0 \tag{1}$$

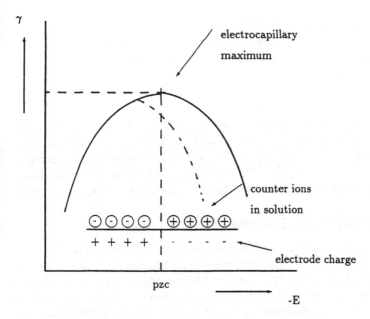

FIG. 3 Electrocapillary curve.

in which $\sigma\,dE$ denotes the electrical work for a change dE in the electrode potential at a charge density of σ and $d\gamma$ the mechanical work for unit area. If the capacitance of the double layer per unit area is assumed constant (C'), the charge density at potential E can be expressed as

$$\sigma = C'(E - E_m) \tag{2}$$

where E_m is the potential of the electrode at which the surface tension of the mercury reaches its maximum value γ_m. In 1 M KCl this potential lies at -0.56 V versus the saturated calomel electrode (SCE). Combining (1) and (2) and integrating produces

$$-\int_{\gamma}^{\gamma m} d\gamma = C' \int_{E}^{E_m} (E - E_m)\,dE \tag{3}$$

or

$$\gamma_m - \gamma = \tfrac{1}{2}\,C'(E - E_m)^2 \tag{4}$$

The parameters that determine the position of maximum of this parabola are E_m and γ_m.

2. Double-Layer Capacitance

In the simple model of the electrocapillary curve given above the specific capacitance of the double layer C' can be found from

$$C' = -\frac{d^2\gamma}{dE^2} \tag{5}$$

The capacitance of the double layer obtained in this way is called the *differential double-layer capacitance*. In aqueous solutions its value is about 20 to 40 $\mu F/cm^2$. It is dependent on the ions that constitute the electrical double layer at a given potential. On the positive side of the electrocapillary maximum the anions determine its value, on the negative side the cations.

C. Adsorption in the Electrical Double Layer

Ions, as well as electrically neutral substances, can be bound to the electrode surface by forces of interaction between the metal of the electrode and the adsorbed species such as van der Waals forces or by chemical bonds (chemisorption). For ionic species, coulombic forces will also play a part if the potential of the electrode is not at its zero charge value. They can support, decrease, or nullify the attraction of the species by the electrode. Anions, in particular, show a tendency for adsorption at the electrode surface due to van der Waals forces, and lose that part of their solvation sphere that is directed toward the metal surface. The stability of these adsorbates is a function of the solvent dielectric constant: a lower dielectric constant favors the adsorption process since it reduces the competition from the solvation process.

Figure 4 shows the situation near the electrode surface, in which specific adsorption of ions on the electrode has influenced the charge distribution and thereby the potential in the vicinity of the electrode. This changes the electrocapillary curve. The presence of adsorptive cations, such as tetraalkylammonium, shifts the maximum on the electrocapillary curve to the positive side and causes a deformation of the negative descending part of the parabola (dashed line in Fig. 3).

Specific adsorption will also have its influence on the double-layer capacitance. The presence of organic compounds at the electrode–electrolyte interface will change the distance of the two plates in the parallel-plate model of the double layer, as well as the dielectric constant of the solvent between these plates.

Adsorption also occurs for electrically neutral compounds. In aqueous solutions the polar water molecules on the electrode surface are displaced by molecules with a lower dielectric constant. If the electrode has a potential that is different from its potential of zero charge, the water dipoles

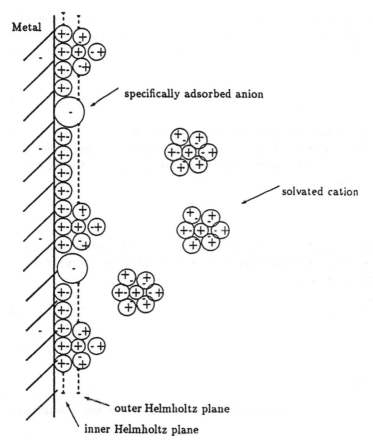

Metal

specifically adsorbed anion

solvated cation

outer Helmholtz plane

inner Helmholtz plane

FIG. 4 Structure of electrical double layer with specific anion adsorption.

are lined up by the high field strengths near the electrode and attracted toward its surface. This process counteracts the adsorption process of the neutral species that generally have a much lower dipole moment. Therefore, the adsorption of these neutral species is most pronounced at the potential of zero charge of the electrode.

1. Adsorption Isotherms

A theoretical description of the adsorption phenomena, based purely on thermodynamic relationships, is only possible for very dilute solutions. For more realistic situations one has to consider the competition between absorbing molecules for the adsorption sites on the surface and interactions between the adsorbed molecules on the surface, and so on. Various models exist from which adsorption isotherms can be derived. The simplest model describes the adsorption process as the formation of a monolayer in which the adsorbed molecules show no interaction and leads to the Langmuir isotherm:

$$\Gamma = \Gamma_s \frac{\beta c}{1 + \beta c} \tag{6}$$

where Γ is the surface concentration of adsorbing species, Γ_s is Γ at full coverage of the electrode, β is the adsorption coefficient, and c the bulk concentration of adsorbing species. The degree of coverage of the electrode can be found from

$$\Theta = \frac{\Gamma}{\Gamma_s} \tag{7}$$

Equations (6) and (7) can be combined to give an expression for the degree of coverage of the electrode as a function of the bulk concentration of the adsorbing species:

$$\Theta = \frac{\beta c}{1 + \beta c} \tag{8}$$

or

$$\beta c = \frac{\Theta}{1 - \Theta} \tag{9}$$

For low coverages this expression can be simplified to the linear isotherm,

$$\Theta = \beta c \tag{10}$$

If interactions of the adsorbed species in the adsorption layer occur, the adsorption can be described by the Frumkin isotherm:

$$c = \frac{\Theta}{1 - \Theta} \exp\left(\frac{\Delta\mu + A\Theta}{RT}\right) \tag{11}$$

The term $A\Theta$ represents the change in interaction of the species in the boundary layer due to adsorption via an energy term that is proportional to the coverage Θ. A negative value for A describes the formation of clusters in the adsorbed state. $\Delta\mu$ is the difference in chemical potential of the adsorbing species in the solution and in the adsorbed state under standard conditions ($c = 1$ and $\Theta = 0.5$) if there is no change in interaction. The value of $\Delta\mu$ is dependent on the potential, not only for ionic species but also for electrically neutral species. Experimental results can be described rather well with

$$\Delta\mu = \Delta\mu_0 + \mu'E + \mu''E^2 \tag{12}$$

which can be used to write Eq. (11) in a potential dependent form as

$$c = \frac{\Theta}{1 - \Theta} \exp \left[\frac{\Delta\mu_m + \mu''(E - E_m)^2 + A\Theta}{RT} \right] \tag{13}$$

where $E_m = -\mu'/2\mu''$ and $\Delta\mu_m = \Delta\mu_0 - \mu''E^2_m$. The physical background behind these equations [7] is that the introduction of a permanent dipole (or ion) in the electrical field of the electrode–electrolyte interface either gives or takes an amount of energy that is proportional to the field strength. The field strength is more or less linear with the potential in the double layer. Moreover, a dipole can be induced. The induced dipoles are proportional to the field strength, and this explains the quadratic term in Eq. (12). At the same time, the amounts of energy that correspond to the removal of solvent molecules from the interface have to be taken into account. Therefore, the values of μ' depend on the difference between the dipole moments of adsorbate and solvent: μ'' depends on the polarizability of both species.

If several different species are adsorbed simultaneously, the Langmuir–Frumkin isotherm cannot be used to describe the situation adequately. In this situation the differences in molecular size have to be taken into account. The way to do this can best be described by starting with a system that has one adsorbing species and that takes into account the difference in surface demand of the solvent and adsorbate molecules. The chemical potential for the adsorbed species is given by

$$\mu_{ads} = \mu_{ads,0} - RT \ln \left[\chi_1 \left(1 - \Theta_1 + \frac{1}{\chi_1} \Theta_1 \right)^{1-\chi_1} \right.$$

$$\left. \times (1 - \Theta_1)^{\chi_1} \Theta_1^{-1} \right] \tag{14}$$

In this equation χ_1 represents the ratio of surface demands of the adsorbed species (P_1) and the solvent molecule (P_0):

$$\chi_1 = \frac{P_1}{P_0} \tag{15}$$

At equilibrium this can be equated to the chemical potential of the free species in the bulk of the solution:

$$\mu_L = \mu_{L,0} + RT \ln c \tag{16}$$

If an interaction term similar to the one in the Frumkin isotherm $(A_{11}\Theta_1)$ is taken into account, this produces the following isotherm:

$$c_1 = \frac{\Theta_1[1 - \Theta_1 + (1/\chi_1)\Theta_1]^{\chi_1-1}}{(1 - \Theta_1)^{\chi_1}\chi_1} \exp \left(\frac{\Delta\mu_1 + A_{11}\Theta_1}{RT} \right) \tag{17}$$

For the adsorption of two competing species, Holleck and Kastening [8,9] give the following expression:

$$c_1 = \frac{\Theta_1[1 - \Theta_1 - \Theta_2 + (1/\chi_1)\Theta_1 + (1/\chi_2)\Theta_2]^{\chi_1 - 1}}{(1 - \Theta_1 - \Theta_2)^{\chi_1}\chi_1}$$

$$\times \exp\left(\frac{\Delta\mu_1 + A_{11}\Theta_1 + A_{12}\Theta_2}{RT}\right) \tag{18}$$

A_{12} denotes the interaction between the two competing species. A similar expression for the isotherm of the second species can be obtained by exchange of the indices.

2. Adsorption Kinetics

The preceding sections dealt with situations in which the adsorption equilibrium at the electrode–electrolyte interface is rapidly established. When this is not the case, then quantities such as Γ and Θ become time dependent as well, and kinetic quantities such as rate of adsorption and rate of diffusion enter the relationships and sometimes completely govern the adsorption process.

For quite a number of cases the diffusion rate of the adsorbing species toward the electrode is the rate-determining step [i.e., for those compounds that already show adsorption at very low concentrations (large values of β and small concentration gradient)]. For the growing mercury drop electrode, Ficks' second law has an extra term that describes the contraction of the diffusion layer during drop growth:

$$\frac{\delta c}{\delta t} = D \frac{\delta^2 c}{\delta x^2} + \frac{2}{3} \frac{x}{t} \frac{\delta c}{\delta x} \tag{19}$$

The solution of this well-known equation for the concentration gradient at the electrode surface,

$$\left(\frac{\delta c}{\delta x}\right)_{x=0} = -\frac{\Delta c}{(\frac{3}{7}\pi Dt)^{1/2}} \tag{20}$$

can be used to derive the Ilkovic equation to calculate the diffusion-limiting current for depolarizers, but it can also be used, according to Koryta [10], to calculate the function $\Gamma = f(c,t)$ for diffusion-limited adsorption. The number of moles that diffuse to the electrode per unit area and per unit time is then given by

$$\frac{dn}{dt} = Dc \left(\frac{3}{7} \pi Dt\right)^{-1/2} \tag{21}$$

The area of the growing mercury drop is given by

$$q_t = 0.85m^{2/3}t^{2/3} \tag{22}$$

in which m represents the mercury flow rate. Multiplication of Eq. (21) with the expression for q_t and integrating over the time the drop has grown gives the total number of moles adsorbed on the electrode:

$$N_t = 0.627D^{1/2}cm^{2/3}t^{7/6} \tag{23}$$

From this the surface concentration of the adsorbate at time t can be calculated:

$$\Gamma_t = \frac{N_t}{q_t} \tag{24}$$

or

$$\Gamma_t = 0.736D^{1/2}ct^{1/2} \tag{25}$$

where D is the diffusion coefficient, c the concentration in the bulk, and t the time.

3. Change of Double-Layer Capacitance Due to Adsorption

The double-layer capacitance C of an electrode partly covered by a single adsorbate can be found if it is considered as a parallel arrangement of the capacitor formed by the uncovered part with a specific capacitance of C_0 and the covered part with a specific capacitance of C_s:

$$C = C_0(1 - \Theta) + C_s\Theta \tag{26}$$

This equation can easily be extended to a situation with more than one adsorbate by including more parallel capacitors.

If there is only one adsorbing species, every adsorbing molecule contributes the same amount to the change of the double-layer capacitance and a *molar capacitance depression* can be defined:

$$\Delta C_n = \frac{\Delta C}{\Gamma} \tag{27}$$

Experimentally, this molar capacitance depression can be found from the capacitance depression at full coverage of the electrode (ΔC_s) and the surface concentration of the adsorbate at full saturation (Γ_s).

The expression for the capacitance decrease per unit area of an electrode can be obtained by combining Eqs. (25) and (27):

$$\Delta C_t = 0.736\Delta C_n D^{1/2}ct^{1/2} \tag{28}$$

and for the whole electrode area given by Eq. (22) one finds that

$$\Delta C_{E_t} = \Delta C_t q_t = 0.627 \Delta C_n D^{1/2} cm^{2/3} t^{7/6} \tag{29}$$

4. Desorption Maxima in the Capacitance Curve

If the differential capacitance of the electrical double layer between mercury and an electrolyte solution containing surfactants is measured against the potential of the electrode, one generally finds a curve exhibiting two more or less pronounced maxima as drawn in Fig. 5.

Around the potential of zero charge the capacitance is lower than that of the corresponding pure supporting electrolyte solution (dashed line in Fig. 5) and can be calculated with (26). The maxima in the curve delimit the potential range where the surfactant is adsorbed on the electrode surface. At potentials more anodic than the left peak, the capacitance curve follows the curve for the supporting electrolyte. This can also be noticed for potentials more cathodic than the right-hand-side maximum. In the vicinity of the maxima, also called tensammetric peaks, Eq. (26) should be extended with an extra term $(q_s - q_0)(d\Theta/dE)$ which is called a pseudocapacitance because it does not represent a pure capacitor, but has ohmic

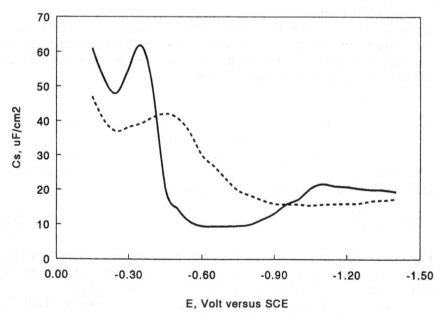

FIG. 5 Adsorption–desorption peaks in differential capacitance curve (dashed curve: no surfactants, supporting electrolyte only).

components as well as being strongly frequency dependent (via the adsorption kinetics). Equation (26) now becomes

$$C = C_0(1 - \Theta) + C_s\Theta + (q_s - q_0)\frac{d\Theta}{dE} \tag{30}$$

The highest values of $d\Theta/dE$ occur in the desorption peaks. In contrast to the situation near the potential of maximum adsorption, a small change in the potential here changes the adsorption equilibrium significantly. Just as in case of Eq. (26), Eq. (30) can be extended to more than one adsorbing species:

$$C = C_0 (1 - \Theta_A - \Theta_B) + C_A\Theta_A + C_B\Theta_B$$
$$+ (q_A - q_0)\frac{d\Theta_A}{dE} + (q_B - q_0)\frac{d\Theta_B}{dE} \tag{31}$$

with C_0, C_A, and C_B the double-layer capacitances at $\Theta = 0$, $\Theta_A = 1$, and $\Theta_B = 1$, respectively, and q_0, q_A, and q_B the charge of the electrical double layer at these coverages. The heights of these tensammetric peaks are a function of the concentration of the adsorbate. However, this function is a linear one only over a limited concentration range [11,12].

The position of the tensammetric peaks on the potential axis is much less dependent on the concentration of the adsorbate [13]. Sometimes this position can be used for identification of the adsorbing compound, but generally it is only used for quantitative purposes in the higher concentration ranges. Theory based on adsorption that can be described by the Frumkin isotherm predicts a variation of the potentials of desorption that can be described by

$$E = a + b \log c \tag{32}$$

in which a and b are constants specific for the adsorbate/supporting electrolyte combination [14].

For cationic detergents the tensammogram often shows only one tensammetric peak. The reason for this is that then there is no potential range at which the cations are replaced by the cations of the supporting electrolyte, even at a very negative charge of the electrode. Adsorption takes place across the whole cathodic branch as well as in part of the anodic branch of the capacitance–potential curve. Only a single peak due to adsorption–desorption is visible on the anodic branch of the curve.

5. Adsorption and the Complex Impedance of the Electrical Double Layer

In the vicinity of the potential of maximum adsorption, the electrical double layer with its adsorbed species can be regarded as a pure condenser with

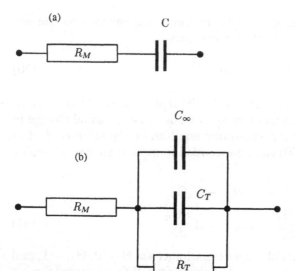

FIG. 6 Equivalent circuits for electrochemical cell: (a) near PZC; (b) at adsorption–desorption potentials.

a capacitance that is independent of frequency. The complete electrochemical cell can be represented by an equivalent circuit consisting of a series connection of a capacitor (C) and a resistor (R_M, the ohmic resistance of the solution) (Fig. 6a).

At the adsorption–desorption potentials a more complicated equivalent circuit is needed to explain the complex impedance of the electrochemical cell [15]. The circuit now incorporates an extra capacitor C_T and resistor R_T, both components of the frequency-dependent kinetic impedance for which Lorenz et al. [16] have given quantitative relationships with parameters of the adsorption and diffusion processes. The extra admittance caused by the adsorption–desorption processes is given as

$$Z_T^{-1} = (\omega^2 C_T^2 + R_T^{-2})^{1/2} \tag{33}$$

Its associated loss angle δ can be found from

$$\tan \delta = (\varphi C_T R_T)^{-1} \tag{34}$$

The kinetic extra admittance, Z_T^{-1}, decreases with frequency due to the changes in C_T and R_T with frequency. It disappears altogether at high frequencies, and the capacitance measured in this situation, C_∞, corre-

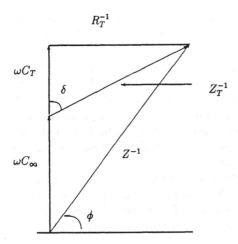

FIG. 7 Vector diagram of cell impedance at adsorption–desorption potentials.

sponds to the pure double-layer capacitance given in Eq. (26). The relatively slow adsorption–desorption processes cannot keep up with the electrical field at these frequencies. From the vector diagram representing the total admittance of the electrochemical cell (Fig. 7) one can see that knowledge of C_∞ enables separate determination of C_T and R_T from measuring the total admittance of the cell Z^{-1} and its associated phase angle ϕ. Mathematically, the total cell admittance at the adsorption–desorption potentials is given by

$$Z^{-1} = [\omega^2(C_\infty + C_T)^2 + R_T^{-2}]^{1/2} \tag{35}$$

III. TENSAMMETRIC MEASUREMENT TECHNIQUES

Tensammetry is not a strictly defined measurement technique. Some consider it as the collection of all electrochemical measurements on nonfaradaic processes, whereas others restrict the use of the term to polarography applied to the study of the electrical double-layer capacitance in relation to adsorption phenomena [15]. In this chapter it is used in its broadest sense, so most of the measurement techniques that can be used to study adsorption phenomena at the electrode–electrolyte interface will be covered here.

Modern electroanalytical instrumentation offers a multitude of possibilities in this respect, ranging from the application of standard techniques that are normally used for the study of faradaic processes, such as differential pulse polarography, to specialized techniques that exploit measurement protocols that discriminate against the here unwanted faradaic response.

Although studies of adsorption phenomena have been performed on numerous electrode materials among which are platinum [17], gold [18], and bismuth [19], the most popular still is, and probably will remain so for a long time, the mercury electrode. Only with mercury electrodes can reproducible results suitable for quantitative treatment be obtained with relative ease; this is due to the simplicity with which a clean and reproducible electrode surface can be ensured.

Automation has played an important role in the development of equipment suited for tensammetry. Modern computerized equipment allows assessment of the complete tensammogram of a sample in a few minutes, together with elaborate data processing techniques such as multivariate calibration.

Basically all modern measurement setups for tensammetry consist of an electrochemical cell with a polarizable small mercury drop electrode (either static or dropping), a nonpolarizable reference electrode, and a counter electrode. The cell is driven by a three-electrode potentiostat that controls the polarization voltage of the working electrode and that measures the current flowing through it. The measurement techniques differ in the polarization–time profiles applied to the cell and in the protocols used for sampling the current. Instruments in which the potentiostat can be controlled by computer can be programmed to run almost any of the techniques of interest.

A. AC Techniques

Very precise measurements of the capacitance of the electrical double layer can be performed with the use of an ac impedance bridge [20]. The measurements, however, are tedious to perform and require considerable skill and have now mostly been superseded by ac polarographic techniques.

1. Phase-Selective AC Polarography

In phase-selective ac polarography the electrochemical cell is polarized with a slowly changing dc ramp voltage onto which a small-amplitude (10 to 30 mV) sine-wave ac voltage is superimposed. The measured quantity is the ac current flowing through the cell with a fixed phase relationship to the applied ac excitation. A lock-in amplifier can be used to obtain the required phase selectivity. The choice of the phase relationship is up to the experimenter. This form of polarography can be used to determine the impedance characteristics of an electrochemical cell as a function of the dc polarization and the frequency of the applied ac modulation. A typical experimental setup is given in Fig. 8.

If two measurements are performed at different phase angles (generally, 0 and 90°), the full complex impedance of the electrochemical cell can be found by using vector calculations, but in the absence of substances undergoing faradaic processes, it suffices to carry out the measurement solely at a 90° phase angle. The measured ac current is then a direct measure for the double-layer capacitance. This is the method of choice if one wants to exploit Eq. (26) for quantitative purposes using the capacity depression caused by the adsorption phenomena.

Phase-selective ac polarography can also be used to determine the double-layer capacitance in the presence of faradaic processes. Using a phase angle of 135° nullifies the ac current component that is caused by the reduction–oxidation of the depolarizers present in the samples and still leaves $\frac{1}{2}\sqrt{2}$ of the ac current through the double-layer capacitance, as can be seen from the equivalent circuit given in Fig. 9 (the Randles circuit [21]) and its corresponding vector diagram, where it is assumed that the charge transfer resistance ($R\Theta$) and ohmic resistance of the solution ($R\Omega$) can be neglected.

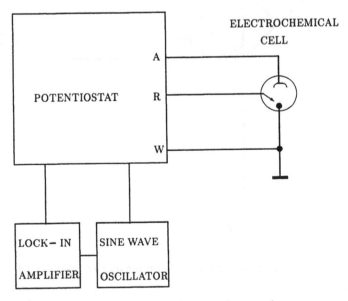

FIG. 8 Setup for phase-selective ac polarography.

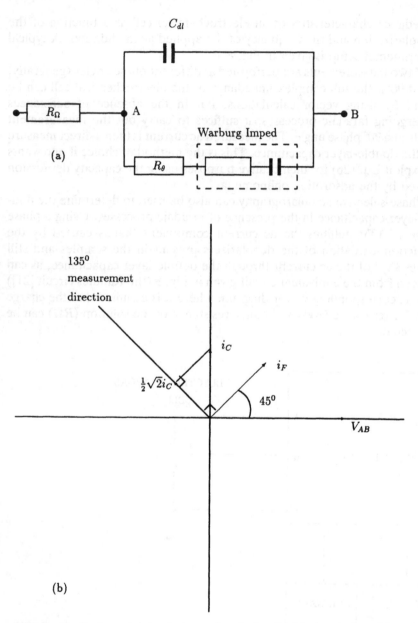

FIG. 9 Equivalent circuit of electrochemical cell with depolarizer (a) and associated vector diagram (b).

2. Second Harmonic AC Polarography

In second harmonic polarography, the experimental setup is identical to
that shown in Fig. 8. The measured quantity, however, is now the ac current
flowing through the cell at twice the frequency of the modulation voltage
[22]. The resulting tensammograms now show a negative and positive peak
separated by a zero crossing near the adsorption–desorption potentials of
the adsorbate [23] if measured phase selectively or the normal peak shape
if total second harmonic current is measured [24]. In both cases the baseline
is much less curved, which makes the peaks easier to evaluate. However,
the technique can only be used for systems showing adsorption–desorption
peaks. In the potential range of maximum adsorption the capacitance of
the cell is only weakly dependent on the potential, and thus the current
will not contain harmonics.

3. The Use of Network Analyzer Equipment

In network analyzer equipment, the process of the determination of the
complex cell impedance at various frequencies by the method described
in Sec. III.A.1 is fully automated. A computer controls the dc polarization
voltage of the electrochemical cell as well as the frequency of the small-
amplitude ac modulation. Also, phase-selective measurement of the resulting ac current is performed by computer. Although this type of equipment may be somewhat too sophisticated (and expensive) for the routine
determination of surfactants, it certainly is very useful for finding the optimal tensammetric procedure [25].

4. Resonant Frequency Measurements

In a method proposed by Bos and Bruggink [26], a very simple experimental
setup is combined with computerized data processing to obtain information
on the double-layer capacitance of the electrochemical cell. In the circuit
that provides the polarization voltage for the cell, an inductor is present,
as shown in Fig. 10. Together with the electrochemical cell this forms a
resonant circuit of which the resonant frequency is determined by the
equivalent capacitance of the cell and the fixed value of the inductance of
the coil. The cell excitation voltage consists of a linear dc ramp onto which
a 5-μs pulse is applied at a fixed time during the drop life of the electrode.
Meanwhile, the resulting ac voltage is sampled at a high rate, from which
the frequency is obtained by fast Fourier transformation, which, in its turn,
serves to calculate the equivalence capacitance of the cell. Apart from the
simplicity of the equipment, the technique has the advantage that it can
determine the double-layer capacitance in the presence of faradaic processes.

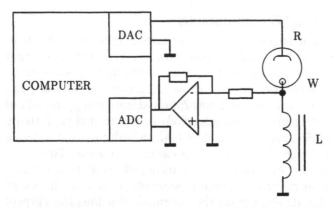

FIG. 10 Circuitry for resonance frequency measurements to determine double-layer capacitance.

B. Pulse Techniques

1. Normal Pulse Polarography

In normal pulse polarography, the excitation voltage program of the DME has the form given in Fig. 11. With each new drop of the DME, the voltage is kept at a prepolarizing value for a fixed time, after which a pulse is applied whose height changes linearly with each new drop. The current is sampled during this pulse. For the study of faradaic processes, the sampling starts after the capacitive response has decayed to a low level, but for tensammetric studies the sensitivity of the method can be much increased if current sampling is started immediately after the start of the pulse.

For tensammetric purposes the duration of the prepolarization step previous to the pulse can be very important, especially if the potential selected

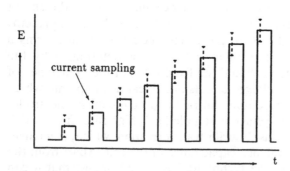

FIG. 11 Voltage program and timing of current sampling for normal pulse polarography.

is in the range where maximum adsorption of the surfactant occurs. If adsorption is strong and rapid, the concentration of the adsorbate at the electrode surface will be governed by diffusion, and longer prepolarization times will favor the sensitivity of the method [27].

2. Differential Pulse Polarography

In differential pulse polarography, most of the capacitive contribution to the cell current is removed. In the absence of faradaic contributions, however, the differential pulse polarogram, in fact, still constitutes a differential capacity curve [28]. Canterford and Taylor [11] have shown that this technique can be used successfully in the analysis of surfactants, especially if a simple modification is applied to the timing of the current measurement with regard to the pulses superimposed on the dc voltage ramp that is applied to the cell [29]. Figure 12 shows a timing diagram of the voltage waveform that normally is used in differential pulse polarography, together with the current response of the cell that is due to the charging of the double-layer capacitance. In the normal situation the signal of interest is the difference in the current measured just before application of the pulse and the current at the end of the applied pulse, where the capacity current has decayed almost completely. In tensammetry it is advantageous to

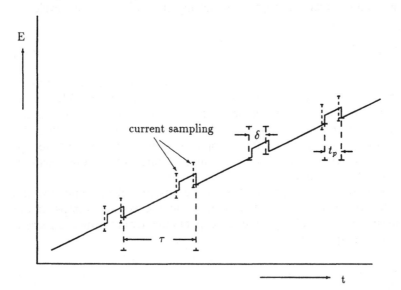

FIG. 12 Timing diagram for differential pulse polarography and charging current: τ is drop time, t_p is pulse duration, and δ is delay between the two current measurements

shorten the delay, δ, so that a more significant difference is obtained. Using this technique, well-developed tensammetric adsorption/desorption peaks can be observed of which the heights show a linear relation with the surfactant concentration up to a certain limit, after which the peak current gradually becomes independent of the concentration.

3. Kalousek Polarography

Generally, the Kalousek measuring technique is used for mechanistic studies, especially for reversibility studies [30]. Cosovic and Branica [31] and Bos [32] have shown that it can also be used for study of the adsorption phenomena in respect to the double-layer capacitance. The characteristics of the Kalousek technique are that (1) a series of square-wave voltage pulses are applied to the electrochemical cell during the lifetime of the mercury drop, and (2) the current measurement takes place during either the even or odd half-cycles of this excitation voltage. In the form most suited for the study of adsorption phenomena, the excitation voltage has a fixed base value that is near the potential of zero charge during one half of the square wave, whereas the voltage value during the other half-cycle changes linearly with each subsequent mercury drop. Figure 13 shows this waveform and the timing of the current measurement. According to Cosovic, the mean current, $\bar{i}$, registered during the pulse-top half-cycles is given by

$$\bar{i} = f \, \Delta E \, C \overline{A} \tag{36}$$

where f is the frequency, ΔE the amplitude of the polarizing potential, C the specific capacitance of the electrode, and $\overline{A}$ its mean surface. Measuring this current at a potential near the desorption range for a surfactant sample and for pure supporting electrolyte without the sample and taking the difference produces a signal $\Delta \bar{i}$ that is proportional to the degree of coverage of the electrode [via Eq. (26)], which can be related to its concentration via the adsorption isotherm.

With computerized equipment and in the absence of faradaic currents it is possible to measure the current decay at the voltage jumps of the excitation signal of the cell and find the capacitance from fitting the current response to the equation

$$i_t = \frac{\Delta V}{R} e^{-t/RC} \tag{37}$$

in which C is the double-layer capacitance and R is the ohmic cell resistance.

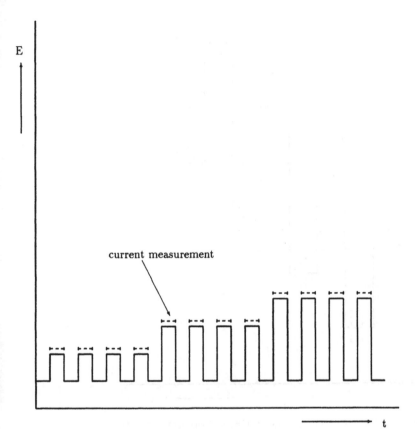

FIG. 13   Kalousek waveform and timing of current sampling.

4. Normal and Double-Pulse Staircase
Sweep Polarography

Britz has developed a fast-scan method with which a complete tensammogram can be obtained at the DME during the last part of its growth
[33]. The waveform of the voltage applied to the DME is a double potential
step sweep as drawn in Fig. 14. The current is sampled at the indicated
positions and summed per positive and negative step. The tensammogram
is constructed by plotting the difference of these two sums versus the
average dc potential during the double pulse (dashed line).

In the same paper Britz showed that differential tensammetric peaks
could be produced from a normal staircase sweep program for the cell
voltage, combined with current measurements at each step level and fol-

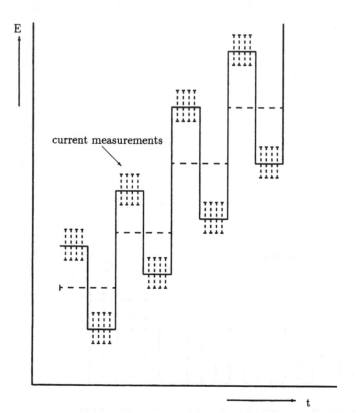

FIG. 14 Voltage waveform used in single-drop tensammetry.

lowed by subtracting the sum of the measured current values of the previous step level from the sum of the next step level. This is particularly applicable to the use of the linear relationship between the adsorption–desorption potential (which in this technique corresponds to a zero crossing between a positive and a negative peak of the signal) and the logarithm of the concentration of the surfactant.

5. Square-Wave Polarography

Square-wave polarography resembles ac polarography in that an alternating signal of small amplitude (5 to 15 mV) superimposed on a linear voltage ramp is used as the excitation signal for the DME. The superimposed signal is not sinusoidal, however, but has a square waveform. The current is usually measured by sampling at short intervals of time during each cycle of the square wave. A comparison with the conventional ac technique by Okamoto [34] shows that it is particularly suited to exploiting the linear relationship between the adsorption–desorption summit potentials and the logarithm of the surfactant concentration.

6. Digital Synchronous Detection

A technique that fully utilizes the advantages of digital signal processing was developed by Seelig and de Levie [35]. The excitation voltage for the DME is a linear dc ramp onto which a special waveform is superimposed that consists of the algebraic sum of a number of sine waves with frequencies that differ by a power of 2 from a base frequency. The current is sampled at short fixed intervals during the application of this waveform and processed by the Hadamard transform, which produces the in-phase and quadrature components of the current at all the frequencies used in the excitation signal.

The advantage of this method clearly lies in the use of various frequencies simultaneously. The cell response is obtained for all these frequencies in the same time that is needed in ac polarography to determine the cell response at a single frequency. In tensammetry, this advantage could be used to discriminate between substances adsorbing at different rates.

C. Indirect Tensammetric Techniques

Adsorption of surfactants in the electrical double layer influences the faradaic processes of electroactive compounds in various ways. First, it can change the standard exchange current density, i_0, and the transfer coefficient, α, of the transfer reaction of the electroactive compound. A quantitative description of this phenomenon by Rek [36] shows that it can be used in a calibration procedure for the concentration of the adsorbing species.

Furthermore, surfactants have a strong influence on the height of polarographic streaming maxima that are caused by an increased transport of the electroactive compound toward the electrode by a streaming motion of the solution [30]. In fact, these substances are used in the polarographic determination of electroactive compounds to suppress these maxima. Faradaic processes that lend themselves well to use this maximum suppression by surfactants quantitatively are the polarographic reduction of oxygen and of mercury ions [37].

D. Hyphenated Tensammetric Techniques

Tensammetry can be used as a detection technique in various other analytical methods. The reason to do so is often the lack of selectivity that hampers some direct applications of tensammetry or the need to operate the analysis in a continuous or semicontinuous mode.

1. Tensammetric Flow-Through Detectors in HPLC and FIA

If selectivity is the issue, a tensammetric detector can be used in combination with high-performance liquid chromatography (HPLC) [38]. For production control, (semi-)continuous methods such as flow injection analysis are more appropriate than the batch operation of polarographic methods [39]. Both techniques require the use of a flow-through detector, the design of which strongly influences the performance of the method. Design rules based on a theoretical description of the mass transfer–controlled adsorption of an electroinactive compound in a flow-through system are given by de Jong et al. [40]. The theory given can be used to derive the optimal electrode radius for a given type of detector and shows that a detector cell in which the fluid flow is directed perpendicular to the mercury flow is optimal in case of the use of a DME. Furthermore, the theoretical expressions can be used to evaluate the signal-to-noise ratio and the linearity of the detector.

An unorthodox design of a tensammetric flow-through detector, in which the role of the mercury and the sample solution are reversed, was given by Scholz et al. [41]. In this design the solution to be investigated flows through a cylinder of glassy carbon that forms the auxiliary electrode, then through a silver cylinder which constitutes the reference electrode, and finally, passes into a mercury pool, where it forms bubbles. As long as the solution bubble is still connected with the main stream of sample, the inner surface of the bubble acts as the working electrode. The main advantage of this bubble electrode is that it circumvents the difficulty of handling a DME in a flow-through manifold.

2. Tensammetric Titrations

Titrations are still popular for the determination of surfactants because of their superior accuracy and stoichiometric nature. The use of tensammetry for the detection of the titration endpoint can circumvent the tedious two-phase titrations, which require mixing and separation of the two phases after each titrant addition [42].

3. Tensammetry with Accumulation on the Hanging Mercury Drop Electrode

The detection limits of the tensammetric determination of surfactants can be improved considerably by preconcentration of the compounds on the mercury electrode [43]. In this process a stationary hanging mercury drop electrode (HMDE) is used. The electrode is kept at the potential of maximum adsorption of the compound. Due to the slow transport of low concentrations of surfactant to the electrode surface, the system is not in an adsorption equilibrium state. Preconcentration is accomplished by keeping the electrode at this potential under stirring, which increases the rate of transport of the compound. After a fixed time the potential is changed in such a way that a desorption peak can be measured. By standardizing all conditions, the height of this desorption peak can be calibrated with the concentration of the surfactant.

E. Practical Aspects in the Choice of a Suitable Technique

In 1988, Bersier and Bersier [44] published a fine review of polarographic adsorption analysis and tensammetry in which they concluded that tensammetry generally needs more attention and care than do other instrumental techniques, and that a specialist is required for the development of new tensammetric techniques. There is no general "best" tensammetric technique. Much depends on the analytical problem to be solved; the main issue in this respect is whether or not selectivity is needed. According to their paper, selectivity can be improved in three ways:

1. *The right choice of the drop time of the DME.* From the theory given in Sec. II.C.2 it can be seen that a difference in the rate of adsorption of two surfactants causes a time dependency in the ratio of their surface concentrations. If the measurements are performed at a relatively short drop time, the faster-adsorbing component will be responsible primarily for the measured signal, whereas at a longer drop time the slower compound will be measured, as it will have displaced the first compound.
2. *The choice of the preconcentration potential.* In tensammetry with accumulation on the hanging mercury drop electrode, Batycka and Lukaszewski have shown that the preconcentration potential is an extra parameter than can be used to control the selectivity of the method [45].
3. *The use of HPLC to bring about the desired separation followed by tensammetric detection.* This is described in Sec. III.D.

If selectivity is not needed (e.g., when the *total* concentration of a number of surfactants is to be determined), the method of choice is the measurement of the depression of the double-layer capacitance near the potential of zero charge of the electrode. The technique required for this, using unmodified, commercially available equipment, is phase-selective ac polarography with ac current measurement at a phase angle of 135° to discriminate against faradaic response (see Sec. III.A.1).

In more demanding situations, one of the more specialized techniques described above can be used. With some programming effort, all the pulse techniques can be performed with a general-purpose computerized electrochemical equipment system consisting of a computer-controlled potentiostat and current measuring system, provided that the equipment is supplied with open software. If a low detection limit is very important, the techniques that use accumulation of the surfactant or the indirect methods that use the polarographic maximum depression of oxygen or mercury are good choices.

IV. APPLICATIONS

The aforementioned review by Bersier and Bersier [44] covers the practical applications of tensammetry and polarographic adsorption analysis from 1974 to 1987, not only for cationic surfactants, but for anionic and nonionic surfactants as well. Two reviews about the analysis of cationic surfactants in general make no mention at all of electrochemical methods [46,47]. Since these reviews, a number of papers have appeared that clearly show the feasibility of the use of tensammetric techniques in the determination of cationic surfactants. These applications can be divided broadly into the following categories: (1) routine quality control, (2) determinations at use level, and (3) determinations in the environment.

A. Routine Quality Control

For the routine quality control of cationic surfactants, titration methods have been developed based on a tensammetric pulse technique for determination of the endpoint [42] and on ac polarography [48]. In the system devised by Bos [42], the sample compounds were diisobutylphenoxyethoxyethyldimethylbenzylammonium chloride monohydrate (Hyamine 1622), N-cetyltrimethylammonium bromide, benzalkonium chloride, and diisobutylcresoxyethoxyethyldimethylbenzylammonium chloride monohydrate (Hyamine 10X). The amounts of sample were between 0.5 and 2.5 μmol and the titrant used was sodium dodecyl sulfate. With proper choice of the measuring potential, differential titrations were possible for some binary mixtures of these compounds. The accuracy of the titrations ranged from 1 to 5%.

Shen et al. [48] used sodium tetraphenylborate as the titrant for the sample compounds cetyltrimethylammonium bromide, tetradecylpyridinium bromide, and cetylpyridinium chloride. Titrations were performed at pH 5.3 in an acetic acid/sodium acetate buffer. Jehring studied the mixed adsorption of the cationic surfactant dodecylpyridinium bromide and the anionic surfactant potassium isooctylsulfonate with ac polarography and found that the depression of the double-layer capacitance by the mixture exceeds the sum of the depressions caused by the separate components [49]. This phenomenon was explained by the interaction forces between the cationic and anionic compounds.

B. Determinations at Use Level

Kinoshita et al. compared the indirect method, in which depression of the polarographic maximum of the oxygen reduction is used, with direct determinations by means of measurement of the depression in double-layer capacitance for a large number of surfactants, including cationics [50]. Suppression of the maximum of oxygen reduction was measured by normal pulse polarography, whereas the double-layer capacitance depression was measured by phase-selective ac polarography. Table 1 shows a comparison of the sensitivities that were obtained for the two methods for a number of cationic surfactants. The values are expressed as the slopes of the following calibration equations:

$$\frac{I}{I_0} = 1 - kc \tag{38}$$

$$\frac{C}{C^0} = 1 - Kc \tag{39}$$

where Eq. (38) is for the suppression of the oxygen maximum and Eq. (39) for suppression of the double-layer capacitance.

Bonivert et al. developed a hand-held device to measure the concentration of surfactants in a plating solution containing metal ions [51]. The measurement comprises the polarization of a working electrode with a dc

TABLE 1 Comparison of Sensitivity for Tensammetric Determination by Suppression of Maximum of Oxygen Reduction and by Measurement of Double-Layer Capacitance Depression

Surfactant	Molecular Mass	k ($\mu g/cm^3$)	K ($\mu g/cm^3$)
Dodecylpyridinium chloride	284	0.92	0.031
Hexadecylpyridinium chloride	358	0.65	0.022
Dodecyltrimethylammonium chloride	264	0.63	0.031
Tetradecyltrimethylammonium chloride	292	0.61	0.030
Hexadecyltrimethylammonium bromide	365	0.60	0.023

bias modulated with a small-amplitude ac signal and determination of the phase angle of the ac current response. The technique was also applied in flow-through cells.

Thermodynamic data on the adsorption of tetradecylammonium chloride, n cetyltrimethylammonium bromide, and cetylpyridinium chloride were obtained by Matysik et al. [52] by double-layer differential capacitance measurements, and by drop-time measurements at a potential of -0.5 V versus SCE from which surface tension data were calculated. The curves were useful analytically within the concentration range from 0.008 to 0.1 mmol/dm^3. The Frumkin isotherm showed the best agreement with the experimentally determined values of the surface tension as a function of the surfactant concentration.

C. Determinations in the Environment

Ćosović et al. compared the use of ac polarography with measurement of the suppression of polarographic maxima for the determination and characterization of surface-active substances in fresh waters [53]. Results were expressed as equivalents of Triton-X-100 and showed that the ac polarography shows more specific differences than the maximum suppression method.

An indirect method for the determination of traces of surfactants in the environment based on their suppression of adsorptive accumulation of the bis(dimethylglyoximato)nickel(II) complex [Ni(DMG)$_2$] was described by Pihlar et al. [54]. They found a linear relationship between the suppression of the height of the Ni (DMG)$_2$ peak and the concentration of various surfactants. The quaternary compounds tested were tetramethylammonium hydroxide, tetramethylammonium tetrafluoroborate, and cetrimide. For the latter substance they found that a much wider linear calibration range is obtained at a preconcentration potential of -0.8 V than at -0.7 V. In their conclusions they stress the importance of correct selection of this preconcentration potential. They claim that the precision that can be obtained with these indirect methods is usually below 2% if sufficient care is taken to keep the surface renewal of the HMDE and the stirring constant.

Sawamoto and Uga described the determination of surfactants in piped and underground water by means of differential capacitance measurements at the mercury–electrolyte interface with ac polarography [55]. Expressed in equivalents of Triton X-100, they were capable of determinations in the ppb range.

Kalousek polarography was applied by Kozarac et al. to estimate the surfactant activity of polluted seawater [56]. Accumulation was used to enhance the sensitivity of the method. The analytical signal was the height of the desorption wave. Less than 1 ppm of N-cetyltrimethylammonium bromide could be determined with a 5-min accumulation period and measurement of the desorption wave at -1.4 V.

The influence of various pretreatment procedures, such as dilution, centrifugation, and filtration, on the results of the determination of surface-active substances in natural waters by capacitance measurements with ac polarography was studied by Ćosović and Vojvodić [57].

V. CONCLUSIONS

From the brevity of the part of this chapter that deals with specific applications, it can be concluded that tensammetry has not seen a major breakthrough in the analysis of cationic surfactants since the review by Bersier and Bersier in which they posed the question: *"Polarographic adsorption analysis and tensammetry: toys or tools for day-to-day routine analysis?"* [44]. Their conclusion that it can be a valuable tool when used properly is confirmed by subsequent publication of the new applications. Still, one wonders why the ratio of papers dealing with the *development* of the tools to those concerning their *application* is still so high. First, this can be caused by the multitude of variants of electrochemical measuring techniques that are suited for tensammetry and the difficulty of making a motivated choice between them. Hopefully, the guidelines given above will provide the necessary information.

Second, quite a number of tensammetric measuring techniques require modification of commercial equipment, which certainly has hindered the widespread use of these techniques. In modern electroanalytical equipment these types of modifications can now be brought about by a change in the software of the computer that controls the instrument. All specialized measuring techniques described in this chapter can be performed by commercially available general-purpose electrochemical equipment with the right software.

Considering the fact that tensammetry can be carried out with relatively simple equipment and that it is a rapid method, it deserves a more prominent place in the analysis of cationic surfactants than it currently has. Maybe this place will be won by the recognition that many analytical problems concerning cationic surfactants can be solved with either phase-selective ac polarography or Kalousek polarography combined with modern chemometrical data processing methods.

REFERENCES

1. B. Breyer and S. Hacobian, *Aust. J. Sci. Res. Ser. A 5*: 500 (1952).
2. B. Breyer and S. Hacobian, *Aust. J. Chem. 6*: 186 (1953).

3. A. Gouy, *J. Phys.* 9: 457 (1910).
4. D. Chapman, *Philos. Mag.* 25(6): 475 (1913).
5. O. Stern, *Z. Elektrochem.* 30: 508 (1924).
6. D. C. Grahame, *Z. Elektrochem.* 62: 264 (1958).
7. B. Kastening and L. Holleck, *Talanta 12*: 1259 (1965).
8. L. Holleck, B. Kastening, and R. D. Williams, *Z. Elektrochem.* 66: 396 (1962).
9. B. Kastening, *Ber. Bunsenges. Phys. Chem.* 68: 979 (1964).
10. I. Koryta, *Collect. Czech. Chem. Commun. 18*: 206 (1953).
11. D. R. Canterford and R. J. Taylor, *J. Electroanal. Chem.* 98: 25 (1979).
12. D. R. Canterford, *Anal. Chim. Acta 94*: 377 (1977).
13. H. Jehring, *Geol. Biol. 6*: 197 (1966).
14. A. N. Frumkin and B. B. Damaskin, in *Modern Aspects of Electrochemistry*, Vol. 3 (J. O'M. Bockris and B. Conway, eds.), Butterworth, London, 1964, p. 149.
15. H. Jehring, *J. Electroanal. Chem. 21*: 77 (1969).
16. W. Lorenz, F. Möckel, and W. Müller, *Z. Phys. Chem. N. F. 25*: 145 (1960).
17. J. J. McMullen and N. Hackerman, *J. Electrochem. Soc. 106*: 341 (1959).
18. G. M. Schmid and N. Hackerman, *J. Electrochem. Soc. 109*: 243 (1962).
19. U. Palm and A. Alumaa, *J. Electroanal. Chem.* 90: 219 (1978).
20. G. J. Hills and R. Payne, *Trans. Faraday Soc. 61*: 316 (1965).
21. J. E. B. Randles, *Discuss. Faraday Soc. 1*: 11 (1947).
22. D. J. Curran, *Int. Lab. 11*: 28 (1981).
23. A. K. Shallal and H. H. Bauer, *Anal. Lett. 4*: 205 (1971).
24. E. Bednarkiewicz and Z. Kublik, *Anal. Chim. Acta 176*: 133 (1985).
25. C. P. M. Bongenaar, M. Sluyters-Rehbach, and J. H. Sluyters, *J. Electroanal. Chem. 109*: 23 (1980).
26. M. Bos and W. H. M. Bruggink, *Anal. Chim. Acta 152*: 35 (1983).
27. J. Flemming, *J. Electroanal. Chem. 75*: 421 (1977).
28. J. H. Christie and R. A. Osteryoung, *J. Electroanal. Chem. 49*: 301 (1974).
29. D. R. Canterford and R. W. Brown, *J. Electroanal. Chem. 119*: 355 (1981).
30. J. Heyrovskey and J. Kuta, *Principles of Polarography*, Academic Press, New York, 1966, p. 477.
31. B. Cosovic and M. Branica, *Electroanal. Chem. Interface Electrochem 46*: 63 (1973).
32. M. Bos, *Anal. Chim. Acta 122*: 387 (1980).
33. D. Britz, *Anal. Chim. Act 115*: 327 (1980).
34. K. Okamoto, *Bull. Chem. Soc. Jpn. 37*: 293 (1964).
35. P. F. Seelig and R. de Levie, *Anal. Chem. 52*: 1506 (1980).
36. J. H. M. Rek, Adsorption and electrokinetic parameters, Ph.D. thesis, Utrecht, The Netherlands, 1963.
37. V. Zutic, B. Cosovic, and Z. Kozarac, *J. Electroanal. Chem. 78*: 113 (1977).
38. J. Lankelma and H. Poppe, *J. Chromatogr. Sci. 14*: 310 (1976).
39. M. Bos, J. H. H. G. van Willigen, and W. E. van der Linden, *Anal. Chim. Acta 156*: 71 (1984).
40. H. G. de Jong, W. Th. Kok, and P. Bos, *Anal. Chim. Acta 155*: 37 (1983).

41. F. Scholz, M. Kupfer, J. Seelisch, G. Glowacz, and G. Henrion, *Fresenius Z. Anal. Chem. 326*: 774 (1987).
42. M. Bos, *Anal. Chim. Acta 135*: 249 (1982).
43. R. Kalvoda, *Anal. Chim. Acta 138*: 11 (1982).
44. P. M. Bersier and J. Bersier, *Analyst 113*: 3 (1988).
45. H. Batycka and Z. Lukaszewski, *Anal. Chim. Acta 162*: 207 (1984).
46. L. D. Metcalfe, *J. Am. Oil Chem. Soc. 61*: 363 (1984).
47. H. Klotz, *Tenside Deterg. 24*: 370 (1987).
48. Y. Shen and H. Gong, *Fenxi Huaxe 19*: 680 (1991); *C.A. 115*: 282495g (1991).
49. H. Jehring, *Z. Phys. Chem. 246*: 1 (1971).
50. H. Kinoshita, M. Kobayashi, and A. Matsuoka, *Bull. Chem. Soc. Jpn. 60*: 1195 (1987).
51. W. D. Bonivert, J. C. Farmer, and J. T. Hachman, U.S. patent 4,812,210 (1989); *C.A. 111*: 89471k (1989).
52. J. Matysik, H. Kroszka and A. Persona, *Adsorption Sci. Technol. 4*: 53 (1987).
53. B. Ćosović, V. Vojvodic, and T. Plese, *Water Res. 19*: 175 (1985).
54. B. Pihlar, B. Gorenc, and D. Petric, *Anal. Chim. Acta 189*: 229 (1986).
55. H. Sawamoto and K. Uga, *Bunseki Kagaku 38*: 288 (1989).
56. Z. Kozarac, B. Cosovic, and M. Branica, *J. Electroanal. Chem. 68*: 75 (1976).
57. B. Ćosović and V. Vojvodic, *Marine Chem. 22*: 363 (1987).

8

Analysis of Low Concentrations of Cationic Surfactants in Laboratory Test Liquors and Environmental Samples

JAMES WATERS Ecotoxicology Section, Unilever Research Port Sunlight Laboratory, Wirral, Merseyside, England

I. INTRODUCTION

Cationic surfactants are important synthetic chemicals that are produced in high tonnages and are widely distributed in their use. Surfactants are conveniently classified by the nature of the ionic charge on the molecule in solution. In the case of cationic surfactants, they are characterized by a positively charged group such as a quaternary ammonium nitrogen. How-

235

ever, molecules such as long-chain amines and amine oxides that can be protonated in aqueous solution to give a positive charge are also considered as types of cationic surfactant. A high proportion of all surfactants used, such as detergent ingredients, will be used in conjunction with water and will eventually be disposed of in sewers in domestic and industrial wastewaters.

Even though the use of cationic surfactants is less than that of anionic and nonionic surfactants, approximately 70,000 tons of these actives are used annually in Western Europe. The quaternary ammonium compounds containing at least one long hydrophobic alkyl chain, used commercially as disinfectants, fabric softening agents, foam depressants, and antistatic agents, account for a large proportion of this total. Until recently the single most important surfactant type was the dialkyldimethylammonium compounds, typically with 14 to 18 alkyl carbons, used in fabric conditioning products (representing up to 70% of total cationic use). However, in Europe this type of surfactant is undergoing replacement by cationic surfactants which contain ester functions in the long hydrophobic groups. These new molecules biodegrade more rapidly than do the dialkyldimethylammonium compounds, due to the presence of the "weaker" ester moiety. Other quaternary ammonium compounds with alkylaryl, imidazoline, and pyridinium structures are also used in significant quantities. There is also a range of high-molecular-weight aliphatic and heterocyclic amine–amine oxides that can be positively charged in aqueous solution and will have properties analogous to those of the quaternary ammonium compounds. The importance and widespread use of cationic surfactants have necessitated the development of analytical methods to determine them in their manufacture and in the many products in which they are used. Although there is a comprehensive range of analytical methods detailed in this volume for the foregoing purpose, those available for use in environmental testing applications are more limited.

Since the introduction in the early 1950s of the first major synthetic surfactants used in household detergents, there has been an ever-increasing awareness within both the detergent industry and regulatory authorities of the importance of laboratory testing of new synthetic organic molecules, such as surfactants, to establish their biodegradability, aquatic toxicity, and possible effects on sewage treatment processes, and so on, prior to widescale use. For large-volume materials that can reach the environment, it is now generally accepted that confirmation of laboratory testing should be obtained by environmental monitoring programs, which establish their distribution and fate in sewage treatment plants, rivers, and so on. Such laboratory and environmental testing can only be performed with the aid of appropriate analytical methodology.

In this chapter we cover those analytical methods that have either gained a degree of acceptance or offer improved means for the determination of cationic surfactants in environmental testing applications. Although it is very apparent from the published literature that many researchers claim methodology suitable for the determination of low concentrations of cationic surfactants, the majority of the methods will not readily be applicable to support studies for environmental testing of these substances.

Analytical measurement techniques based on the use of radio-labeled compounds (e.g., ^{14}C, ^{35}S), physicochemical properties (foaming, surface tension, etc.), and gross parameters that summate the total organic content of test systems (biochemical oxygen demand, BOD; chemical oxygen demand, COD; total organic carbon, TOC, CO_2 evolution, and the like) have been used successfully to support laboratory testing of organic chemicals, particularly for biodegradability. These techniques have been well documented by Swisher [1]. However, for the analysis of environmental samples, they will generally have limited value. Therefore, for the purposes of this chapter, only those chemical techniques capable of determining a specific cationic surfactant or this class of surfactant are considered.

A. Analysis of Mixtures

Compared to the major anionic and nonionic surfactants, many cationic surfactants have simple, well-characterized compositions. The commercial long-alkyl-chain quaternary ammonium compounds will be composed of a limited number of homologous structures and contain typically only small amounts of impurities, such as long-chain amines. In this respect, the analysis of cationic surfactants in environmental testing can be easier compared to other major surfactants. However, the presence of a positively charged group in these molecules means that they have a strong affinity to associate with negatively charged sites on a wide range of substrates and surfaces found in nature. The fact that many cationic surfactants, certainly those used in fabric conditioning applications, also have a very low water solubility will further result in these molecules being removed from solution by complexation (with anions) and adsorption onto solid matrices. This will obviously be a major consideration for analyses of such surfactants in laboratory test liquors and environmental samples where opportunities for adsorption and complexation arise. Further consideration of this topic is given in the next section.

For the purpose of laboratory testing of cationic surfactants, methodology is typically required for their determination at mg/L concentrations of lower, whereas for the analysis of environmental samples such as effluents, river waters, and so on, even lower concentrations (μg/L) must be quantified. This analytical capability is available for a number of major cationic surfactants, with recent advances in analytical techniques particularly exemplified by high-performance liquid chromatography with conductivity detection and postcolumn solvent extraction–ultraviolet/fluorescence detection, fast atom bombardment mass spectrometry, and so on. Further work is required, however, to extend this capability to other surfactants.

B. Representative Sampling and Sample Preservation

The analysis of cationic surfactants at the low concentrations encountered in environmental testing and monitoring can be influenced significantly by the tendency of these materials to concentrate and adsorb at interfaces. Surfactants discharged to sewers will distribute between the liquor and solid fractions of sewage and as a result will undergo quite different forms of treatment (aerobic and anaerobic) in sewage works. An assessment of their subsequent distribution and fate in the environment will therefore necessitate the use of appropriate sampling and analytical techniques for determining these materials in "adsorbed" as well as "soluble" forms.

Should foam be formed on samples, it will contain a much higher concentration of surface-active substances than that in the bulk solution, and thus difficulty may be encountered in obtaining a representative sample for analysis. For this reason, the formation of foam should be avoided during sample preparation. Adsorption onto the walls of sample containers and analytical equipment may result in losses of material before and during analysis for samples containing low concentrations of surfactant, particularly cationic surfactants. Special precautions can be employed to minimize the problem, including the rinsing of emptied sample bottles with solvent to ensure quantitative recovery of cationics, the use of sample stabilizers to hold surfactants in solution, and glassware preconditioned to reagents and test substances to reduce losses. For example, Osburn [2] recommended that environmental liquor samples containing low concentrations of cationic surfactant should be stabilized by the addition of 1% v/v of formalin (40% v/v formaldehyde solution) and 5 mg/L of nonionic surfactant, to prevent sizable losses of material through biodeterioration and/ or adsorption onto the sample vessel. Both laboratory testing and environmental liquor samples can contain suspended solids of varying character upon which surfactants can adsorb. For an analysis of total surfactant content a representative sample will be required. Where this is difficult or impossible to achieve, the alternative is to remove the particulates, preferably by centrifugation, and to analyze the liquor and solids separately. Solids on which surfactant could be adsorbed, such as sewage sludge, soil, and river sediments, will require very careful preparation to ensure that they can be sampled representatively [3]. It is usual to dry, grind, and particle size such samples prior to analysis for organics.

In a somewhat different context, the ready degradation of some cationic surfactants may result in an underestimation of their concentration in environmental testing samples if they are not analyzed within a short time of being sampled (i.e., a few hours). This problem can be overcome by the use of sample preservatives [1]. An addition of 1% v/v of formalin, a 40% formaldehyde solution, has been found to be very effective for holding the concentration of all major surfactant classes in environmental liquor samples for extended periods (i.e., months) if stored in a refrigerator at 4°C. The use of 5% v/v of formalin has been recommended for preserving samples with high concentrations of biomass, such as sewage sludges [1]. However, it should not be assumed without further investigation that the loss of a cationic surfactant from solution during a biodegradability study is attributable to its degradation alone. For the reasons already discussed above, a mass balance for the material in the test system should be established to ascertain the amount in solution, amount adsorbed on the test vessel, and by difference that biodegraded.

C. Nonspecific Versus Specific Analysis

The need to determine routinely low concentrations (0 to 20 mg/L) of surfactants in laboratory test liquors and environmental samples has led in the first place to the development of analytical methods capable of assessing the important members of the main surfactant classes (e.g., colorimetric disulfine blue procedure for cationics, colorimetric methylene blue procedure for anionics, and the Wickbold potentiometric titration method for alkoxylated nonionic surfactants). The chemistry of these summary methods is such that determinations are not specific for synthetic surfactant alone (hence the term nonspecific analysis). For the analysis of a known synthetic surfactant in clean water solutions these methods can give an accurate determination of the surfactant present. Their use for determining intact surfactant in liquors from biodegradability tests are generally considered satisfactory for assessing primary biodegradability; control liquors are usually available to correct for any interfering materials in test liquors. (Primary biodegradation is defined as the oxidation or otherwise modification of a molecule by bacterial action to the extent that its characteristic properties are no longer evident or when it no longer responds to analytical procedures more or less specific for determining the original surfactant.) Such analytical methodology has been adopted for use in legislative biodegradation test protocols, in particular the EEC directive for determining the biodegradability of anionic and nonionic surfactants

[4], because of their relative simplicity, convenience, and freedom from interferences. However, the lack of specificity and sensitivity of these non-specific procedures means that their application to the determination of surfactants in environmental samples, where concentrations are generally lower and many more interferences are encountered, is less satisfactory. At best they provide only conservative estimates of surfactants. In addition, these methods are usually unable to cope with the complete range of environmental matrices (e.g., sludge, sediment, and soil samples).

The need to demonstrate the distribution and fate of major chemicals in the environment is increasingly important for assessments of their environmental safety. For this purpose, specific and sensitive analytical methods are required for determining the actual concentrations of a given molecule in all relevant matrices. Major advances in the development of such methodology for cationic surfactants are described in later sections.

D. Concentration and Separation Techniques

The success of analytical methodologies used for the environmental testing of surfactants is usually dependent on having techniques for concentrating low concentrations of material free of interfering compounds in addition to a sensitive analytical detection. When analytes are present at ppb to ppm concentrations in samples, large sample sizes (e.g., liter volumes or gram quantities) are generally required to allow enough material (as little as a few micrograms) to be concentrated for an accurate analytical determination. The physical form of the sample will very much determine the type of techniques that can be readily employed for the concentration and separation of cationic surfactants.

Solvent extraction steps employing direct reflux and Soxhlet techniques are used to recover and concentrate cationic surfactants from solid samples (e.g., sludge, sediment, and soil) [5,6], and are invariably followed by further cleanup of the resulting extracts. However, there is a need to develop alternative techniques for recovering the new "ester" cationic surfactant molecules from solid substrates to avoid their chemical breakdown; new approaches are being sought such as the use of supercritical fluid extraction (SFE). Although trace enrichment [or solid-phase extraction (SPE)] techniques based on the use of disposable minicolumns are increasingly preferred for concentrating polar organics, typified by surfactants, at low concentrations from large volumes of liquor, to date, there are few documented examples for the cationic surfactants [7]. A wide range of absorbent packings such as octadecyl-bonded silica, ion-exchange resins, macroreticular resins, and so on, have been used to concentrate surfactants selectively on these columns, followed by their elution in very small volumes of solvent. Direct evaporation of aqueous samples has found considerable use as a first step for concentrating cationic surfactants [2,8]. Liquid–liquid solvent extraction has also been employed successfully for concentrating cationic surfactant from aqueous samples [9]. In addition, the Wickbold solvent sublation technique has been found effective by some researchers for concentrating and separating cationic surfactant free from nonsurfactant interfering material [10,11] from dilute aqueous solutions. With this technique "soluble" surface-active substances are transported from the aqueous solution into an overlying layer of solvent by bubbling gas through it.

Many of these concentration techniques result in a separation of the analyte free from possible interfering materials. The complex nature of environmental matrices, however, generally require further cleanup of the resulting concentrates to free the cationic surfactants from most interfering materials. Additional separational steps may typically include trace enrichment onto disposable columns, solid–liquid extraction, liquid–liquid extraction, and so on. Ion-exchange chromatography is a particularly important technique for separating cationic surfactants from anionic surfactants, which can cause serious interferences. Examples of all these concentration/cleanup techniques will be given at appropriate points in the text. Most specific methods employed for determining surfactants have a number of sample preparation steps and are consequently involved and time consuming. Only in this way can the necessary accuracy and precision be assured for trace analyses in environmental testing matrices. The use of the separational capability of modern chromatographic techniques together with highly specific detection systems has further enhanced the ability to determine cationic surfactants. In this respect, HPLC is particularly effective for this purpose.

II. NONSPECIFIC ANALYTICAL METHODS: COLORIMETRIC TECHNIQUES

Colorimetric techniques are used routinely to determine low concentrations of cationic surfactants in aqueous solutions [1,12] and have found application in the laboratory testing and environmental monitoring of these materials. Molecules containing a quaternary ammonium nitrogen moiety as well as amines and amine oxides that can be protonated under appropriate aqueous conditions can be analyzed. The colorimetric (fluorescence) methods generally have the same common analytical basis, that is, the formation of solvent extractable compounds between the cationic surfactant and an intensely colored or fluorescent anionic species [12]. The most popular anionic reagent for this purpose is probably disulfine blue [2,8,13]. This same determination principle has also been used extensively as the basis of many other procedures employed for the determination of anionic surfactants. Some authors have referred to this as paired-ion extraction (PIX) analysis.

The anionic reagents used are not appreciably extracted from water by organic solvents but can form solvent-extractable, stoichiometric ion-association compounds with cationic surfactant by the neutralization of their negative charge with that on the cationic polar group, (i.e., quaternary ammonium nitrogen, protonated amine, etc.). The ion pair is readily extractable into organic solvents such as chloroform provided that there are sufficient carbon atoms in the surfactant structure to impart enough hydrophobicity. The intensity of the color (fluorescence) of the anion–cation ion pair in the extractant gives a direct measure of the surfactant present. The sensitivities of the procedures are such that microgram amounts of cationic surfactant can readily be analyzed.

PIX reactions are generally applicable to all the major classes of cationic surfactants. Hence their application to mixtures of cationic surfactants will result in an estimate of the total surfactant present, and reference to an appropriate standard will have to be made to express the result. Differentiation between quaternary ammonium compounds and amine/amine oxides can be achieved by the use of an appropriate anionic reagent at two suitable solution pH values. Under "acidic" conditions both quats and amine/amine oxides can form extractable ion-association complexes, whereas at a higher pH, where the amine/amine oxide is predominantly unprotonated, only the quat will respond.

The simple chemistry of the anion–cation reaction means that it cannot be specific for synthetic cationic surfactants, since any natural molecules containing a single cationic grouping and a hydrophobic moiety should also be capable of forming extractable compounds with anionic reagents. All procedures are therefore susceptible to positive interferences. Negative interferences can also occur as a result of the direct competition from other anionic materials, such as organics containing sulfonate and sulfate groups, with the anionic reagent for cationic surfactant. Colorimetric procedures can therefore be expected to show similar interferences, the degree of interference being dependent on the anionic reagent and the procedure used. The interference of some species can be partially or completely removed by the correct choice of reaction conditions or by separation. However, not all interferences can be eliminated, and the results obtained for unknown samples cannot be taken as representing cationic surfactant alone.

A. Disulfine Blue Analysis

The disulfine blue molecule is used as a pH indicator dye (pK_a 2.63) and has a very high molar extinction coefficient (96,200 at 640 nm in water)

[13]. Biswas and Mandal [13] have demonstrated that only the blue base form of the dye forms chloroform-extractable 1:1 stoichiometric compounds with cationic surfactants from both neutral and acidic solutions. Klotz [14] found that the use of citrate buffer (pH 3) allowed quaternary ammonium compounds and tertiary amines to be extracted as their disulfine blue ion association compounds, whereas a phosphate buffer (pH 7) resulted in extraction of the quaternary compounds alone. He also showed the use of a chloroform/n-butanol (95:5) extractant in place of chloroform resulted in a more efficient extraction of the disulfine blue ion-association compounds particularly those formed with ethoxylated cationic surfactants.

Waters and Kupfer [8] and Klotz [14] have determined the structural requirements for some major cationic surfactant classes to respond to disulfine blue. Monoalkyltrimethyl- and dialkyldimethylammonium compounds with greater than 11 and 7 carbons, respectively, in the alkyl groups were efficiently extracted as their disulfine blue ion-association compounds into chloroform. On this basis, the commercially important cationic surfactants should generally respond to the disulfine blue reaction. When applied to the analysis of cationic surfactants in "clean" water samples, accurate trace determinations can be made without any interference [15]. However, the analysis of cationic surfactants in aqueous solutions, such as aquatic toxicity test media, containing high concentrations of certain inorganic anions (e.g., chloride, nitrate, etc.) may require standard calibrations to be performed in the same matrix as the samples, to ensure the most accurate results. The presence of such inorganic anions can result in a somewhat reduced but still linear response for the test substance compared to a distilled water calibration. The sensitivity of the disulfine blue reaction is such that at least 0.01 to 0.02 mg/L of cationic surfactant can be determined in a 100-mL sample.

Use of the direct disulfine blue reaction, however, is unsuitable for the determination of cationic surfactants under more realistic testing conditions in the presence of anionic surfactants or in complex media where other anionic interferences may occur. Procedures fail because the anionic interferences interact more strongly with cationic surfactants than does the anionic dye reagent. This is particularly true of environmental samples, where synthetic and natural anionic material will generally be in great excess over the cationic surfactant present.

Waters and Kupfer [8] reported the first method that attempted to determine cationic surfactants in biodegradation test liquors (0.1 to 10 mg/L) containing "anionics." They concentrated the cationic surfactant in unfiltered aqueous samples (<200 mL) by direct evaporation; linear alkylbenzene-sulfonate (LAS) was actually added to those samples low in anionic to facilitate the overall recovery of cationics. The resulting solid cation–anion association compounds formed were quantitatively separated from the sample matrix by extraction with hot methanol. The methanol cationic extract was anion-exchanged (BioRad AGI-X2, chloride form) under nonaqueous conditions to remove all interfering anionic components. The isolated cationic surfactant was dissolved in chloroform and partitioned with an aqueous solution containing disulfine blue reagent (pH 5, acetate buffer) to form the chloroform-extractable disulfine blue-cationic complex. A single extraction technique gave a simple, convenient, and quantitative determination of the cationic surfactant. The absorbances of extracts were measured at 628 nm and the samples quantified with the aid of a standard calibration (0 to 50 µg) for the test material or a suitable reference standard. The estimated limit of detection of the disulfine blue reaction was about 1 µg of di(hardenedtallow)dimethylammonium chloride (DHTDMAC), corresponding to 0.025 mg/L in a 200-mL sample. The recovery of added cationic surfactant from biodegradation test liquors, containing up to a tenfold molar excess of anionic surfactant, was generally better than 95%.

The method has been adopted as the official French and U.K. method for the determination of cationic surfactant in biodegradation test liquors. Waters and Kupfer were careful to point out that the disulfine blue reaction was not specific for cationic surfactant. Any natural long-chain amines and other compounds that can be protonated in weakly acidic conditions or which contain a quaternary nitrogen group may form extractable disulfine blue compounds [i.e., disulfine blue active substances (DSBAS)] that can interfere. Control liquors are generally used to correct for low concentrations of interfering materials in biodegradability test liquors. The concentrations of natural substances in waste and surface waters are likely to be so high that the procedure would not readily be applicable to environmental monitoring.

Osburn [2] modified the Waters and Kupfer disulfine blue (DS) procedure extensively to extend its applicability to a broad range of samples (sewage, effluent, sludge, river water, and sediment) for use in establishing the fate and distribution of DHTDMAC in the environment. Osburn found it necessary to preserve and stabilize samples containing low concentrations of cationic surfactant with formaldehyde/nonionic surfactant to prevent sizable losses of material through biodeterioration and/or adsorption onto the sample vessel. As a first step, he also evaporated to dryness an appropriate volume or weight of sample (containing 5 to 500 µg of DSBAS). The dried residue was extracted with hot methanolic hydrochloric acid (0.4 M) to recover all DHTDMAC quantitatively, even that bound strongly to matrices such as clays. The cationic surfactant in the resulting extract was removed from inorganic salts by extraction into chloroform from an acidified aqueous phase. The removal of chloride by this solvent extraction step eliminated the possibility of the partial elution of anionic surfactant during the subsequent anion-exchange step. This anion-exchange step was performed using a methanol eluant; particular care was taken to ensure a high anion-exchange efficiency to avoid negative interferences in the final colorimetric estimation stage. The dried eluate residue was dissolved in chloroform and shaken with an aqueous disulfine blue reagent (pH 1) to form the chloroform-soluble DS–cation complex. Two further chloroform extractions were used to ensure complete recovery. The evaporated extracts were redissolved in methanol for an absorbance measurement at 625 nm. The quantity of DSBAS in a sample was determined using a standard calibration chosen to correspond with the concentration present. The use of methanol for the final colorimetric measurement overcame the problems observed by Waters and Kupfer [8] of chloroform of poor color stability and the adsorption of DS complex onto optical cells.

The general applicability of the procedure was demonstrated for a range of environmental samples. The recovery of standard additions of DHTDMAC from these samples was acceptably high (ca. 90%), while the reproducibility was good for duplicate determinations. The sensitivity of the method was such that low-ppb concentrations of DSBAS could be determined in environmental liquor matrices. Osburn recognized that the modified DS procedure was still nonspecific for the cationic surfactant of interest and subsequently developed an additional TLC analysis step to enable the actual contribution of DHTDMAC to the measured concentration of DSBAS in samples to be estimated (see Sec. III.A).

The work of German researchers [16–18] has led to the adoption by the German Standards Organization (DIN) of a disulfine blue procedure for determining DSBAS in aqueous samples [11]. The method estimates the concentrations of cationic surfactants of the long-alkyl-chain quaternary ammonium, imidazoline quaternary, and amine types in sewage treatment liquors and surface waters. A solvent sublation technique (see Sec. D) was used to concentrate the DSBAS (50 to 1000 μg) from 1- to 4-L volumes of unfiltered sample to which sodium chloride (100 g/L) and LAS (2 mg/L) had been added. The latter facilitated the recovery of the cationic surfactant through the formation of solvent-sublatable ion-association compounds. The resulting sublation extract was dissolved in methanol and ion-exchanged on a strong anion-exchange resin, such as Dowex 1 × 2 (chloride form), to remove all interfering anionic components. The isolated cationic material was dissolved in chloroform/n-butanol solvent mixture (95:5) and then partitioned with an aqueous citrate buffer, pH 3, containing disulfine blue reagent to form the solvent-extractable disulfine blue–cationic complex. The absorbance of extracts was measured at 628 nm and samples were quantified with the aid of a DHTDMAC standard calibration (0 to 100 μg). In common with the Osburn method, the procedure allowed ppb concentrations of DSBAS to be determined in environmental liquor samples. However, many analysts believe that the procedure is more likely to assess only the "dissolved" DSBAS rather than the total concentration in samples, particularly those containing high concentrations of suspended solids (e.g., raw and settled sewage) on which cationic surfactant may be strongly adsorbed. Hellman [5] has also reported the use of a disulfine blue colorimetric determination for the quantification of DSBAS recovered from sludges and sediments by reflux extraction with methanol and cleaned up by alumina chromatography.

B. Other Anionic Reagents

Many other anionic dyes and metal chelates have been described as colorimetric (and fluorometric) PIX reagents for the determination of low concentrations of cationic surfactants in clean water samples: for example, Orange II [19–21], bromothymol blue [22], bromophenol blue [23], alizarin sulfonate [24], picric acid [25], eosine yellowish [26], methyl orange [27,28], 9,10-dimethoxyanthracene-2-sulfonate [29], bis[2-(5-bromo-2-pyridylazo)-5-(n-propyl-n-sulfopropylamino)phenolato]cobalt(III) [30], tetrabromo-phenolphthalein ethyl ester (TBPE) [31], ternary ion association compounds of diprotic acid dyes,and quinidine [32], to name but a few. However, with only a few notable exceptions, these reagents have not found any useful application in environmental testing situations.

  Picric acid, 2,4,6 trinitrophenol, forms stable intensely colored ion-association compounds (λ_{max}375nm) with cationic surfactants of commercial importance in aqueous solution (pH 4 to 12) that can be extracted into either dichloromethane or chloroform [12,25]. In contrast to some anionic dyes, high solution ionic strengths have no significant effect on the extraction of picrate–cation ion-association compounds. In the experience of the author, picric acid can be used to determine cationic surfactants (0 to 20 mg/L) in biodegradation test liquors, such as those recommended by the Organization of Economic Co-operation and Development (OECD) [33] in the absence of anionic surfactant and other anionic interferences. Similarly, Orange II [19,20] p(2-hydroxy-1-napthylazo)benzenesulfonate has gained some application in supporting simple tests of primary biodegradability and aquatic toxicity with cationic surfactants when well-defined test liquors are used. Orange II forms chloroform extractable 1:1 stoichiometric ion-association compounds (λ_{max} 485 nm) with commercially important cationic surfactants. The reagent can be used to determine total cationic surfactant or just quaternary ammonium compounds in samples containing mixtures through the use of an appropriate extraction pH. The applicability of the Orange II reaction for determining certain amphoteric surfactants has also been demonstrated [20].

Bonilla Símon et al. [28] have reported a simple colorimetric procedure for estimating the concentration of cationic surfactant (used as preservative) in the tissue of a marine organism (squid). The cationic surfactant, recovered by extraction of the tissue with ethanol, is partitioned from an aqueous phase, pH 4, into chloroform as its methyl orange ion-association compound (λ_{max}418nm). The procedure had a detection limit of approximately 1 mg/kg.

PIX reactions have provided the basis for the automatic determination of cationic surfactant in aqueous samples employing both air-segmented continuous-flow techniques exemplified by the Technicon Autoanalyzer (Orange II [21]) and unsegmented/segmented flow injection analysis (FIA) systems (Orange II [34] and TBPE [35]). These automatic systems introduce samples into a continuous-flow system providing for the addition of reagent, mixing, extraction with immiscible chlorinated solvent, and measurement of the color intensity of the solvent stream. Automatic techniques are particularly successful in determining cationic surfactants in clean water situations, with high analysis rates (60 samples/h) and significantly better precisions than manual procedures. As yet, no automatic application has gained wide acceptance for environmental testing purposes. The use of anionic metal chelates in PIX reactions has also led to the development of sensitive atomic absorption end determinations for indirect determination of cationic surfactants in natural waters [36,37]. Le Bihan and Courtet-Coupez [38] used the cobaltothiocyanate anion, $Co(CNS)_4^{2-}$, with a benzene extractant, and determined the extracted cobalt by flameless atomic absorption.

III. SPECIFIC ANALYTICAL METHODS

A. Thin-Layer Chromatography

Thin-layer chromatography (TLC) has proven to be a useful and comparatively simple means for specifically determining low concentrations of cationic surfactants in "environmental" samples. Michelsen [10,39] would probably claim the first application of TLC for the specific determination of cationic surfactants in aqueous environmental samples. He concentrated the cationic surfactant in samples (1- to 5-L volumes) into ethyl acetate via the solvent sublation technique described earlier. The residue from the sublation extract was dissolved in methanol and the cationics concentrated on a strong cation-exchanger resin to free them from other surface-active materials, particularly anionic surfactants, prior to their elution with methanolic hydrochloric acid. The isolated cationic surfactant was redissolved in chloroform (1 mL) and an aliquot spotted (10 to 50 μL) alongside appropriate standards (generally 0.5 to 5.0 μg) on a ready-made silica gel type 60 plate and an ethyl acetate/glacial acetic acid/water (typically, 4:3:3) solvent system was used to run the cationic surfactant as a single spot to its characteristic R_f value. The plate was visualized by spraying with a dilute Dragendorff reagent [barium chloride–potassium tetraiodobismuthate (III)] to form red-brown spots on a yellow plate background. The intensity of sample spots in transmitted light (λ_{max} 525 nm) was assessed using a scanning densitometer and then quantified by reference to suitable standard spots. The applicability of the procedure was demonstrated for a range of commercial cationic surfactants: dialkyldimethylammonium, monoalkyl-trimethylammonium, arylalkyldimethylammonium, and alkylpyridinium compounds. Michelsen claimed that the procedure had a range of application of 0.002 to 1.0 mg/L in aqueous samples, including waste and surface waters. The success of the methodology, however, was heavily dependent on the standardization of the steps for carrying out the TLC end determination, especially the visualization of the plates. The use of the sublation concentration step was also considered by many to restrict the procedure to assessing the "dissolved" cationic surfactant rather than the total content in samples, particularly those containing high concentrations of suspended solids. Because of these shortcomings, the simpler TLC technique described by Osburn [2] has gained more acceptance for determining cationic surfactants (in particular, DHTDMAC) in a range of environmental liquor and solid matrices [40,41].

Osburn [2] recognized that his modified disulfine (DS) colorimetric procedure (Section II.A) was nonspecific for cationic surfactant. He therefore incorporated an additional TLC analysis step into the procedure to make it possible to semiquantitatively estimate (precision ±20%) the actual contribution of DHTDMAC to the measured concentration of DSBAS in samples. The methanol DS extract from the colorimetric estimation was transferred to a macroreticular cation-exchange resin (Amberlyst 15) to concentrate selectively the cationic surfactant, which was then eluted with methanolic hydrochloric acid. The isolated DHTDMAC was spotted alongside standards (1 to 5 μg) on a silica gel G plate and a chloroform/methanol/water (75:23:3) solvent system was used to run the DHTDMAC as a single spot ($R_f \sim 0.5$). The plate was visualized by spraying and charring with 25% sulfuric acid and the sample spots were visually quantified with the aid of the standard spots. Using this technique, Osburn showed that between 20 and 80% of the DSBAS in environmental samples (ppb to ppm range) could be attributed to DHTDMAC surfactants.

Sherma et al. [7] have reported a specific TLC method for the determination of the long-alkyl-chain amine, octadecylamine (ODA), in boiler waters. ODA was isolated from the boiler water either by solvent extraction into ethylene dichloride or by concentration onto a micro-Chromosorb column followed by elution with acetone. The recovered ODA was spotted alongside standards on a high-performance silica gel layer, and a n-butanol/ glacial acetic acid/water (3:1:1) solvent system was used to run it as a single spot (R_f − 0.70). The plate was visualized by spraying with ninhydrin and the sample spots quantified by reflectance densitometry. The procedure incorporating the column concentration resulted in reproducible sample results and gave higher recoveries of standard additions (94.2% at 0.3 mg/ L ODA) than the solvent extraction step while encountering no serious interferences. The use of convenient TLC techniques should continue to have a role to play in the determination of surfactants in environmental testing situations; the wider application of modern sophisticated scanning densitometers will certainly provide improved quantification for this form of chromatography.

B. High-Performance Liquid Chromatography

To date, trace enrichment high-performance liquid chromatography (HPLC) has provided the most effective means for specifically and sensitively determining cationic surfactants, both quaternary ammonium compounds and amines, in a range of environmental testing applications. HPLC is particularly well suited for the separation and quantification of these involatile charged molecules. The method of Wee and Kennedy [9] provided the first practical, fully quantitative, and specific means for determining long-alkyl-chain quaternary ammonium compounds in environmental samples. An interesting feature of this analytical method was a chromatography system based on HPLC with conductivity detection. The particular form of ion chromatography employed by Wee and Kennedy eliminated the use of suppressor columns, separated without ion pairing, utilized a nonaqueous medium and for the first time, enabled the determination of submicrogram quantities of non-UV absorbing quaternaries. This was made possible because quaternaries were soluble and ionized in organic solvents. Quaternaries could be separated on a cyanoamino bonded-phase column using an organic mobile phase and detected by their conductivity.

The procedure was applied to the determination of four commonly used quaternaries, including DHTDMAC, in river waters. A simple and efficient extraction scheme was used to recover the cationics from samples. The quaternaries were concentrated from unfiltered river water samples (up to 200 mL), to which LAS and hydrochloric acid had been added, by extraction into dichloromethane (three 50-mL portions). The LAS enhanced the extraction of quaternary through the formation of a more readily ex-

tractable anion–cation complex while the acid (0.5 *M*) displaced any adsorbed material from suspended solids and facilitated its recovery. The
dried residue from the extraction step was redissolved in dichloromethane
and extracted three times with deionized water to eliminate ionic species
(including LAS) that would contribute to the conductivity background of
the sample chromatogram. The solvent/phase was evaporated to dryness
and the cationic material solubilized in the mobile phase (500 μL) for an
HPLC analysis (50-μL injections). The quaternaries were separated on a
"preconditioned" column (Partisil PAC 10) using isocratic conditions and
a chloroform/methanol (92:8 v/v) solvent mixture. The four quaternaries
were eluted within 20 min, but a total analysis time of 1 h was needed for
samples to remove unwanted organic ions (including LAS) from the column. The separations obtained were determined by the hydrophilicity of
the quaternaries: the more hydrophilic the cationic, the greater its retention
on the column. The separated quaternaries were detected by conductivity
detection and quantified by peak height measurements with the aid of the
appropriate external standards. Linear calibrations with correlation coefficients greater than 0.99 were obtained for all four quaternaries. The
detection limit for an environmental sample was approximately 0.02 μg
injected or 2 μg/L based on a 50-μL injection and a 100-mL sample volume.
The reproducibility for duplicate river water analyses were all within an
acceptable range of ±20%. With this procedure, standard additions of the
four cationics (10 to 20 μg/L) were quantitatively recovered from river
waters. In addition to its high sensitivity and specificity for cationics, the
method had the added advantage of simplicity and convenience.

Wee [41] subsequently demonstrated the wider applicability of the
HPLC procedure for the determination of DHTDMAC in sewage treatment liquors and at the same time compared its performance with those
of the Osburn DSBAS and TLC methods [2]. Only slight modifications to
the HPLC method were necessary to accommodate the analysis of sewage
treatment samples. The reproducibility of the three procedures was assessed by repeat determinations on sewage, final effluent, and river water
samples. The precision of the HPLC and DSBAS determinations was
±10% or less, whereas that for TLC was much larger, about ±30%. A
minimum of 5 μg of DHTDMAC was required for a satisfactory DSBAS
and TLC determination, but only 0.02 μg was needed for an HPLC injection. The HPLC procedure gave consistently higher recoveries of standard additions of DHTDMAC with less analytical variability than found
for either DSBAS or TLC for all samples investigated.

No serious interferences were encountered for the HPLC analysis of
DHTDMAC in sewage treatment samples, as had previously been established for river waters [9]; the cationic surfactant was eluted as a single

sharp peak at a retention time of about 6 min. The data reported by Wee showed the very high specificity of the HPLC procedure for the analysis of DHTDMAC in environmental samples compared to that of the DSBAS and TLC methods. It was concluded that HPLC provided by far the most convenient, sensitive, and specific means for determining cationic surfactants in environmental testing applications.

Because of these many advantages, the joint industry AIS/CESIO Task Force "Environmental Surfactant Analysis" has developed further the HPLC approach for the determination of DHTDMAC in sludge, sediment, and soil matrices in addition to environmental liquor samples [42]. This was readily achieved by combining the concentration/cleanup steps of the Osburn DSBAS procedure (i.e., methanolic hydrochloric acid extraction, liquid–liquid extraction, and anion exchange) with the aqueous back-extraction step and the HPLC-conductivity analysis of Wee and Kennedy's method. Only minor modifications were required to optimize the combined procedure for the determination of DHTDMAC in liquors (μg/L) and solids (μg/g): for example, use of 1 M methanolic hydrochloric acid extractant (instead of 0.4 M) and up to five extraction steps for solid samples, the addition of LAS at the liquid–liquid extraction stage to facilitate the recovery of cationic, the use of a more efficient PAC 5 column, and a chloroform/methanol/glacial acetic acid (89:10:1) HPLC mobile phase.

This procedure was applicable to the determination of the very low concentrations of DHTDMAC found in samples from sewage treatment plants, rivers and soils. The detection limit of the procedure for environmental liquor and soil samples were estimated to be 2.5 μg/L and 0.5 μg/g, respectively. The HPLC procedure was validated on an interlaboratory basis and was demonstrated to give accurate and reproducible results for samples and high recoveries of standard additions, generally $\geq 90\%$ from all matrices. The methodology as outlined is suitable for DHTDMAC but can with adaptation be used to determine other cationic surfactants, including long-alkyl-chain amines [43].

Matthijs and De Henau [44] have applied such a combined procedure to the analysis of monoalkyltrimethylammonium quaternary salts (MAQs) in domestic wastewater, river water, and sewage sludge. The sensitivity of the method for MAQ was in the region of 1 μg/L of influent sewage, 0.5 μg/L for final effluent and river waters, and 0.5 μg/g for sludges. The mean recovery rate of MAQ from liquor samples (2 to 70 μg/L standard additions) was $88 \pm 7\%$ and for sludges (7 to 100 μg/g) was $97 \pm 25\%$.

De Ruiter et al. [29] have also described an analysis of cationic surfactants, both quaternary ammonium compounds and amines, based essentially on the combined methods of Osburn and Wee, except that they employed their own HPLC separation and detection techniques. They developed a postcolumn detector system that mixed the HPLC column eluant with an aqueous stream containing an ion-pair reagent, either methyl orange or 9,10-dimethoxyanthracene-2-sulfonate sodium salt, to form the cationic surfactant-ion pairs in the organic phase. After phase separation the methyl orange and dimethoxyanthracenesulfonate-cationic ion pairs were monitored by ultraviolet adsorption or fluorescence detection, respectively. A gradient solvent system using a chloroform/methanol/acetonitrile mobile phase with a PAC 10 column produced fast separations for the cationic surfactant, with better resolution of peaks than obtained by Wee and Kennedy's isocratic system. De Ruiter et al. claimed that the 20-min analysis time was three times shorter; the work of the AIS/CESIO researchers [42] showed, however, that provided an anion-exchange step was included to remove anionic interferences, the original isocratic separation was completed within 10 min for all environmental samples. The detection limits for cationic surfactant in river water were about 2 μg/L and an impressive 10 ng/L for the methyl orange and dimethoxyanthracenesulfonate ion pairs, respectively. The reproducibility of river water determinations was very good [i.e., 4.2% (RSD,n = 20)]. The procedure was also successfully validated for sewage treatment liquors, sludges, and river sediments. The only obvious drawback of the method was the need to use a laborious standard addition technique to quantify the DHTDMAC in environmental samples (based on peak height measurements). The essential component of the continuous-flow ion-pair extraction detection system was a newly developed sandwich phase separator which was reliable and robust. Schoester and Kloster [45] have recently reported modifications to this postcolumn extraction system, which they claim will result in better system stability and improved peak shape for cationic surfactants.

Emmerich and Levsen [46] have investigated an alternative trace enrichment–HPLC procedure to those described above for the determination of cationic surfactants in wastewater and activated sludge with only limited success. These workers preferred an alumina chromatography step in place of anion exchange to clean up extracts and the use of a Diol 100 analytical column instead of the usual PAC 5/10 system to separate the cationics.

C. Gas Chromatography

Gas chromatography (GC) has found only limited application in the environmental testing of cationic surfactants. Long-chain quaternary ammonium salts are not sufficiently volatile for their direct analysis by GC. However, they can be converted to the corresponding tertiary amines (in situ in the GC injector port), which can then readily be gas chromatographed [47]. The resultant tertiary amines in the GC traces can give useful information on the chain length distribution in the quaternary ammonium compounds; quats with two or more long chains will, however, break down in a complex manner. Consequently, such techniques are more appropriate for their qualitative identification. Nevertheless, long-chain amines can be analyzed quantitatively by GC. Valls et al. [48–50] have described trace enrichment–GC procedures for determining very low concentrations of long-chain tertiary amines, particularly mono- and dialkylamines with 14 to 18 carbons in the alkyl chains, in environmental samples including sewage treatment liquors, sludges, seawater, marine sediment, and biota. The amines were recovered from solid and liquor samples by Soxhlet extraction [dichloromethane/methanol (2:1)] and liquid–liquid extraction (dichloromethane), respectively. The acid fraction of the resulting extract residues was removed by a further solvent extraction step before the amines were selectively concentrated (in hexane) and cleaned up (diethyl ether eluant) by combined alumina and silica column chromatography. The isolated amines were analyzed by high-temperature capillary GC (OVI column) with either flame ionization or nitrogen-specific detection and the peaks quantified by reference to suitable external standards. GC-MS was used to confirm the identity of the individual peaks in the chromatograms. The procedure was claimed to have a sub-ppb capability for determining tertiary amines in aqueous samples. The recovery of standard additions of tertiary amine was in the range 60 to 90%, while the relative standard deviation for determinations averaged 5%. GC will continue to provide a convenient analytical means for specifically determining a wide range of surface-active amines in support of their environmental testing.

D. Mass Spectrometry

A recent paper by Simms et al. [51] clearly demonstrates the great potential of mass spectrometry techniques for the quantitative determination of surfactants in environmental matrices. Mass spectrometry (MS) has emerged as a powerful tool for the characterization of nonvolatile organic compounds, including surfactants in environmental situations [52,53]. This has resulted primarily from the recent development of techniques such as desorption ionization (DI) and fast atom bombardment (FAB) mass spectrometry. In addition, to the usual qualitative and structural information given, DI and FAB-MS have also been used to quantify surfactants in complex mixtures [54].

Simms et al. developed two FAB-MS methods for the quantification of μg/L concentrations of cationic surfactants in aqueous environmental matrices, including sewage influents, effluents, and river waters. Only a simple sample preparation scheme was needed to concentrate/cleanup cationic surfactant for an accurate FAB-MS analysis. An essential first step was the addition of stable-isotope deuterium-labeled internal standards, [$^2H_3$]di(hardenedtallow)dimethylammonium iodide (DHTDMAI) and [$^2H_3$]dodecyltrimethylammonium iodide (DTMAI), to the samples for the purposes of accurate quantification. The cationic surfactant was recovered from the dry residue of the evaporated samples by solvent extraction techniques similar to those already described above but employing mixtures of methanol/chloroform as the extractant. The resulting extracts were cleaned up by passage through an alumina minicolumn (Alumina B cartridge). The isolated cationic material was taken up in glycerol for analysis by FAB-MS. Quantification was achieved using external standards containing a fixed amount of stable-isotope-labeled standard and varying amounts of the unlabeled reference surfactants to prepare calibrations from which unknown samples could be quantified. Intensity ratios were monitored at m/z 228 and 231 for DTMAC and m/z 522 and 525 for DHTDMAC. Low-resolution full-scanning FAB-MS allowed simultaneous quantification of targeted components. A higher-resolution signal-averaging technique, however, was more accurate and reproducible than full scanning but allowed quantification of only a single component at a time. Average recoveries for DHTDMAC and DTMAC varied from 86 to 107% over the concentration range 10 to 500 μg/L depending on the method used. The average coefficient of variation for environmental determinations was 5 to 14%. Simms et al. claimed a limit of detection for FAB-MS that was one to two orders of magnitude lower than for HPLC-conductivity analysis (ca.100 ng/L). The minimal sample preparation, high sensitivity, and specificity of the FAB-MS procedure make it attractive for the environmental monitoring of cationic surfactants; however, the need for a sophisticated MS system and suitable stable-isotope-labeled internal standards may detract to some extent from these advantages.

IV. CONCLUDING REMARKS

The rapid rate of progress in the development of new and improved methodology for use in the environmental monitoring of cationic surfactants is expected to continue. The powerful capability of modern chromatographic techniques, particularly HPLC, has meant that all the major surfactant types may be separated for analysis under the correct conditions. The main limitations in effectively using this separational capability are (1) the need to develop concentration/separational schemes to isolate sufficient material of acceptable purity for analysis; (2) the requirement for suitable detection systems capable of sensitive and specific determination of separated surfactant components; and (3) the ability to assign the peaks in chromatograms correctly to different surfactant types. The use of powerful combined techniques such as HPLC/MS, GC/MS, and SFC/MS and the direct application of mass spectrometry in the form of FD-MS and FAB-MS should in the future provide the necessary means to overcome the shortcomings of existing methodology.

REFERENCES

1. R.D. Swisher, *Surfactant Biodegradation,* Surfactant Science Series, Vol. 18, Marcel Dekker, New York, 1987.
2. Q.W. Osburn, *J. Am. Oil Chem. Soc. 59*: 453 (1982).
3. *Methods for the Examination of Waters and Associated Materials, the Sampling and Initial Preparation of Sewage and Waterworks Sludges, Soils, Sediments and Plant Materials, and Contaminated Wildlife Prior to Analysis,* 2nd ed., HMSO, London, 1986.
4. *EEC Directive 4311/82,* Annex, Brussels, Jan. 1982, Chap. 3.
5. H. Hellman, *Z. Wasser Abwasser Forsch. 22*: 4 (1989).
6. H. Klotz, *Tenside Deterg. 24*: 370 (1987).
7. J. Sherma, K. Chandler, and J. Ahringer, *J. Liquid Chromatogr. 7*: 2743 (1984).
8. J. Waters and W. Kupfer, *Anal. Chim. Acta 85*: 241 (1976).
9. V.T. Wee and J.M. Kennedy, *Anal. Chem. 54*: 1631 (1982).
10. E.R. Michelsen, *Tenside Deterg. 15*: 169 (1978).
11. Deutsche Einheitsverfahren zur Wasser-, Abwasser- und Schlammuntersuchung, *Summarische Wirkungs und Staffkenngrößen (GruppeH): Bestimmung der disulfinblau-aktiven Substanzen (H 20),* DIN 38409, Teil 20, July 1989.
12. J.T. Cross, in *Cationic Surfactants,* Vol. 4 (E. Jungermann, ed.), Marcel Dekker, New York, 1970, pp. 440–452.
13. H.K. Biswas and B.M. Mandal, *Anal. Chem. 44*: 1636 (1972).
14. H. Klotz, Kongreßberichte, *Welt-Tenside-Kongreß, Munich, Bd. 3; 1984, p. 305.*
15. A. Arpino and C. Ruffo, *Riv. Ital. Sostanzo Grasse 53*: 395 (1976).
16. E. Kunkel, *Die Analytik der Tenside,* Chemische Werke Hüls, 1976.
17. E. Kunkel, *Microchim, Acta 11*: 227 (1977).
18. W. Kupfer, *Tenside Deterg. 19*: 158 (1982).
19. A. V. Few and R.H. Ottewill, *Colloid Sci. 11*: 34 (1946).
20. G.V. Scott, *Anal. Chem. 40*: 768 (1963).
21. J. Kawase and M. Yamanaka, *Analyst 104*: 750 (1979).
22. M.E. Auerbach, *Anal. Chem. 15*: 492 (1943).
23. M.E. Auerbach, *Anal. Chem. 16*: 739 (1944).
24. B. Baleux, P. Caumette, J.-F. Coite, and J.-C. Lebecq, *J. Franc. Hydrol. 13*: 41 (1974).
25. I. Sheiham and T.A. Pinfold, *Analyst 94*: 387 (1969).
26. G.E. Batley, T.M. Florence, and J.R. Kennedy, *Talanta 20*: 987 (1973).

27. L.K. Wang and D.F. Langley, *Ind. Eng. Chem. Prod. Res. Dev. 14*: 210 (1975).
28. M.M. Bonilla Símon, A. De Elvira Cózar and L. Maria Polo Diez, *Analyst 115*: 337 (1990).
29. C. De Ruiter, J.C.H.F. Hefkens, U.A. Th. Brinkmann, R.W. Frei, M. Evers, E. Matthijs, and J.A. Meijer, *Int. J. Environ. Anal. Chem. 31*: 325 (1987).
30. I. Kasahara, M. Kanai, M. Taniguchi, A. Kakeba, N. Hata, S. Taguchi, and K. Goto, *Anal. Chim. Acta 219*: 239 (1989).
31. T. Sakai, *Bunseki Kagaku 21*: 199 (1975).
32. T. Sakai and N. Ohno, *Bunseki Kagaku 32*: 302 (1983).
33. *Pollution by Detergents: Determination of the Biodegradability of Anionic Synthetic Surface Active Agents,* Organisation for Economic Co-operation and Development, Paris, 1972.
34. J. Kawase, *Anal. Chem. 52*: 2124 (1980).
35. T. Sakai, *Analyst 117*: 211 (1992).
36. P.M. Gonzalez, C.C. Rica, and L.P. Diez, *J. Anal. Atomic Spectrom. 2*: 809 (1987).
37. P. Martinez Jimenez, M. Gallego, and M. Valcarcel, *Anal. Chim. Acta 215*: 233 (1988).
38. A. Le Bihan and J. Courtot-Coupez, *Analusis 4*: 58 (1976).
39. E.R. Michelsen, *Seifen Oele Fette Wasche 104*: 93 (1978).
40. B.W. Topping and J. Waters, *Tenside Deterg. 19*: 164 (1982).
41. V.T. Wee, *Water Res. 18*: 223 (1984).
42. P. Gerike, H. Klotz, J.G.A. Kooijman, E. Matthijs, and J. Waters, *Water Res. 28*: 147 (1994).
43. H. Klotz, lecture to the *Bavarian Water Research Institute's 44th Wastewater Biology Course,* Munich, Oct. 16–19, 1989.
44. E. Matthijs and H. De Henau, *Vom Wasser. 69*: 73 (1987).
45. M. Schoester and G. Kloster, *Vom Wasser. 77*: 13 (1991).
46. M. Emmrich and K. Levsen, *Vom Wasser. 75*: 343 (1990).
47. L.D. Metcalfe, *J. Am. Oil Chem. Soc. 61*: 363 (1984).
48. M. Valls, J.M. Bayona, and J. Albaigés, *Nature 337*: 722 (1989).
49. M. Valls, P. Fernández, and J. M. Bayona, *Chemosphere 19*: 1819 (1989).
50. P. Fernández, M. Valls, J. M. Bayona, and J. Albaigés, *Environ. Sci. Technol. 25*: 547 (1991).
51. J.R. Simms, T. Keough, S.R. Ward, B.L. Moore, and M.M. Bandurraga, *Anal. Chem. 60*: 2631 (1988).
52. S.J. Burlingame, T.A. Baillie, and P.J. Derrick, *Anal. Chem. 58*: 165R (1986).
53. S.J. Pachuta and R.G. Cooks, *Chem. Rev. 87*: 647 (1987).
54. H. Shiraishi, A. Otsuki, and K. Fuwa, *Bull. Chem. Soc. Jpn. 55*: 1410 (1982).

9

Mass Spectrometry of Cationic Surfactants

HENRY T. KALINOSKI Analytical Chemistry Section, Unilever Research U.S., Edgewater, New Jersey

I. INTRODUCTION

Properties that render cationic surfactants commercially valuable have, until recently, limited the utility of mass spectrometry for their characterization. In fact, of over 1100 references in Volume 34 of this series [1], only one is to the use of mass spectrometry. But recent advances in the field are beginning to allow the observations of J. J. Thomson to be realized for this diverse class of compounds: "There are many problems in Chemistry which could be solved with far greater ease by this than by any other method. The method is surprisingly sensitive . . . requires an infinitesimal amount of material and does not require this to be specially purified . . ." [2].

In this chapter we describe the basic principles of mass spectrometry and the specific advances that have allowed the technique to be applied routinely to cationic surfactant samples. This includes some pertinent details of the instrumental techniques but will not supply all of the information necessary for the uninitiated to produce meaningful results from mass spectral investigations. Examples of the various techniques specifically applied to cationic surfactant mixtures are given. The examples include some specific references to the extension of mass spectrometry as a tool to support chromatography and environmental investigations, but these are covered briefly here, in deference to more extensive coverage elsewhere in this volume. Limitations of the various approaches are also discussed.

The chapter is based on a review of the recent literature in mass spectrometry, including development of new techniques and applications to cationic surfactants. The abstracts from recent American Society for Mass Spectrometry Conferences on Mass Spectrometry and Allied Topics were also valuable in identifying initial investigations of emerging techniques. The work does not exhaustively list every citation to the mass spectrometry of cationic materials, but the resulting bibliography provides those interested with a valuable resource for a more detailed understanding of the mass spectrometry of cationic surfactants. This bibliography includes a number of recent reviews and reference works [3–5] which illustrate an even greater scope of this powerful analytical technique.

Three recent developments most responsible for extension of mass spectrometry to cationic surfactant materials are covered in some detail. Two areas, fast atom bombardment (FAB) [6] and electrospray [7], involve introduction of samples to the mass spectrometer and the third, tandem mass spectrometry (MS/MS) [3], relates to analysis of surfactant ions. The detection of ion signals, recording of mass spectral data, and the overwhelming influence of computers in mass spectrometry are not covered. Chromatographic separations, often important steps in sample introduction for mass spectrometry, are covered elsewhere. The chapter closes with some indications on the future of mass spectrometry and how the analyses of cationic surfactants are likely to be affected.

II. BASIC MASS SPECTROMETRY

The simplest description of mass spectrometry is the measurement of ion intensity versus mass-to-charge (m/z) ratio. From measurements of m/z and ion intensity, an understanding of the molecular structure of the analyte can be gained. Less simple are the instruments required to provide this information. Mass spectrometer systems are comprised of six basic components: sample inlet, ion source, ion analyzer, signal detector, data recording device, and vacuum system. A number of different principles have been explored, resulting in different combinations of components to reach these ends. The most common instrument types are based on the quadrupole mass filter or a combination of electric and magnetic sectors. The basic procedures, however, apply to all mass spectrometers. A sample is introduced to the ion source, where ions directly related to the intact structure of the analyte are formed. This is followed by separation based on or directly related to ion mass of intact ions and any related fragment ions (through the influence of a combination of electrical, magnetic, radiofrequency and/or gravitational fields), with each ion producing a signal in proportion to the abundance of the species present.

Information from mass spectrometry is in the form of mass-to-charge ratio and ion intensity. In most forms of mass spectrometry, the charge (z) on the ions is unity, so the measured m/z value is a direct indication of ionic mass. Much of organic mass spectrometry, particularly for cationic surfactants, is performed in the positive ion mode. Charge on the ions comes either from stripping one electron from an organic molecule, to form a radical cation (M^+), or through ion-molecule chemistry, to add a singly charged species (proton, ammonium, or alkali metal ion), forming an adduct ion $(M + I)^+$. Materials may also be presented to the mass spectrometer ion source already carrying a formal charge, as is the case with cationic surfactants. The ionization process then transports the performed ions to the gas phase for analysis. All of these ionization processes impart additional energy to the analyte molecule, often causing bond cleavage and formation of fragment ions. If fragment ions are formed in the source of the mass spectrometer, they will be mass separated along with the analyte parent ion.

Both the absolute intensity of the ions and their relative abundance normalized to the most abundant ion formed are generally dependent on the structure and stability of the parent species. Conditions used to form radical cations are energetic enough to break any bond in an organic molecule, usually producing little information on intact molecules and abundant information on structural fragments. Processes leading to adduct formation are of lower energy, yielding more information on the molecular weight of the intact parent. Fragmentation of the adduct ions is usually through the loss of small, neutral molecules (H_2O, CH_3OH, HCl, etc.), yielding some information on functional groups. From an understanding of the chemistry of these high-energy ionic species, insight into the nature of the parent molecule can be gained [8,9].

A. Sample Introduction

To produce ions for mass spectrometry, the analyte must be present in the ion source in the gas phase. For most mass spectrometers and many analytes, this is accomplished simply through introduction of the sample to the high vacuum of the ion source, usually in the range 10^{-4} to 10^{-6} torr. The most direct introduction method is having a gaseous sample in a vessel connected via inert tubing (usually glass) directly to the ion source. Unfortunately, this approach is useful only for permanent gases or materials that readily volatilize at room temperature.

For a larger variety of materials, the sample can be introduced on a probe that can be placed near or in the ion source. This type of probe can usually be heated to assist volatilization of analyte species. The need to heat materials for volatilization often partially or totally degrades the analyte prior to sample ionization. A modification of the probe inlet method, where the sample is inserted on a wire filament capable of rapid heating (>1000°C/s), overcomes thermal degradation for some samples.

B. Gas Chromatography/Mass Spectrometry

Gas sampling and probe inlet introduction, however, are generally limited to samples of high purity, to ensure that only one analyte species is present in the ion source at one time. Most problems requiring mass spectrometry involve mixtures of materials, necessitating inlet systems that allow for some form of separation. The combination of separation and mass analysis has been met most elegantly through the coupling of gas chromatography and mass spectrometry (GC/MS). This was realized partially through the complementary nature of sample requirements for the two techniques. Gas chromatography is accomplished for materials that are volatile (or readily made volatile) and thermally stable from ambient through relatively high temperatures (up to about 400°C). Gas chromatography can deliver "isolated" fractions of samples in the gas phase directly to the mass spectrometer. A variety of interfaces coupling the two techniques have been developed [10–13], but currently the most common system is the direct inlet for capillary GC. In this interface, the fused-silica open-tubular separation column (generally 250 to 320 μm ID) is terminated in close proximity to the mass spectrometer ion source. Much of the column effluent is swept directly into the ion source for analysis. The basics of GC/MS and its application can be found in Refs. 10–13.

C. Liquid Chromatography/Mass Spectrometry

Unfortunately, the vast majority of organic compounds, including intact cationic surfactants, are not amenable to direct analysis via GC/MS due to lack of volatility or to thermal instability. This has prompted wide-ranging efforts to couple other forms of chromatography, particularly liquid chromatography (LC), with mass spectrometry. In fact, when considering the ideal detector for LC [14], some would argue that mass spectrometry satisfies the requirements quite well [4]. Conversely, liquid chromatography provides most of the characteristics, including minimum sample preparation, separation, and sequential delivery of samples, needed for an "ideal inlet to mass spectrometry" [4].

Some of the techniques that have been developed, or are now finding applications, for liquid chromatography–mass spectrometry (LC/MS) have been found to be ideal for the analysis of cationic surfactants. The LC/MS interfaces now finding wide utility (e.g., thermospray, flow FAB, electrospray) were developed specifically to allow direct mass spectral analysis of nonvolatile, polar or ionic, higher-molecular-weight materials. The interfaces function to separate analytes from the bulk of the chromatographic mobile phase and deliver the analyte to the high-vacuum mass spectrometer ion source. Through this approach, a highly sensitive, selective, and flexible technique for mass analysis can be obtained.

D. Other Approaches

There are, of course, other means of sample introduction for mass spectrometry. Some of these, planar chromatography/MS and laser desorption, are covered in more detail later, with specific reference to their utility for the analysis of cationic surfactants.

III. FAST ATOM BOMBARDMENT

A variety of techniques were employed to produce molecular weight and structural information from cationic surfactants, including electron ionization (EI) [15], chemical ionization (CI) [16], surface ionization [17], electrohydrodynamic (EHD) ionization [18], and field desorption (FD)

[19]. Most, however, were found to yield limited information, or to be applicable to only some of the materials of interest, or were difficult to use routinely. Fast atom bombardment (FAB) is the ionization technique that finally allowed for simple direct mass spectrometry of a wide range of thermally labile, low-volatility, polar or charged molecules, including cationic surfactants.

The technique now known as FAB was first described by Barber and co-workers in 1981 [20] and was the topic of some controversy regarding its novelty [21]. It is now generally accepted that the approach is all but identical to the surface analysis technique secondary ion mass spectrometry (SIMS). The novel aspect of FAB is the use of a low-volatility, viscous matrix to introduce the sample to the mass spectrometer ion source. This allows for analysis of samples with no preparation other than dissolving the analyte. The viscous liquid matrix, usually glycerol, permits constant replenishment of the sample surface and sample lifetime in the mass spectrometer is from a few minutes to an hour or more. This is particularly useful for a range of advanced mass spectrometric techniques, enhancing the signal-to-noise ratio, allowing acquisition of accurate mass data and facilitating tandem mass spectrometry experiments. Further, the surface activity of a sample is related directly to its ease of ionization in FAB, and the importance of surface activity has been the subject of numerous studies [22–25]. A variety of reviews and texts have been written regarding the mechanism and applications of FAB [6,21,26–28], so only some basic fundamentals and applications to cationic surfactants are presented here.

A. Mechanism

A schematic representation of the sample introduction/ionization mechanism operable under FAB conditions is given in Fig. 1. The sample is present in the drop of matrix on the tip of a direct-insertion probe. Probe tips are generally made of a conducting material (copper, stainless steel, molybdenum, gold plate) to allow the tip to be grounded, avoiding charge buildup in the sample: they may also need to be relatively inert to prevent reaction (i.e., redox, metal exchange) with some analytes. The primary particle beam, usually composed of energetic xenon or argon atoms or cesium ions, is directed at the insertion probe target, perpendicular to the axis of the mass spectrometer analyzer. Interaction of the primary particle beam, at energies of 6 to 10 kV for gases and up to 35 kV for cesium ions, with the matrix–sample mixture results in a sputtering of ionic and neutral species from the sample target. A fraction of the ions of appropriate polarity are electrically extracted, and this secondary ion beam is focused by the ion source and optics for subsequent mass analysis. Some portion of the

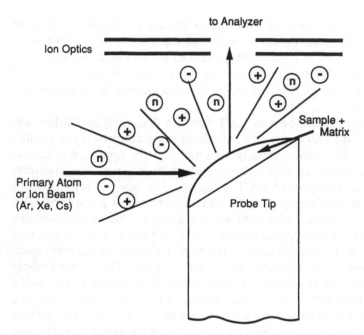

FIG. 1 Schematic illustration of the ionization process in FAB mass spectrometry. The primary particle beam initiates an emission of a variety of secondary particles (cations, +; anions, −; neutrals, n) from the sample-matrix mixture. Some ions may also be formed in the gas phase above the liquid matrix. The type of ions sampled (positive or negative) depends on the tuning of the ion source optics.

high-energy neutral materials travel the axis of mass analysis and may contribute to "noise" on the instrument detector. This is particularly a problem in systems with the FAB source and instrument detector in a straight, "line-of-sight" geometry.

The energy distribution of the primary beam is generally quite broad and the sputtering/ionization processes also yield a broad distribution of ion energies. The relatively high intensity of the secondary ion beam available for analysis compensates for the broad distribution, allowing for significant discrimination by the ion source optics to permit high-resolution mass analysis. The high intensity and long lifetime of this secondary ion beam were immediately found to be an improvement over field desorption for the analysis of cationic surfactants.

FAB was originally classed as a "soft" ionization technique, seen as imparting little energy to analyte ions. This was reinforced by the tendency to form proton attachment $[(M + H)^+]$ or proton abstraction $[(M - H)^-]$ ions, similar to those found in CI. The ease with which the technique forms gas-phase ions from materials present as ions in the matrix (such as quaternary ammonium cationic surfactants) suggested that the proton attachment ions are present as preformed ions as well. FAB particle bombardment then serves only to desorb the ions from the matrix [29,30]. Further study into the ionization process found that during ionization, energy on the order of 30 eV was transferred to analyte molecules [31] capable of inducing some bond cleavage and fragmentation. Fragmentation found for proton attachment ions is similar to that found in CI, with the loss of polar substituents as small neutral molecules. Additionally, depending on analyte structure, a range of charge-site-initiated cleavages, rearrangements, and charge-remote cleavages are also observed [32–34].

It is also clear that the FAB matrix plays a role in the ionization of various analyte species. Of initial importance in choosing a matrix for FAB is that the analyte of interest must show some solubility but should not react with it. The matrix may also contribute ions to the mass spectral background, possibly interfering with sample characterization. Ions can also be formed from adducts of analyte and matrix or analyte and matrix contaminants (e.g., alkali metals). Some references list the variety of materials that have been used as FAB matrices and their physical properties [6,35]. Both glycerol and m-nitrobenzyl alcohol have been found to be effective matrices for cationic surfactants in the author's laboratory.

B. FAB of Cationic Surfactants

The physical and chemical properties of cationic surfactants have generally limited the utility of more "classical" mass spectrometry techniques (EI, CI) in their characterization. These same properties—low volatility, high surface activity, formal ionic charge—make cationic surfactants ideal for analysis using FAB ionization. Figure 2 shows the positive-ion FAB mass spectrum of a commercial ditallowdimethylammonium chloride sample (Arquad 2HT-75, Armak Industries), illustrating the characteristics of such spectra. Intense molecular cations are found, corresponding to the alkyl chain distribution of the tallow fatty acid starting materials (C_{14}, C_{16}, and C_{18}). The spectrum contains ions corresponding to the trialkylmethylammonium materials also present in this commercial mixture (m/z 700 to 810).

Positive-ion FAB is a very sensitive technique for cationic surfactants, with a detection limit of 0.05 μg/mL of matrix reported [36]. The ease with which ions can be formed from cationic surfactants has led to the proposal that the materials be used as calibration compounds for FAB mass spectrometry [37].

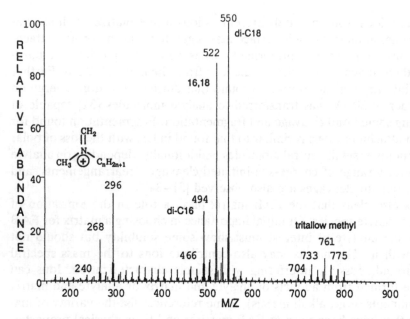

FIG. 2 FAB mass spectrum of a commercial ditallowdimethylammonium chloride sample produced using xenon atom bombardment from a *m*-nitrobenzyl alcohol matrix. Along with the molecular cations from the C_{14}, C_{16}, and C_{18} components are ions corresponding to the tritallowmethyl cations and the C_{14}, C_{16}, and C_{18} methylimine fragment ions (*m/z* 240, 268, and 296).

The other significant feature of the spectrum in Fig. 2 is the presence of fragment ions, particularly those at *m/z* 240, 268, and 296. These ions correspond to alkylmethylimine ions of the structure shown in Fig. 2, with *n* equal to C_{14}, C_{16}, and C_{18}, respectively. Fragments of this type are believed to be formed via the mechanisms illustrated in Scheme 1. Some of the fragments found may not be due to gas-phase ion decomposition, as

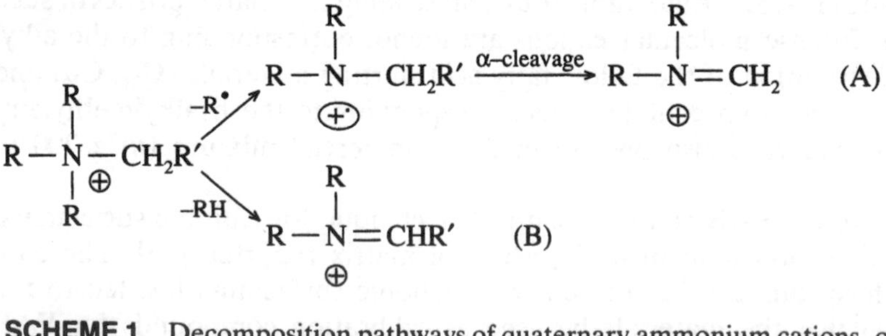

SCHEME 1 Decomposition pathways of quaternary ammonium cations, observed in positive-ion FAB mass spectra.

the gas-phase species should be of low internal energy, and have been attributed to high-energy, liquid-phase processes [38].

The intact molecular cation is not always the most abundant ion (base peak) in the FAB mass spectrum of a quaternary ammonium cationic surfactant. Figure 3 shows the FAB mass spectra of an ethoxylated di(tallowamido)ammonium cationic surfactant (Varisoft 110, Sherex) produced using krypton (Fig. 3a) and xenon (Fig. 3b) as the bombarding gases. For both krypton and xenon bombardment, the fragment ions at m/z 282 and 310 are of greater abundance than are the intact molecular cations. The major fragment ions (m/z 282, 310) result from simple cleavage of the carbon–quaternary nitrogen bond with charge retention on the alkyl amide portion, as shown in Fig. 3a. The minor fragments (m/z 357, 385) also result from cleavage of the carbon–nitrogen bond, with hydrogen transfer to the tertiary amine. Both these processes appear different from those found for the dialkyldimethyl type of quaternary ammonium materials (Scheme 1).

Although both spectra in Fig. 3 were produced at the same primary particle beam accelerating voltage (9 kV), the spectrum produced using xenon (Fig. 3b) is of greater absolute abundance, owing to the higher momentum (energy) of the bombarding particles. Less clear are the reasons for the greater relative abundance of intact molecular cations produced using the xenon bombarding gas.

The pair of spectra in Fig. 3 also illustrate the utility of FAB in identifying other components in complex mixtures of cationic surfactant materials. At 14 Da lower than each of the ions corresponding to a quaternary ammonium cation are ions corresponding to the protonated molecules of the di(tallowamido) ethoxy amines used to form the cationic materials. FAB is generally useful for producing this qualitative information on components in mixtures but is not as effective in providing quantitative information on such mixtures. A number of studies indicated the selective desorption/ionization of components in mixtures of cationic surfactants [22,25,39,40] and it appears that this selectivity is correlated with the surface activity of the various mixture components [22,24,25]. One proposed solution was the addition of a strong anionic surfactant to the sample–matrix mixture to equalize desorption of mixture components [22]. Negative-ion FAB mass spectrometry has been used for cationic surfactants but is characterized by adduct ions of the quaternary cation and two anions. This can be useful in identifying the counterion present in cationic surfactant mixtures.

C. Applications of FAB Mass Spectrometry

One distinct advantage of FAB is the ability to obtain detailed mass spectral information with little or no sample pretreatment. Figure 4 shows the

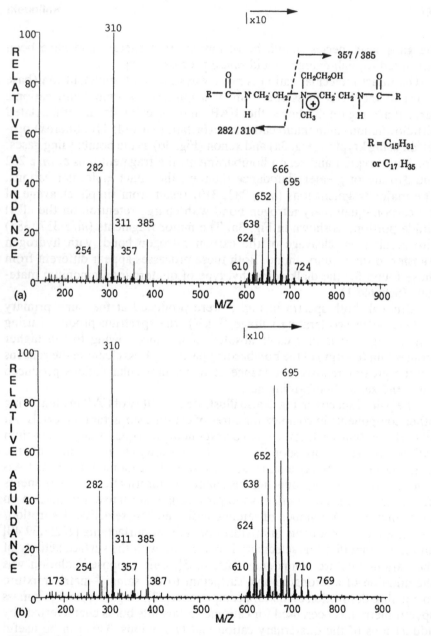

FIG. 3 FAB mass spectra of an ethoxylated di(tallowamido)methylammonium surfactant mixture produced using krypton (a) and xenon (b) atom bombardment. Both spectra were produced from a *m*-nitrobenzyl alcohol matrix with the FAB source operating at 9 kV and 3 mA of ion current. Absolute intensity scales for the two spectra are different. Cleavage of the carbon-quaternary nitrogen bond shown would yield the fragment ions found.

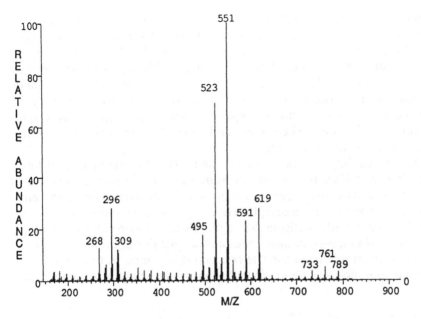

FIG. 4 FAB mass spectrum of a commercial liquid fabric softener product. Spectrum was produced from a mixture of the product with the m-nitrobenzyl alcohol matrix using xenon atom bombardment. Nominal masses differ from those shown in Fig. 2 due to mass assignment and rounding routines in the computer data system.

positive-ion FAB mass spectrum of a commercial liquid fabric softener product. The spectrum was obtained by mixing the fabric softener with the FAB matrix, m-nitrobenzyl alcohol, and subjecting a drop of the mixture to xenon bombardment. Comparison with the spectrum in Fig. 2 indicates the presence of a ditallowdimethylammonium mixture (m/z 551, 523, 495, 296, and 268) in this product. Nominal, unit mass assignments of ions in Fig. 4 appear one mass higher than those in Fig. 2 due to the calibration and mass assignment routines used by the computer data system. The accurate mass assignment for the species appearing at m/z 550 in Fig. 2 and at m/z 551 in Fig. 4 (dioctadecyldimethyl ammonium ion, $C_{38}H_{80}N^+$) is 550.627 Da based on ^{12}C equal to 12.000. This type of mass assignment rounding in low-resolution mass analyzers (such as quadrupoles) with computer data systems must be considered when interpreting data obtained in mass spectrometry. As is also found in the ditallowdimethylammonium raw material (Fig. 2), the product analyzed for Fig. 4 contains the tritallowmethylammonium materials. Distinct from the raw material spectrum are the ions at m/z 619, 591, 310, 309, 282, and 281. When considered in combination with other information on this product, a tentative identification as a tallow imidazoline mixture is possible from these mass spectral data. To confirm the tentative identification, the FAB mass spectrum of an authentic tallow imidazoline (Varisoft 516, Sherex) can be obtained (Fig. 5). The m/z 282 and 310 ions are likely to be of the same alkyl amide structure as those found for the di(alkylamido) cationic surfactant (Fig. 3). The ions at m/z 309 and 281 correspond to protonated $C_{17}H_{35}$- and $C_{16}H_{33}$-substituted imidazoles. These ions can be explained via the carbon–nitrogen bond cleavage seen in Fig. 3.

The applicability and limitations of the FAB technique in a quantitative mode is exemplified by spectral measurements made in the presence of anionic surfactants [41]. Quantification of cationic surfactants in an environmental matrix was accomplished using two FAB-mass spectral approaches, both with particular strength. Both a low-resolution scanning approach and a high-resolution continuum acquisition approach were found to yield comparable results in the concentration range 10 to 500 ppb. The full-scan approach was more advantageous in the analysis of samples con-

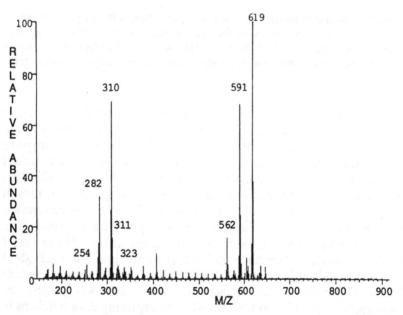

FIG. 5 FAB mass spectrum of an authentic sample of a tallow imidazoline fabric softener active, believed to be a component of the mixture shown in Fig. 4. The spectrum was produced from a m-nitrobenzyl alcohol matrix using xenon atom bombardment under conditions equivalent to those used for the sample in Fig. 4.

taining a broad distribution of homologous materials. The continuum acquisition was more accurate and reproducible, particularly at lower sample concentrations. The approach was conservatively judged to be five times faster than existing HPLC methods. The FAB-MS method was regarded as the method of choice for work requiring precision, accuracy, and limits of quantitation but may be limited by the need for suitable (preferably stable-isotope-labeled) internal standards.

D. Dynamic or Flow FAB

FAB in the direct-insertion probe approach is limited by the use of a high concentration of viscous matrix. These limitations are manifested in ion suppression effects, generally poor sensitivity, and high background. In an effort to address limitations of the direct insertion probe (static) method, modifications were made to allow FAB to be used with flowing (dynamic) systems [42–44]. Further, these modifications allow FAB to be used online in conjunction with liquid separations techniques.

Figure 6 shows a simple diagram of a system developed to perform flow-FAB. Much like a GC/MS interface, the probe for flow FAB allows a fused silica capillary tube (75 to 100 μm ID) to be introduced directly into the mass spectrometer ion source. A solvent, typically 5% glycerol in water, flows through the tube to the FAB target, for bombardment under typical FAB conditions. The system can be used in a flow injection mode, with samples injected into the flowing carrier solvent, or in conjunction with a

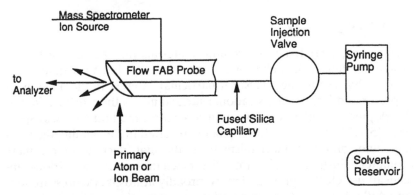

FIG. 6 Schematic representation of the apparatus required to perform dynamic or flow FAB. A separation column can be in place between the valve and FAB probe tip to allow the technique to be used as an on-line LC detection method. Numerous modifications of the basic design have been made to apply or improve the technique.

liquid separation technique, such as HPLC [42,44] or capillary electrophoresis [45–47]. One aspect of the union with separations techniques is that flow FAB operates at low liquid flow rates, microliters per minute, so that separations must be conducted on that scale (capillary or packed capillary HPLC or capillary electrophoresis) or the separation column effluent must be split prior to introduction to the flow FAB system.

An excellent recent application [48] highlights the fundamentals of the technique in the analysis of cationic surfactants. In contrast to nearly all reported flow FAB efforts, normal-phase HPLC was used, with chloroform, hexane, methanol, and acetic acid in the mobile-phase and packed capillary separation columns. To permit stable operating conditions and to be compatible with the nonaqueous HPLC mobile phase, the matrix solution was 75:25 glycerol/methanol. The FAB matrix was added to the column effluent following the separation via an in-line "T" (postcolumn, coaxial addition) [49]. The analytes studied were a commercial ditallow-dimethylammonium chloride (Akzo Chemicals) and a commercial fabric softener product. The analysis was conducted on a double-focusing sector mass spectrometer, under both moderate resolution and accurate mass conditions. Low-resolution spectra, for both the standard material and the fabric softener product, were comparable to that shown in Fig. 2, with the added advantage of separating the ditallowdimethyl from the tritallow-methyl substituents. Accurate mass data consistently had mass errors of less than 10 ppm. Flow FAB has been demonstrated to be of practical value in the analysis of cationic surfactants and its use in all areas of mass spectrometry should increase significantly in the future.

IV. ELECTROSPRAY

The desire to get relatively nonvolatile materials into the gas phase for spectroscopic studies, including mass spectrometry, led to a number of significant advances in the 1980s. Of particular interest in the analysis of large, preformed ions, such as cationic surfactants, is the work leading to the development of electrospray as an ionization method for mass spectrometry [5,7,50,51]. These developments resulted in decoupling the sample inlet and ionization mechanism from the high vacuum of the mass spectrometer. Further studies into the mechanisms of the various approaches to use electrospray, and more broadly atmospheric pressure ionization (API), have led to a wide variety of interesting applications.

The diagram in Fig. 7 shows the basic components of an electrospray ionization source patterned after the apparatus of Yamashita and Fenn [7]. Liquid samples are introduced through the syringe needle, the end of which is specifically prepared to permit stable operation. The needle is at elevated

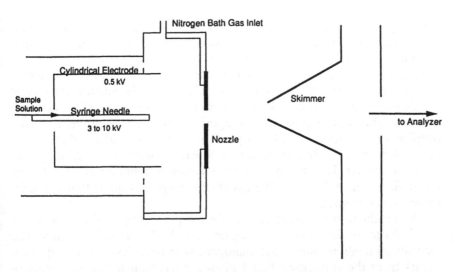

FIG. 7 Schematic diagram of the apparatus to perform electrospray ionization mass spectrometry, generally based on the description from Ref. 7. The diagram is not to scale and does not illustrate extra pumping regions that may be required to ensure proper operation. Numerous modifications have been made to the basic design for improvements or specific applications.

voltage relative to ground. The charge on the needle, and resulting field gradient between the needle and the lower voltage nozzle, induce a spray of ionized droplets, directed at the inlet to the mass spectrometer analyzer. In some systems, a flow of dry gas is used to assist in evaporation of the liquid solvent. A variety of methods have been developed to focus the ion beam, to maximize sampling efficiency, to perform gas-phase ion-molecule chemistry and, in general, improve basic system performance or to suit a particular application [5,50,51]. Early efforts at electrospray were made on quadrupole instruments, for ease of initial development and to accommodate the generally low translational energies of API-produced ions. More recently, the technique has been adapted to a wide range of analyzer types [51].

A. Parameters and Characteristics

As with flow FAB, electrospray systems are constructed using small-bore capillary tubing and accommodate relatively low flow rates (1 to 10 µL/min). Some laboratories have developed modifications to address higher flow rates, 40 to 50 µL/min for a pneumatically assisted system [52]. These dimensions and flow rates are compatible with microscale liquid chromatography, capillary electrophoresis, and other, low-flow liquid separation techniques [50–52]. Higher-flow-rate systems can be interfaced by splitting the effluent of the separation column [53].

In general, aqueous-based solvents have been used for electrospray. Use of alcohols or other solvents generally lowers the surface tension of the liquid being sprayed, resulting in more uniform production of small droplets. The particular solvent composition is often determined by the coupling of electrospray with a separation technique and is seldom composed of pure water. In fact, it can be difficult to maintain a stable system using pure water. Use of auxiliary nebulization or addition of a modifying liquid via a "sheath" flow [54] expands the range of liquids that are amenable to electrospray.

The mechanism of ionization in electrospray is not fully understood [7,51]. As ionization occurs in a region of high electric field strength, the mechanism might be somewhat analogous with field desorption [19]. As it occurs from the condensed liquid phase, it has similarities with electrohydrodynamic ionization [18]. Whatever the exact process, the technique is very mild, producing ions of intact molecular species with little or no fragmentation. With some modifications to the ion optics, ions produced in electrospray can be made to undergo some ion-molecule chemistry, increasing the amount of structural information available.

B. Application to Cationic Surfactants

In many of the early studies of electrospray, quaternary cationic compounds were chosen as representative of the polar, involatile materials of greatest challenge for mass spectrometry [7,50–53]. It was found that these compounds ionized quite well under a variety of conditions and sensitivity at attomolar (10^{-18} M) concentrations appears feasible.

The promise of a rapid mass spectrometric method for surfactant analysis, unencumbered by the limitations of FAB, led to the exploration of electrospray of surfactant solutions [55]. In this very recent study, cationic materials were examined along with anionic, nonionic, and zwitterionic surfactants under a variety of electrospray conditions. Mixtures of isopropanol and water were most useful for the surfactant solutions, with the positive-ion detection limits in the high-femtogram (10^{-15}) range. In most cases, no fragmentation was found. It was indicated that quantitative flow-injection and LC/MS applications and trace environmental studies would follow this initial survey.

Figures 8 and 9 are representative electrospray mass spectra of quaternary ammonium cationic surfactant mixtures. Figure 8 is the spectrum of the ditallowdimethylammonium mixture used to produce the FAB mass

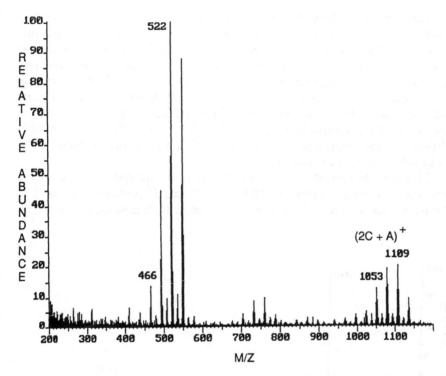

FIG. 8 Electrospray mass spectrum of the ditallowdimethylammonium chloride surfactant mixture used to produce the FAB mass spectrum shown in Fig. 2. The spectrum was produced from an isopropanol–water mobile phase using a flow-injection approach. This spectrum is found to contain the cluster ions over m/z 1000 and to lack the imine-type fragment ions produced using FAB. The tritallow-methyl components are found between m/z 700 and 810.

spectrum shown in Fig. 2. Similar to results from FAB ionization, the most intense ions correspond with the quaternary ammonium molecular cations (m/z 494, 522, and 550). The spectrum also contains the tritallowmethyl-ammonium ions between m/z 700 and 810. Two features of the electro-spray spectrum (Fig. 8) contrast with the FAB spectrum (Fig. 2), the presence of an ion series between m/z 950 and 1150 and the absence of significant fragment ions in the electrospray spectrum. The higher-molecular-weight ion series corresponds with two quaternary ammonium cations complexed with a chloride anion $(2C^+A^-)^+$. Such cluster ions are a common feature of electrospray spectra [7]. Lack of fragment ions support the mildness of electrospray ionization.

Mildness of electrospray ionization is not universal for cationic surfactants, as evidenced in Fig. 9. This di(tallowamido) ethoxylated ammonium cationic surfactant mixture was analyzed using FAB ionization, producing the spectra in Fig. 3. In comparison with the FAB results, the molecular cations are more abundant in the electrospray spectrum. The major fragment ions are still present at m/z 282 and 310, however, as are a great deal of other ions. It is not clear how these ions relate to either the parent species or the fragment ions found in FAB. It is clear that optimum conditions for the electrospray of cationic surfactant mixtures should be developed separately for each sample.

Electrospray ionization is just beginning to be applied as a tool for the characterization of cationic surfactant mixtures. Conditions may need to be developed to allow electrospray to be used with nonaqueous systems,

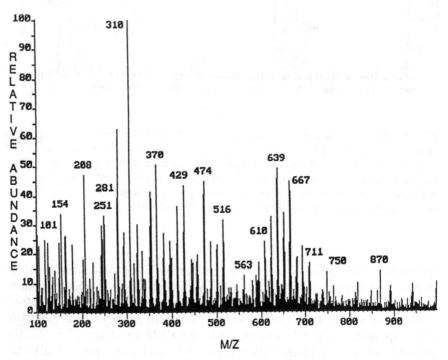

FIG. 9 Electrospray mass spectrum of the ethoxylated di(tallowamido)methyl-ammonium surfactant mixture used to produce the FAB mass spectra shown in Fig. 3. The spectrum was produced from an isopropanol–water mobile phase. Ions related to intact molecular cations are relatively more abundant when produced using electrospray mass spectrometry.

as found for flow FAB [48]. Continued investigation should yield the same exciting results as have been found for other applications of the technique.

V. THERMOSPRAY

Another technique was developed and applied as an LC/MS interface and ionization mechanism for polar, ionic, and low-volatility materials prior to the development of electrospray [56,57]. The thermospray technique utilizes heat to vaporize and nebulize a flowing liquid stream for inlet into a mass spectrometer. In contrast with flow FAB and electrospray, the thermospray interface was developed specifically as an LC/MS inlet and was designed to operate at the flow rates of standard analytical separation columns (1 mL/min). The technique has been the focus of a wide variety of studies to understand the operable mechanisms, to optimize system performance, and to develop applications [4]. These studies have shown the broad applicability of the technique and, particularly when contrasted with electrospray and flow FAB, the potentially severe limitations to the approach. All this being understood, thermospray is still the most successful and widely practiced method for interfacing analytical HPLC with mass spectrometry.

A. Parameters and Characteristics

Thermospray was designed for use with reversed-phase HPLC separations, using aqueous-based mobile phases and ionic buffers. These conditions define both the advantages and limitations to the techniques. The approach requires hardware specifically designed and used exclusively for thermospray. An auxiliary pumping capacity, in addition to that already available on most mass spectrometers, is required. The technique can be used directly in-line with other detectors, such as UV absorption, as long as the mass spectrometer is the final detector. With appropriate control of the thermospray apparatus, the technique can be applied with a wide variety of mobile-phase conditions, including gradient elution. A limitation to the technique is that buffers used for HPLC–thermospray–MS must be volatile under thermospray conditions, to avoid clogging of the interface. This can require modification to the HPLC conditions, not always possible or acceptable to the separations scientist.

Thermospray is also unique in that it is both an inlet technique and an ionization mechanism, and these two processes can be independent from one another. In using thermospray as an inlet technique, the solvent and dissolved samples are heated to vaporize and nebulize the mixture for spraying into the ionization region of the ion source. Gas-phase ion-molecule chemistry, initiated using a high voltage discharge or a filament, is used to produce ions for analysis. These mechanisms are closely associated with chemical ionization and use the mobile-phase components as reactants. As an ionization technique, the ionization mechanism is somewhat analogous with electrospray, using heat and the mobile-phase velocity into a vacuum to affect nebulization. The presence of charged analyte species and ions from mobile-phase buffers produce gas-phase ions for mass analysis. The technique, either as an inlet or ionization, leads to the production of sample-ion adduct species, with the particular form dependent on the materials present during the thermospray operation. Both these descriptions are oversimplifications and more fundamental details are available elsewhere [4]. The technique offered significant advantages to a variety of researchers and applications for the method continue to increase [4].

B. Application to Cationic Surfactants

Along with the myriad of possible uses for the technique, thermospray was applied to the analysis of ionic surfactants, including cationics [58–62]. As with other techniques (FAB, electrospray), quaternary ammonium compounds were seen as "normally intractable compounds" [58] that could demonstrate the potential for the technique. Results of thermospray analyses were found to be erratic, with substantial thermal degradation encountered. The products generally detected were amines, with the most likely pathway related to the Hofmann reaction [63,64], as shown in Scheme 2. In one report [61], it was indicated that the intact cation could "generally be made to be one of the principal peaks" through manipulation of temperatures in the thermospray source. Other evaluations of the technique determined that due to the thermal decomposition, thermospray was unsuitable for quantitative analysis of mixtures of quaternary ammonium salts. These results, combined with the availability of more suitable tech-

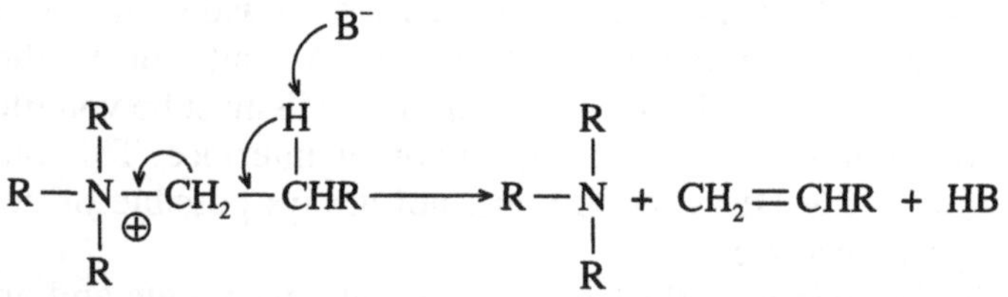

SCHEME 2 Possible route, based on the Hofmann reaction, for the formation of ions from quaternary ammonium compounds found under thermospray conditions. "B" is usually the counterion under classic Hofmann conditions but may be hydroxyl ion or a buffer anion under thermospray conditions.

niques, such as FAB [62] and possibly electrospray, indicate a limited utility for thermospray as a routine technique for analyses of cationic surfactants.

VI. OTHER APPROACHES

A. Field Desorption

Prior to the development of FAB, the most effective means of producing mass spectral information on cationic surfactant materials was through field desorption (FD) ionization [19,65–69]. In this technique, a sample solution is placed on a specially treated wire emitter and the solvent evaporated. The emitter containing a small amount (<1 μg) of sample is inserted into the mass spectrometer ion source. The emitter is subjected to a high-voltage electric field (10 kV potential difference), inducing desorption and ionization of sample materials. The emitter can also be heated, by passing current through the wire, to assist sample desorption. The technique is quite mild, little fragmentation is found in FD spectra. Producing ions in FD appears to be opposite that in EI or CI, quaternary ammonium compounds are among the easiest to ionize [70,71]. The technique is quite sensitive, with claims of as little as 10^{-13} g of material required to obtain spectra suitable for interpretation.

The technique suffers from being more difficult to optimize and use, particularly when compared with FAB. The use of high voltage for ionization is more applicable to magnetic sector instruments as opposed to quadrupole systems. Most important, ion currents produced in FD for most samples are quite unstable and last only seconds. The technique is still being studied for its ability to produce information on intact molecular species [72]. Finally, the approach offers the opportunity to analyze mixtures of all three types of surfactants (cationic, anionic, and nonionic) simultaneously from a single sample (such as environmental samples), as each type desorbs at a distinct emitter heating current [73].

B. Laser Desorption/Ionization

The analysis of surfactants has also been attempted using lasers as the energy source for ionization [74–78]. The advantages attributed to laser ionization include high sensitivity, minimum sample preparation, elimination of a liquid matrix, small sample requirements, rapid analysis, and high-resolution mass measurement of parent and fragment ions. The use of laser ionization and the lack of a liquid matrix allows mass spectrometry to be conducted on systems requiring very high vacuum, such as Fourier transform-ion cyclotron resonance (FT-ICR) spectrometers [77,78].

Laser ionization can yield spectra from cationic surfactants that contain only intact cations or the amount of fragmentation can be increased by increasing the power on the ionizing laser. This allows information on both molecular identity and structural composition to be obtained in a single analysis. Fragment ions from laser desorption appear similar to those found from other ionization techniques. The conditions of the laser desorption experiment can be adjusted to allow electron ionization to be performed on the desorbed surfactants. This may allow library-searchable EI spectra to be obtained.

The use of laser desorption as a routine tool for cationic surfactant analysis has been limited, however, in part by the complexity of systems used with laser ionization. Recent developments in the use of laser ionization, with time-of-flight analyzers and other systems [79,80], may lead to an increase in the use of this approach in the characterization of cationic surfactants.

C. Electron and Chemical Ionization

Both of these techniques require the analyte be in the gas phase for ionization. Approaches to use EI and CI for cationic surfactant analysis have generally focused on the means to place the sample in the gas phase while avoiding sample decomposition. Another approach is to degrade (pyrolyze) the sample intentionally for introduction using gas chromatography [81]. As this is more appropriately a chromatographic technique, it will not be covered.

A number of techniques have been tried to vaporize cationic samples for EI and CI mass spectrometry [15,16,77,82]. One approach is to heat the sample rapidly from a surface [15] or filament [16] inside the ion source of the mass spectrometer. Desorption can also be affected using a laser, as noted previously [77]. The desorbed species can then be ionized using 70-eV electrons for EI or a CI gas plasma.

Results from EI of tetraalkylammonium salts indicated production of the molecular cations [15] and fragment ions similar to those found with other ionization techniques. Ions were also found corresponding with clusters of two cations and one anion. As opposed to other techniques giving rise to thermal fragmentation (i.e., thermospray), ions were found corresponding with alkyl halides, as expected in thermal decomposition (Scheme 2).

Ammonia chemical ionization of quats produced ions corresponding with the protonated ion of the intact salt $[(C^+ + A^-) + H^+]^+$ and fragments related to that species [16]. It should be noted that the analytes in this study were heterocyclic, quaternary ammonium compounds and decomposition pathways previously noted for quaternary ammonium surfactants were unavailable. No indications on the applicability of long-chain ammonium compounds were given.

A recent approach to allow conventional EI or CI to be conducted in conjunction with liquid chromatography is the particle-beam interface [83,84]. In this system, a flowing liquid stream (column effluent or flow-injection stream) is nebulized in a slightly heated chamber to affect desolvation. A two-stage momentum separator pumps evaporated solvent and nebulizing gas away. The mechanism produces a "beam" of solid particles or low-volatility liquid drops for introduction into an EI or CI source. The system can handle flow rates of 0.1 to 0.5 mL/min and reproducible EI spectra have been obtained from injections of 100 ng of sample. When applied to the analysis of cationic surfactants [82] the EI and CI spectra are characterized by ions related to the tertiary amine (from thermal degradation) and a small signal from the intact quaternary cation. It was found that the abundance of the intact quat relative to the tertiary amine increases with increasing ion source temperature. It remains to be seen whether further applications for this approach will be made.

With the difficulties in generating intact samples for gas-phase ionization techniques, it seems unlikely at EI and CI will be extensively used for cationic surfactant analysis. This may change with improvements such as particle-beam interfaces and laser desorption.

VII. TANDEM MASS SPECTROMETRY

Most techniques for the mass spectrometric characterization of cationic surfactants were developed to produce an abundant signal from the intact molecular cation. This usually precludes producing structural information through ion fragmentation. Such a situation sacrifices an important advantage of mass spectrometry. This has been addressed by coupling ionization techniques for cationic surfactants with more sophisticated tandem mass spectrometric techniques [3]. In tandem mass spectrometry, one mass analyzer is used to select an ion for further mass analysis in a second analyzer. Through these techniques, mass spectrometry is used as a separation and sample purification step for mass spectrometry. This has led to the use of the term *mass spectrometry/mass spectrometry* (MS/MS), analogous to GC/MS or LC/MS. The fact that spectra of cationic surfactants are comprised predominantly of intact molecular species satisfies a number of desirable criteria for tandem mass spectrometry (i.e., a unique, single ion for each analyte species, with reasonable magnitude and persistence).

A. Characteristics and Requirements

All types of mass spectrometers have been used to perform MS/MS. Some systems are limited in the type or amount of information that can be generated or the types of experiments that can be performed. Basically, however, the techniques in tandem mass spectrometry are the same. Two analyzers are operated in concert with one another to allow additional structural information to be produced for a sample or a series of samples. These are referred to as *tandem-in-space experiments*. Certain types of instruments, ion traps, and FT-ICR systems can perform MS/MS experiments in a single analyzer, separating the functions (mass selection and mass analysis) in time rather than in space. Each system type has particular benefits for particular types of experiments.

In either regime, the information desired dictates which MS/MS experiment should be performed. Three basic experiments define the types most widely practiced. The first, the product (or daughter) ion scan, involves using the first analyzer to select the single ion of interest and a subsequent analyzer to separate the fragments from the selected parent. There is a reaction region between the first and second analyzers to induce parent ion decomposition, usually through collision with a neutral gas. This type of experiment is particularly useful for analysis of complex mixtures or when background may be a problem (such as in FAB or thermospray). The approach yields information on the structure of the parent selected, in a manner similar to the use of electron ionization for structural characterization of pure compounds. The second experiment type, the precursor (or parent) ion scan, uses the second analyzer to select for some common fragment ion and scans the first analyzer. This approach allows all ions in a mixture with a common decomposition product ion to be detected in a single scan. This type of information is useful in characterizing classes of compounds in mixtures. The third experiment, the neutral loss scan, is similar to a parent scan in that it allows characterization of groups of compounds based on the presence of some common functional group. In the neutral loss scan, the first and second analyzers are scanned at a fixed offset related to mass, such that the second analyzer will pass ions that differ from ions that left the first analyzer by some predetermined molecular mass. As many ions decompose through loss of small neutral molecules (e.g., H_2O, CH_3OH, HCl, NO, etc.) of known molecular mass, the neutral loss scan allows identification of all materials in a mixture undergoing the specified loss.

Instruments composed of electric and magnetic sectors are capable of very high energy (1000 eV or more) operation in the collision region, yielding more ion fragmentation and structural information. Systems with a quadrupole filter as the second analyzer (triple quadrupoles or hybrid instruments) operate at low collision energy (10 to 500 eV), usually producing limited fragmentation.

These are, of course, very brief descriptions of the types of experiments available in tandem mass spectrometry. More detailed descriptions of the reactions, their general utility, and requirements to perform the experiments on specific instruments are available in an excellent review of the technique [3].

Figure 10 shows the FAB mass spectrum of a commercial tallow imidazolinium surfactant mixture (Varisoft 445, Sherex) produced on a triple quadrupole instrument, with a series of molecular cations around m/z 600. A second series of ions is present around m/z 300. Tandem mass spectrometry can be used to determine the relationships between the various ions in this spectrum. Figure 11 shows the product ion mass spectra from two selected ions from Fig. 10, m/z 604 and m/z 632. The species at m/z 604 (Fig. 11a) is found to decompose (under low-energy conditions) to give four fragment ions (m/z 323, 310, 295, and 282), corresponding to

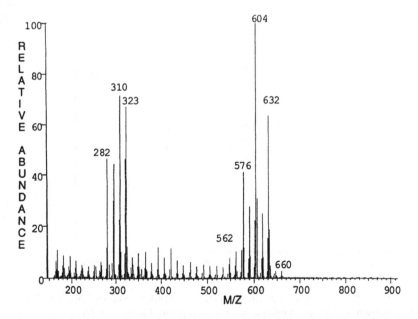

FIG. 10 FAB mass spectrum of a tallow imidazolinium cationic surfactant mixture produced from a m-nitrobenzyl alcohol matrix using xenon atom bombardment. The ions at m/z 604 and 632 were selected for further tandem mass spectrometry for structural characterization.

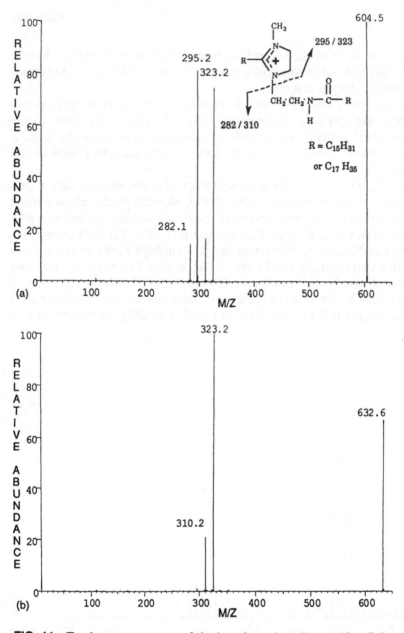

FIG. 11 Tandem mass spectra of the ions from the tallow imidazolinium cationic surfactant shown in Fig. 10. The species producing the ion at m/z 604.5 (a) yields fragment ions indicating the presence of two different alkyl chains (C_{15} and C_{17}) while the same fragmentations produce ions related to only C_{17} species in the spectrum from m/z 632.6 (b). Assignment of fractional masses is due to mass assignment and rounding in the data system and does not imply mass resolution ($m/\Delta m$) greater than unity.

those found in the FAB/MS spectrum in Fig. 10. Cleavage of the carbon–quaternary nitrogen bond leads to all ions found (Fig. 11a), suggesting the parent is composed of alkyl chains (R) of different lengths. The MS/MS spectrum of m/z 632 shows only two fragment ions (m/z 323 and 310), suggesting both alkyl chains in the parent material are of the same length. $C_{15}H_{31}$ and $C_{17}H_{35}$ alkyl chains from tallow fatty acid starting materials would lead to the fragment ions found in these tandem mass spectra.

B. Applications to Cationic Surfactants

Cationic surfactants were among the materials to be studied in one of the earliest applications of MS/MS to an industrial problem [19]. This work used field desorption to produce the molecular cations for subsequent structural analysis by MS/MS. An effort published shortly after this [36] used FAB to produce the cations of quaternary ammonium salts and probed the ion decomposition mechanisms using MS/MS. Field desorption and FAB as ionization methods for MS/MS of cationic surfactants were then compared in another study [69]. The most comprehensive study of MS/MS of cationic surfactants appeared in 1984 [85] when FAB was used to ionize a wide range of commercial materials (including amines, amine oxides, quaternary amines, polyethoxylated quaternary compounds, and amphoteric materials) and probe cation decomposition mechanisms. Cationic surfactants remain among the challenging problems to be addressed using tandem mass spectrometry [48,62,73,86].

It is clear that the various forms of mass spectrometry will continue to provide analytical information in the characterization of this commercially important class of materials.

REFERENCES

1. J. M. Richmond, ed., *Cationic Surfactants: Organic Chemistry,* Surfactant Science Series, Vol. 34, Marcel Dekker, New York, (1990).
2. J. J. Thomson, *Rays of Positive Electricity and Their Applications to Chemical Analysis,* Longman, London, (1913).
3. K. L. Busch, G. L. Glish, and S. A. McLuckey, *Mass Spectrometry/Mass Spectrometry,* VCH, New York, (1988).
4. A. L. Yergey, C. G. Edmonds, I. A. S. Lewis, and M. L. Vestal, *Liquid Chromatography/Mass Spectrometry: Techniques and Applications,* Plenum Press, New York, (1990).
5. J. B. Fenn, M. Mann, C. K. Meng, S. F. Wong, and C. M. Whitehouse, *Mass Spectrom. Rev. 9:* 37 (1990).
6. R. L. Cochran, *Appl. Spectrosc. Rev. 22:* 137 (1986).

7. M. Yamashita and J. B. Fenn, *J. Phys. Chem. 88*: 4451 (1984).
8. F. W. McLafferty, *Interpretation of Mass Spectra*, 3rd ed., University Science, Mill Valley, Calif., (1980).
9. A. G. Harrison, *Chemical Ionization Mass Spectrometry*, CRC Press, Boca Raton, Fla., (1983).
10. F. W. Karasek and R. E. Clement, *Basic Gas Chromatography/Mass Spectrometry: Principles and Techniques*, Elsevier, Amsterdam, (1988).
11. W. McFadden, *Techniques of Combined Gas Chromatography/Mass Spectrometry: Applications in Organic Analysis*, Wiley, New York, (1973).
12. B. J. Guizinowicz, M. J. Guzinowicz, and H. F. Martin, *Fundamentals of Integrated GC/MS*, Part III, *The Integrated GC/MS Analytical System*, Chromatographic Series, Vol. 7, Marcel Dekker, New York, (1977).
13. B. S. Middleditch, ed., *Practical Mass Spectrometry: A Contemporary Introduction*, Plenum Press, New York, (1979).
14. L. R. Snyder and J. J. Kirkland, *Introduction to Modern Liquid Chromatography*, Wiley, New York, (1974).
15. T. D. Lee, W. R. Anderson, Jr., and G. D. Daves, Jr., *Anal. Chem. 53*: 304 (1981).
16. B. Hasiak, G. Ricart, D. Barbry, D. Couturier, and M. Hardy, *Org. Mass Spectrom. 16*: 17 (1981).
17. U. Kh. Rasulev, E. G. Nazarov, K. S. Tursunov, and T. V. Arguneeva, *Zh. Anal. Khim 46*: 1802 (1991).
18. J. H. Callahan, K. Hool, J. D. Reynolds, and K. D. Cook, *Int. J. Mass Spectrom. Ion Processes 75*: 291 (1987).
19. R. Weber, K. Levsen, G. J. Louter, A. J. H. Boerboom, and J. Haverkamp, *Anal. Chem. 54*: 1458 (1982).
20. M. Barber, R. S. Bordoli, D. Sedgwick, and A. N. Tyler, *J. Chem. Soc. Chem. Commun. 7*: 325 (1981).
21. P. A. Lyon, ed., *Desorption Mass Spectrometry: Are SIMS and FAB the Same?* ACS Symposium Series 291, American Chemical Society, Washington, D.C. (1985).
22. W. V. Ligon and S. B. Dorn, *Int. J. Mass Spectrum, Ion Phys. 61*: 113 (1984).
23. W. V. Ligon and S. B. Dorn, *Int. J. Mass Spectrom. Ion Phys. 57*: 75 (1984).
24. M. Barber, R. S. Bordoli, G. Elliott, D. Sedgwick, and A. N. Tyler, *J. Chem. Soc. Faraday Trans. 1 79*: 249 (1983).
25. A. J. DeStefano and T. Keough, *Proceedings of the 31st ASMS Conference on Mass Spectrometry and Allied Topics*, American Society for Mass Spectrometry, Boston, 1983, p. 130.
26. K. L. Busch and R. G. Cooks, *Science 218*: 247 (1982).
27. H. R. Schulten and R. P. Lattimer, *Mass Spectrom. Rev. 3*: 231 (1984)
28. A. Dell and G. W. Taylor, *Mass Spectrom. Rev. 3*: 357, (1984).
29. R. J. Day, S. E. Unger, and R. G. Cooks, *Anal. Chem. 52*: 557A (1980).
30. M. Barber, R. S. Bordoli, G. Elliott, D. Sedgwick, and A. N. Tyler, *Anal. Chem. 54*: 645A (1982).
31. M. Rabrenovic, T. Ast, and J. H. Beynon, *Int. J. Mass Spectrom. Ion Phys. 61*: 31 (1984).

32. A. M. Buko, L. R. Phillips, and B. A. Fraser, *Biomed. Mass Spectrom. 10*: 408 (1983).
33. J. W. Dallinga, N. M. M. Nibbering, J. van der Greef, and M. C. Ten Noever de Braun, *Org. Mass Spectrom. 19*: 10 (1984).
34. G. W. Garner, D. B. Gordon, L. W. Tetler, and R. D. Sedgwick, *Org. Mass Spectrom, 18*: 486 (1983).
35. J. L. Gower, *Biomed. Mass Spectrom. 12*: 191 (1985).
36. E. Schneider, K. Levsen, P. Dähling, and F. W. Röllgen, *Fresenius Z. Anal. Chem. 316*: 277 (1983).
37. A. J. DeStefano and T. Keough, *Anal. Chem. 56*: 1846 (1984).
38. H. J. Veith, *Org. Mass Spectrom. 18*: 154 (1983).
39. R. J. Cotter, G. Hansen, and J. R. Jones, *Anal. Chim. Acta 136*: 135 (1982).
40. T. Keough and A. J. DeStefano, *Proceedings of the 31st ASMS Conference on Mass Spectrometry and Allied Topics,* American Society for Mass Spectrometry, Boston, 1983, p. 128.
41. J. R. Simms, T. Keough, S. R. Ward, B. L. Moore, and M. M. Bandurraga, *Anal Chem. 60*: 2613, (1988).
42. Y. Ito, T. Takeuchi, D. Ishi, and M. Goto, *J. Chromatogr. 346*: 161 (1985).
43. R. M. Caprioli, T. Fan, and J. D. Cottrell, *Anal. Chem. 58*: 2949 (1986).
44. R. M. Caprioli, *Anal. Chem. 62*: 477A (1990).
45. R. D. Minard, D. Chin-Fatt, P. Curry, Jr., and A. G. Ewing, *Proceedings of the 36th ASMS Conference on Mass Spectrometry and Allied Topics,* American Society for Mass Spectrometry, San Francisco, 1988, p. 950.
46. R. M. Caprioli, W. T. Moore, K. B. Wilson, and S. Moring, *J. Liquid Chromatogr. 480*: 247 (1989).
47. J. S. M. de Wit, L. J. Deterding, M. A. Moseley, K. B. Tomer, and J. W. Jorgensen, *Rapid Commun. Mass Spectrom. 2*: 100 (1988).
48. D. L. Lawrence, *J. Am. Soc. Mass Spectrom. 3*: 575 (1992).
49. M. A. Moseley, L. J. Deterding, J. S. M. de Wit, K. B. Tomer, R. T. Kennedy, N. Bragg, and J. W. Jorgensen, *Anal. Chem. 61*: 1577 (1989).
50. R. D. Smith, J. A. Olivares, N. T. Nguyen, and H. R. Udseth, *Anal. Chem. 60*: 436 (1988).
51. E. C. Huang, T. Wachs, J. J. Conboy, and J. D. Henion, *Anal. Chem. 62*: 713A (1990).
52. H. R. Udseth, J. A. Loo, and R. D. Smith, *Anal. Chem. 61*: 228 (1989).
53. J. J. Conboy, J. D. Henion, M. W. Martin, and J. A. Zweigenbaum, *Anal. Chem. 62*: 800 (1990).
54. R. D. Smith, C. J. Barinaga, and H. R. Udseth, *Anal. Chem. 60*: 1948 (1988).
55. G. J. Harvey and J. C. Dunphy, *Proceedings of the 40th ASMS Conference on Mass Spectrometry and Allied Topics,* American Society for Mass Spectrometry, Washington, D.C., 1992, p. 592.
56. C. R. Blakley, J. C. Carmody, and M. L. Vestal, *J. Am. Chem. Soc. 102*: 5931 (1980).
57. C. R. Blakley, J. C. Carmody, and M. L. Vestal, *Anal. Chem. 52*: 1636 (1980).
58. A. L. Yergey, M. L. Vestal, C. R. Blakley, and D. J. Liberato, *Proceedings*

of the 30th ASMS Conference on Mass Spectrometry and Allied Topics, American Society for Mass Spectrometry, Honolulu, 1982, p. 188.
59. D. S. Mitchell, K. R. Jennings, and J. H. Scrivens, *Proceedings of the 35th ASMS Conference on Mass Spectrometry and Allied Topics,* American Society for Mass Spectrometry, Denver, 1987, p. 427.
60. D. Vecchio and A. L. Yergey, *Proceedings of the 36th ASMS Conference on Mass Spectrometry and Allied Topics,* American Society for Mass Spectrometry, San Francisco, 1988, p. 1075.
61. D. Vecchio and A. L. Yergey, *Org. Mass Spectrom. 24*: 1060 (1989).
62. Md. A. Mabud, P. A. Dreifuss, W. E. Killinger, and M. W. Smith, *Proceedings of the 37th ASMS Conference on Mass Spectrometry and Allied Topics* American Society for Mass Spectrometry, Miami Beach, Fla., 1989, p. 929.
63. A. C. Cope and E. R. Trumbull, *Org. React. 11*: 317 (1960).
64. D. A. Archer and H. Booth, *J. Chem. Soc.* 322 (1963).
65. D. A. Brent, D. J. Rouse, M. C. Sammons, and M. M. Bursey, *Tetrahedron Lett. 42*: 4127 (1973).
66. H. R. Shulten and F. W. Röllgen, *Angew. Chem. Int. Ed. Eng. 14*: 561 (1975).
67. M. C. Sammons, M. M. Bursey, and C. K. White, *Anal. Chem. 47*: 1165 (1975).
68. H. J. Veith, *Org. Mass Spectrom. 11*: 629 (1976).
69. H. J. Veith, *Org. Mass Spectrom. 18*: 154 (1983).
70. F. W. Röllgen, U. Giessmann, H. J. Heinen, and S. J. Reddy, *Int. J. Mass Spectrom. Ion Phys. 24*: 235 (1977).
71. R. J. Cotter, *Trends Anal. Chem. 1*: 307 (1982).
72. J. H. Scrivens, K. Rollins, M. J. Taylor, and R. C. K. Jennings, *Proceedings of the 40th ASMS Conference on Mass Spectrometry and Allied Topics,* American Society for Mass Spectrometry, Washington, D.C., 1992, p. 1320.
73. E. Schneider, K. Levsen, A. J. H. Boerboom, P. Kistemaker, S. A. McLuckey, and M. Przybylski, *Anal. Chem. 56*: 1987 (1984).
74. B. Schueler and F. R. Krueger, *Org. Mass Spectrom. 14*: 439 (1979).
75. R. Stoll and F. W. Röllgen, *Org. Mass Spectrom. 14*: 642 (1979).
76. K. Balasanmugam and D. M. Hercules, *Anal. Chem. 55*: 145 (1983).
77. D. Weil and P. Lyon, *Proceedings of the 39th ASMS Conference on Mass Spectrometry and Allied Topics,* American Society for Mass Spectrometry, Nashville, Tenn., 1991, p. 819.
78. D. Weil, P. Lyon, R. Pachuta and C. Moore, *Proceedings of the 40th ASMS Conference on Mass Spectrometry and Allied Topics,* American Society for Mass Spectrometry, Washington, D.C., 1992, p. 842.
79. F. Hillenkamp, M. Karas, R. C. Beavis, and B. T. Chait, *Anal. Chem. 63*: 1193A (1991).
80. A. McIntosh, T. Donovan, and J. Brodbelt, *Anal. Chem. 64*: 2079 (1992).
81. N. J. Haskins and R. Mitchell, *Analyst 116*: 901 (1991).
82. B. H. Solka, *Proceedings of the 39th ASMS Conference on Mass Spectrometry and Allied Topics,* American Society for Mass Spectrometry, Nashville, Tenn., 1991, p. 1661.
83. R. C. Willoughby and R. F. Browner, *Anal. Chem., 56*: 2625 (1984).

84. R. C. Willoughby and F. Poeppel, *Proceedings of the 35th ASMS Conference on Mass Spectrometry and Allied Topics,* American Society for Mass Spectrometry, Denver, 1987, p. 289.

85. P. A. Lyon, F. W. Crow, K. B. Tomer, and M. L. Gross, *Anal. Chem. 56*: 2278 (1984).

86. D. F. Fraley and D. L. Lawrence, *Proceedings of the 33rd ASMS Conference on Mass Spectrometry and Allied Topics*, American Society for Mass Spectrometry, San Diego, 1985, p. 144.

10
Chromatography of Cationic Surfactants: HPLC, TLC, and GLC

BRUCE PAUL McPHERSON and HENRIK T. RASMUSSEN
Analytical Sciences, Colgate-Palmolive Company, Piscataway, New Jersey

I. INTRODUCTION

Reviews of the chromatographic separations of cationic surfactants have appeared in various forms in previous volumes of this series [1,2] and elsewhere [3]. This review summarizes the progress made in high-perform-ance liquid chromatography (HPLC), gas chromatography (GC), and thin-layer chromatography (TLC), as well as in emerging separation techniques such as supercritical fluid chromatography (SFC) and capillary electropho-resis (CE). The content is organized by major compound type, followed by chromatographic technique in an effort to make the information readily accessible. The intention is not to evaluate the work cited herein, but simply to summarize the approaches that have been taken to date. A primary concern has been to specify not only the conditions necessary for analysis, but also (1) the matrix types for which the analysis has shown applicability; and (2) the expected analytical accuracy, reproducibility, and limits of detection.

II. QUATERNARY AMMONIUM SALTS

A. High-Performance Liquid Chromatography

The approach to the analysis of quaternary ammonium salts by HPLC is governed primarily by the substituents. In this review, two classifications are considered: (1) aromatic quaternary ammonium salts such as alkyl-benzyl and alkylpyridinium salts, for which direct ultraviolet (UV) detec-tion is possible; and (2) nonaromatic compounds, for which detection is the primary obstacle.

1. Aromatic Quaternary Ammonium Salts

Analysis of quaternary ammonium salts by HPLC typically involves the use of ion-pairing reagents. A general paper on the practical aspects of ion-pair chromatography has been published by Gloor and Johnson [4]. The use of ion-paired HPLC for the separation of quaternary ammonium salts has been examined extensively by Abidi [5], who studied the influence of stationary phase, mobile phase, counterion, counterion concentration and pH, on separation of n-alkylbenzyldimethylammonium chlorides. Some of this work is summarized in Table 1. No specific applications were noted. However, the summary of data may serve as a useful guideline in manipulating selectivity.

Homologous series of alkylbenzyldimethylammonium chlorides and al-kylpyridinium halides with C_{10} to C_{18} alkyl groups have been separated by Nakae et al. [6,7] using a 500 mm × 4 mm ID column packed with Hitachi

TABLE 1   Capacity Factors of C_{12}, C_{14}, C_{16}, and C_{18} n-Alkylbenzyldimethyl Ammonium Chlorides Using Various Reversed-Phase HPLC Conditions

Percent organic modifier[a]	Component			
	C_{12}	C_{14}	C_{16}	C_{18}
Octadecylsilica				
95% methanol (0.1)	0.51	0.83	1.25	2.08
90% methanol (0.1)	0.80	1.32	2.54	4.85
95% acetonitrile (0.1)	0.40	0.61	1.13	2.12
90% acetonitrile (0.01)	0.75	1.12	2.38	4.62
90% acetonitrile (0.05)	0.82	1.25	2.50	4.88
60% THF (0.1)	1.25	1.90	2.65	4.15
50% THF (0.1)	1.74	2.81	4.59	6.84
Phenylpropylsilica				
90% methanol (0.1)	0.52	0.61	0.75	0.93
85% methanol (0.1)	0.77	0.98	1.31	1.79
80% acetonitrile (0.1)	1.10	1.17	1.50	1.83
70% acetonitrile (0.1)	2.27	2.95	4.10	5.37
70% acetonitrile (0.05)	2.67	3.50	4.83	6.42
60% THF (0.1)	1.21	1.59	2.09	2.71
50% THF (0.1)	2.74	3.43	4.18	5.64
Cyanopropylsilica				
60% methanol (0.1)	1.56	2.06	2.89	4.11
50% methanol (0.1)	5.47	8.76	15.50	29.60
50% acetonitrile (0.2)	3.38	4.13	5.25	6.63
50% acetonitrile (0.05)	4.51	5.75	6.59	9.25
40% acetonitrile (0.1)	8.60	12.20	17.60	25.80
50% THF (0.1)	2.07	2.53	3.07	3.67
40% THF (0.1)	4.06	5.56	7.56	

[a]Values in parentheses are molar concentrations of perchlorate buffers, pH 3.
Source: Ref. 5.

Gel 3011 (porous microspherical styrene/divinyl benzene copolymer particles with an average diameter of 10 to 15 μm). Components of both sample types were eluted in less than 30 min, at a temperature of 30°C, using 0.5 M perchloric acid in methanol as the mobile phase. Improved resolution was achieved, at the expense of analysis time, by increasing the perchloric acid concentration. Other acids and salts, dissolved in methanol, were also investigated as mobile-phase additives, but it was concluded that only the pairing anion influenced retention. Identification of individual homologs is aided by a linear relationship between alkyl chain length and solute capacity factor. Agreement between the molar percentages of each component and HPLC peak areas were typically within 5%. Quantitative reproducibility, as measured from five replicate determinations, resulted in relative standard deviations (RSDs) of less than 5% for all components.

Nakamura et al. [8] utilized a 250 mm × 4 mm ID column packed with 5 μm TSK Gel ODS/silica to separate alkylbenzyldimethylammonium chlorides with alkyl chains ranging from 10 to 18 carbons. The optimized mobile phase consisted of 85/15 (v/v) methanol/water with 0.4 M sodium chloride. UV detection was at 210 nm, the flow was set to 1.5 mL/min, and the column was thermostatted at 50°C. Under the outlined conditions the coefficient of variance (CV) of the relative peak areas of each component was less than 0.6%. The chief advantage of the conditions used, however, was broad applicability. The same conditions also proved viable for the determination of the homologous distributions of N-acylsarcosinates, N-acyl-N-methyl taurates, N-acyl-L-glutamates, fatty acid monoethanolamides, fatty acid diethanolamides, monoisopropanolamides, and alkyl sulfates. Use of the method to determine the homologous distribution of various commercial surfactants gave results that were in excellent agreement with results obtained by GC. Whereas the GC analysis of several of the mixtures required extensive sample separation, the HPLC method required only dilution and analysis times were typically less than 20 min. In subsequent papers [9,10] the same authors demonstrate that in addition to separating the individual homologous series, mixtures of homologous series may be separated simultaneously (see Sec. II.A.2).

Benzalkonium chloride homologs have also been separated by Meyer [11] using a 300 mm × 4 mm ID μBondapak CN (10 μm) column, a 60:40 (v/v) acetonitrile/0.1 M sodium acetate eluant (adjusted to pH 5.0 with acetic acid), and UV detection (254 nm). Homologs with alkyl chain lengths of up to 18 were separated in less than 13 min. The method was used to determine the alkyl chain distribution of benzalkonium chloride (50 ppm) from various sources with good reproducibility. The compositions of two samples from different sources were found to be (for nine replicate determinations):

Sample 1: 53.1% C_{12} (RSD = 2.0%), 33.1% C_{14} (RSD = 3.7%), 13.9% C_{16} (RSD = 8.7%)

Sample 2: 72.0% C_{12} (RSD = 3.6%), 23.1% C_{14} (RSD = 4.2%), 4.9% C_{16}(RSD = 16.4%)

By summation of the peak areas from the three chain lengths, a recovery of 100.0 ± 2.3% was obtained. The latter approach was used to quantitate benzalkonium chloride in an ophthalmic drug at the 40-ppm level.

Subsequent papers [12,13] on the analysis of ophthalmic systems, however, have revealed that polymeric materials, suspended particulate matter such as steroid suspensions, and alkaloids may interfere with the method as described. To circumvent the problem, benzalkonium chloride has been ion paired with methyl orange and isolated by extraction into 1,2-dichloroethane [12]. Alternatively, Ambrus et al. [13] have quantitated benzalkonium chloride in various ophthalmic formulations using one of two mobile phases/diluents:

1. 33:9:58 (v/v/v) acetonitrile/2-propanol/water adjusted to pH 2.2 with perchloric acid.
2. 34:15:51 (v/v/v) acetonitrile/THF/water adjusted to pH 2.2 with perchloric acid.

Recoveries ranged from 91.7 to 101.9% with relative standard deviations of less than 3.3%. The column employed was a 150 × 4 mm ID μBondapak CN (4 μm) and the detection wavelength was changed to 214 nm for increased sensitivity. A typical chromatogram, obtained in mobile phase (1), but illustrative of the separation achieved in each media, is shown in Fig. 1.

Meyer and Takahashi [14] have attempted to apply the conditions used by Meyer [11] for the analysis of cetylpyridinium chloride in chondrotin sulfate. While the method gave acceptable retention, spiked chondrotin sulfate samples provided inadequate recoveries. To increase recoveries, a mobile phase consisting of 60:40 (v/v) methanol/0.36% (w/v) tetramethylammonium hydroxide pentahydrate in water, adjusted to pH 3.5 with acetic acid, was used (other conditions as in Ref. 11). The modified con-

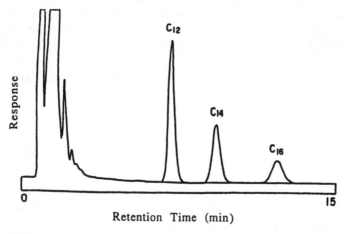

FIG. 1 Separation of benzalkonium chlorides by HPLC. For conditions, see the text. (From Ref. 13.)

ditions gave recoveries for cetylpyridinium chloride, at the 1 ppm level, of 98% with a RSD of 1.3%. The new method was also shown to be applicable to the analysis of cetylpyridinium chloride in a commercially available mouthwash at the 500-ppm level.

The use of ion-paired reversed-phase HPLC for the analysis of the cetylpyridinium ion has also been demonstrated by Linares et al. [15]. Using an endcapped 250 mm × 4.6 mm ID Ultrabase C_{18} (5 μm) column and a mobile phase consisting of 10:50:40 (v/v/v) water/chloroform/methanol with 0.01 M sodium dioctyl sulfosuccinate, analyses were complete in less than 6 min. Detection was performed using a photodiode array detector which allowed for the cetylpyridinium ion and the co-analytes (benzocaine and dextromethorphan) to be monitored at their optimum wavelengths. No interferences were observed from sorbitol, mint flavor, and magnesium stearate. At a wavelength of 244 nm, the detection limit for the cetylpyridinium ion was 175 μg/mL (3 × the signal-to-noise ratio). Reproducibility at the 1000-μg/mL level was better than 3% RSD.

Kawase et al [16] have used a 150 mm × 4.6 mm ID column packed with Develosil ODS-3 (3 μm) to analyze imidazoline-type surfactants. Separation was achieved using 0.1 M sodium perchlorate in 60:60:5 (v/v/v) methanol/acetonitrile/water, at 40°C, and a detection wavelength of 240 nm. The molar alkyl chain distribution and average molecular weight of the sample compared favorably to results obtained by the GC analysis of the fatty acids derived by hydrolysis of the imidazoline ring [17]. Furthermore, the method could be applied to the analysis of commercial products.

Abidi [18] has separated benzyldiisobutylphenoxyethoxyethyldimethylammonium chloride from its cresoxy analog using: a 30 cm × 4.0 mm ID Ultrasphere ODS (5 μm) column; a mobile phase consisting of 70:30 (v/v) acetonitrile/water with 0.01 M sodium pentylsulfonate/0.01 M perchloric acid (adjusted to a pH of 3 with disodium hydrogen phosphate); and UV detection at 215 nm. Limits of detection were 10 ng (2 × the signal-to-noise ratio). Recoveries from spiked natural waters, using cation-exchange chromatography for sample enrichment, averaged 85.2% ± 6.7% for levels of 1 to 2000 μg/L.

In an attempt to eliminate the need for ion pairing with perchloric acid, Abidi [19] has used a 25 cm × 4.6 mm ID column packed with β-cyclodextrin for the analysis of mixtures of n-alkylbenzyldimethylammonium chloride. Separations of C_{12} to C_{18} homologs were achieved in less than 15 min using both 70:30 (v/v) water/acetonitrile and 50:50 (v/v) methanol/water at a detection wavelength of 215 nm. Quantitative analysis using the water/methanol system resulted in RSDs of 3.67 to 6.61% for standards prepared to contain from 10 to 100 μg/mL of each homolog. When the method was applied to the analysis of unknown samples, determinations were in excellent agreement with results obtained by mass spectrometry. It was, additionally, noted that the addition of 0.1 M of certain electrolytes to 30:70 (v/v) acetonitrile/water (adjusted to pH 3.5 with phosphoric acid) improved resolution. The most notable enhancement was achieved using silver nitrate. The use of silver nitrate, however, necessitates the use of ion exchange to convert the alkylbenzyldimethylammonium salts to nitrates to the prevent the precipitation of silver chloride.

2. Nonaromatic Quaternary Ammonium Salts

Tetraalkyl quaternary ammonium materials are used in vastly different applications with alkyl (R) carbon numbers from 1 (methyl) to 18 (stearyl) in most instances. In this section only compounds containing at least one alkyl group with 10 or more carbons will be considered (i.e., the tetrapropyl and tetrabutyl types will not be included).

Helboe [20], building on previously reported techniques [21], used paired-ion chromatography with vacancy or indirect photometric detection to determine four alkyltrimethylammonium bromides: dodecyl, tetradecyl, hexadecyl, and stearyl. Peak shapes and response comparisons were made with Nucleosil 5 C_{18}, C_8, and CN columns as well as Nucleosil 7 C_6H_5. Other key variables revolved around the use of p-toluenesulfonate and naphthalene-2-sulfonate UV-absorbing counterions. It was demonstrated that the CN column (120 × 4.6 mm) with an eluant of 55:45 methanol/water (v/v) containing 5 mM p-toluenesulfonic acid provided optimal results (Fig. 2). Detection was at 254 nm, with linearity established for injected amounts ranging from 0.4 nmol (the detection limit) to 64 nmol. Helboe confirmed previous findings by Parris [22] that the capacity factor decreases for increasing amounts of solute and hence that peak areas must be used for quantitation. It was also confirmed that the change in capacity factor with sample amount is much more pronounced when naphthalene-2-sulfonate is the counterion rather than p-toluenesulfonate. The precision of the method was investigated, with relative standard deviations between 1.1 and 1.7% reported.

Vacancy UV detection was also employed by Larson and Pfeiffer [23] in the determinations of tetradecyltrimethyl-, hexadecyldimethylethyl-, and didodecyldimethylammonium salts. The mode of separation was different from Helboe's in that a Partisil-10 SCX (4.6 × 250 mm) strong cation-exchange column was used. The mobile phase was acetonitrile/water (70:30 v/v) with 1% acetic acid (pH 3.7), containing 0.010 M benzyltrimethylammonium chloride as the UV-absorbing counterion. Calibration runs yielded straight lines showing linear dependence of peak area over a range of 10 to 1000 ppm at the 268-nm detector setting. It was also demonstrated that the peak areas obtained were nearly independent of the quaternary am-

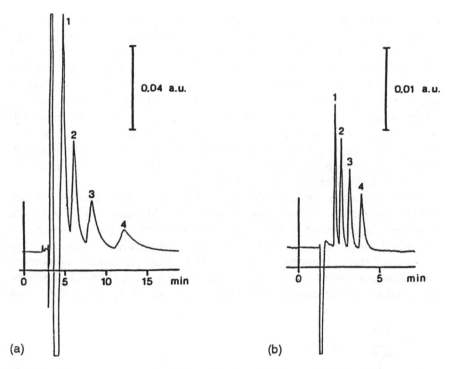

FIG. 2 Chromatograms of ca.16 nmol of each of four alkyltrimethylammonium bromides: (a) Nucleosil 7 C_6H_5 (250 × 4.6 mm) with 75:25 methanol/water containing 0.4 mM sodium naphthalene-2-sulfonate and 10 mM phosphoric acid; (b) Nucleosil 5 CN (120 × 4.6 mm) with 55:45 methanol/water containing 5 mM p-toluenesulfonic acid. Detection wavelength: 254 nm. Flow rate: 1 mL/min. Peaks: 1, dodecyl; 2, tetradecyl; 3, hexadecyl; 4, stearyl. (From Ref. 20.)

monium salts injected indicating that concentrations of unknowns may be estimated from the compounds for which standards are available, as previously reported by Small and Miller [24]. Detection limits for the various quaternary ammonium salts injected were 0.6 to 0.8 μg, calculated as 2.5 times short-term noise.

Spagnolo et al. [25] took Larson and Pfeiffer's work a little further with an application to determine aliphatic and aromatic quaternary ammonium salts simultaneously. A Partisil-10 SCX (2.1 × 250 mm) strong cation-exchange column was used with an eluant of 0.04 M ammonium formate in 100% methanol, but indirect photometric UV detection was not employed. Direct UV (264 nm) and refractive index detectors were used in series to determine methyltrialkyl-, methylbenzyldialkyl-, dimethyldialkyl-, dimethylbenzylalkyl-, and trimethylalkylammonium chlorides present in organoclays. The order of elution follows the order of listing in the preceding sentence with the methyltrialkyl eluting first (4.3 min) and trimethylalkyl(s) eluting last at 12.9 and 13.6 min. In their application, the alkyl substituents were hydrogenated tallow groups that are mixtures of C_{18} and C_{16} chain lengths. It was noted that splitting of the C_{18} and C_{16} chain lengths was only evident in the monoalkyl compounds: dimethyl-benzylalkyl- and trimethylalkylammonium salts. Quantitative analysis was not included because there were no pure standards available, although they reported that rough estimates of the relative amounts can be obtained via peak area normalization with the RI detector.

Refractive index was the sole means of detection for a diverse mixture of nine surfactants as reported by Nakamura and Morikawa [9]. Their work produced fundamental methodology for the simultaneous separation of individual surfactants and their respective homologs from each other re-gardless of their ionogenic properties. The three quaternary ammonium compounds included were alkylpyridinium, alkylbenzyldimethylammon-ium, and alkyltrimethylammonium chlorides (Fig. 3). A 6 mm × 200 mm

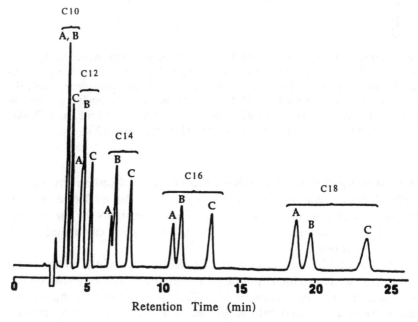

FIG. 3 Separation of alkyltrimethylammonium (A), alkylpyridinium (B), and al-kylbenzyldimethylammonium (C) chloride homologs. Conditions as described in the text with 1.0 M sodium perchlorate at pH 2.5. (From Ref. 9.)

ODS/silica column (TSK Gel LS 410) was used and maintained at 50°C. Several mobile-phase additives (inorganic salts) and pH variables were explored, resulting in two eluants of choice: 15:85 (v/v) water/methanol containing 1.0 M sodium perchlorate (adjusted to pH 2.5 with phosphoric acid), or containing 0.1 M at pH 3.5. A subsequent study [10] with Kanesato and Nakata includes their version of a postcolumn reaction technique [26,27] employing a Technicon Auto-Analyzer II. The system performs an ion-pair extraction with Orange II to detect only the cationics at 484 nm with a pH of 7.3. HPLC is performed just as in the prior Nakamura and Morikawa paper. However, to detect different classes of ionogenic surfactants selectively, the postcolumn chemistry is altered. At a pH of 3.4, with Orange II, both cationics and amphoterics are detected. Using methylene blue to form the hydrophobic ion-pair complex results in detection of only the anionic surfactants.

An additional vacancy UV detection application by Huang [28] deals solely with a dialkyldimethylammonium chloride compound. Arquad 2C-75, in which the alkyl chain distribution ranges from 8 to 18, was detected as a single negative peak. A DuPont Zorbax C_8 (4.6 × 250 mm) column was used and compared to Helboe's findings with the Nucleosil CN. Huang reports twice the sensitivity for Arquad 2C-75 (25 ppm) in rolling oils when using the Zorbax C_8 column. The mobile phase used was 25:75 (v/v) methanol/water containing 5 mM of p-toluenesulfonic acid UV counterion at a wavelength of 260 nm for detection. One of the drawbacks illustrated is that under these conditions other quaternary ammonium salts (dioctadecyldimethyl- and octadecyltrimethyl-) coelute with the 2C-75.

Specific determinations of dialkyldimethylammonium compounds in personal care formulations and raw materials have been done using refractive index detection [29]. Caesar and colleagues describe quantitative determinations of myristamidopropyl-N,N-dimethyl-N-(2,3-dihydroxypropyl) ammonium chloride and oleamidopropyl-N,N-dimethyl-N-2,3-dihydroxypropyl) ammonium chloride with two μBondapak CN columns (150 × 4 mm) in series. A mobile phase consisting of 1100 mL of water and 900 mL of acetonitrile containing 2 mL of trifluoroacetic acid was used to determine the quaternary ammonium chloride content in the raw materials, and also to determine quantitatively the oleyl salt in a skin moisturizer formulation. The myristyl salt was quantitatively determined in a clear conditioning shampoo with a mobile phase consisting of 1140 mL of water, 840 mL of acetonitrile, and 20 mL of THF containing 2 mL of trifluoroacetic acid. Also illustrated are the palmitoleic, linoleic, and palmitic components in the oleyl raw materials. Precision and accuracy data were included for all determinations.

Conductometric detection and nonaqueous ion chromatography are the unique features in an application reported by Wee and Kennedy [30]. They determined two non-UV-absorbing quaternaries, ditallowdimethyl- and dodecyltrimethylammonium chlorides, and two UV-absorbing quaternaries, stearyldimethylbenzylammonium and 1-hexadecylpyridinium chlorides, in river water. A Whatman 25 cm Partisil PAC 10 column (cyanoamino bonded phase) was used with a mobile phase of 92:8 (v/v) chloroform/methanol. All columns were rinsed prior to use with at least 100 mL of methanol to elute residual ions and reduce background conductivity to improve the sensitivity. River water samples were acidified with HCl, and linear alkylbenzenesulfonate was added prior to repeated extractions with methylene chloride and subsequent dryings/concentration steps. Final dissolution was in the HPLC mobile phase. Response linearity for each quaternary ammonium salt was illustrated and the detection limit for the standard solution was approximately 0.01 μg. For an environmental sample where the background interference was higher, the detection limit was approximately 0.02 μg or 2 ppb based on a 50 μL sample size (out of 0.5 mL) injected and a 100 mL sample analyzed. Recovery varied from 93 to 106% with standard deviations no worse than 20%. Environmental concentrations of quaternary ammonium salts found in several river waters were all in the low parts per billion concentrations, ranging from 5 to 30 ppb for ditallowdimethylammonium chloride and less than 2 ppb for the other three.

Normal-phase HPLC, using gradient elution and evaporative light-scattering detection was employed by Wilkes and co-workers [31] for quantitative determinations of low levels of monoalkyltrimethyl- and trialkylmethylammonium chlorides in dialkyldimethylammonium chloride. The method has also been extended successfully to household and cosmetic products and an imidazolinium-type quaternary ammonium salt was included to demonstrate the applicability of the procedure. In their studies a 250 × 4.6 mm column packed with 5 μm RSil polyphenol was used with the gradient elution/mobile phase program summarized in Table 2 (flow

TABLE 2 Gradient Elution Program

Time (min)	Solvent A[a]	Solvent B[b]
00	90	10
20	10	90
25	10	90

[a]5mM Trifluoroacetic acid (TFA) in n-hexane.
[b]5mM Trifluoroacetic acid in THF/methanol (3:1).

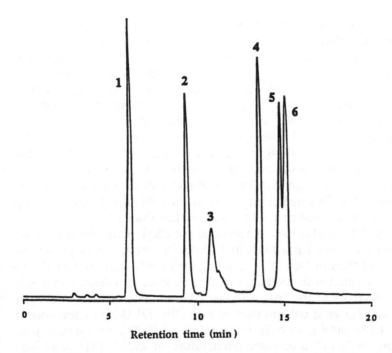

FIG. 4 Separation of trialkylmethylammonium (1), dialkyldimethylammonium (2), imidazolinium (3), cetylbenzyldimethylammonium (4), C_{18} alkyltrimethylammonium (5), and C_{16} alkyltrimethylammonium (6) salts. Conditions as described in text. (From Ref. 31.)

rate of 1.5 mL/min). The chromatogram, illustrating a quaternary ammonium salt mixture separated under their conditions, is shown in Fig. 4. It was noted that hexane with a high percentage of methanol/THF, rather than diethyl ether or chloroform with a much smaller amount of modifier, gives better retention time stability than traditional normal-phase solvent systems, as experienced by Wee and Kennedy and others [32]. An internal standard procedure with dimethylbenzylhexadecylammonium chloride as a reference standard was adopted for quantitative determinations. Detector response/linearity was established over a range of 0 to 100 µg injected. Studies on precision and accuracy were included in their work.

B. Gas Chromatography

Due to the low volatility of quaternary ammonium compounds, direct GC methods have been difficult to implement. The majority of published papers have instead focused on means of derivatization or degradation. Warrington [33], for example, has hydrogenated benzalkonium chloride to yield toluene and alkyldimethylamines. The latter compounds were separated using a 0.75% SE-30 on 100 to 140-mesh Gas Chrom P column, temperature programmed from 60 to 276°C at the rate of 9°C/min. Through the use of alkaline columns, Metcalfe [34] achieved the on-column decomposition of trimethylalkyl-, dimethylbenzylalkyl-, and dimethyldialkylammonium halides to the corresponding tertiary amines. Using Chromosorb-W, 60 to 80 mesh, coated with 10% (by weight) potassium hydroxide and 5 to 20% Apiezon L or Carbowax 6000, compounds containing up to 38 carbons were separated. Using a similar column, Barry and Saunders [35] have determined the alkyl chain distribution of commercial alkyltrimethylammonium bromides. The quaternary ammonium salts, along with impurity bromoalkanes, were analyzed with a RSD of less than 1%.

Takano and Tsuji [17] have determined the alkyl chain distribution of imidazolinium cationic surfactants by hydrolyzing them first in potassium hydroxide and then in HCl. The resulting fatty acids were extracted into ether and esterified with methanol in the presence of concentrated sulfuric acid. Analysis of the fatty acid methyl esters was conducted using a 2 m × 3 mm ID glass column packed with 20% DEGS on Chromosorb WAW (60 to 80 mesh), an oven temperature of 190°C, an injection port temperature of 250°C, a detector temperature of 250°C, FID detection, and helium carrier gas at 40 mL/min.

Abidi [36] has derivatized alkylbenzyldimethylammonium halides as trichloroethyl carbamates and cyanamides to obtain compounds sensitive to electron capture and thermionic nitrogen detection, respectively. Analyses were conducted on 2 m × 2 mm ID coiled glass columns packed with 3% OV-17 or 3% SE-30 on Gas-Chrom Q (80 to 100 mesh), using a nitrogen flow rate of 30 mL/min, detector temperatures of 350°C, and an injection port temperature of 250°C. Separations were achieved isothermally at 250°C for the trichloroethyl carbamate derivatives, resulting in an analysis time of 25 min, and at 220°C for the analyses of the cyanamide derivatives (analysis time = 10 min). Detection limits were in the low-picogram region using electron capture detection and at 0.1 ng using the nitrogen detector. Both detectors showed good linearity for concentrations in the range 0.01 to 10 μg/mL. The coefficients of variation for repetitive injections of the same sample averaged 3.6%.

The derivatization process described by Abidi [36] was used subsequently for the analysis of aryloxyethoxybenzyldimethylammonium salts in natural waters [18]. Using a 25 m × 0.21 mm ID fused silica column coated with OV-101, an isothermal column temperature of 260°C, and an inlet pressure of 10 psi, the detection limits for trichloroethyl carbamate derivatives were less than 2 pg. Cyanamide derivatives had detection limits of 0.1 ng. Recoveries from natural waters spiked with 1 to 2000 μg/L were 80.4 ± 7.1%.

Suzuki et al. [37–39] have used the Hofmann degradation to convert alkyltrimethyl- and dialkyldimethylammonium compounds to the corresponding 1-olefins and alkylamines. The degradation procedure consists of treating a dried sample with potassium tert-butoxide in 80:20 (v/v) benzene/DMSO and washing the resultant with (1) 5% hydrochloric acid and (2) water. Formation of the 1-olefin is favored when the sample is refluxed with the tert-butoxide for 30 min. Under milder conditions, where the reagents are merely shaken and allowed to stand for 10 min at room temperature, alkyldimethylamine formation is favored. Conversion to 1-olefins has been used to quantitate alkyltrimethyl- and dialkyldimethylammonium chlorides in hair rinses containing cetyl alcohol, propylene glycol, methylcellulose, and perfume in water [37]. Recoveries averaged 96.3 to 102% with a CV of less than 3%. Calibration curves were linear from 100 to 800 ppm for all compounds tested, and detection limits were approximately 40 ppm.

Using a 2 m × 3 mm ID glass column packed with 5% SE-30 on Chromosorb W AW DMCS (80 to 100 mesh) at 200°C, with a nitrogen flow rate of 40 mL/min, the method was also applied to the analysis of 3-(trimethoxysilyl)propyloctadecyldimethylammonium chloride in commercial textile products [38]. Recoveries ranged from 85.8 to 91.4% with a CV of less than 4%. The 1-octadecene calibration curve was linear from 40 to 200 ppm but did not pass through the origin. Modifying the outlined GC conditions to use a column temperature program from 140 to 230°C at 5°C/min and converting benzalkonium chlorides to their corresponding alkyldimethylamines [39], wet wipes were analyzed for benzalkonium chloride with a recovery of 92.4 to 104.2% and a CV of 1.4 to 4.5%. Calibration curves showed good linearity from 0.4 to 2.0 mg/mL and the limit of detection was determined to be approximately 0.1 mg/mL.

The Hofmann degradation has also been used by Takano et al. [40]. In their approach, 0.2 g of alkyltrimethylammonium halide, alkylbenzyldimethylammonium chloride or dialkyldimethylammonium halide was refluxed with 0.5 g of sodium methoxide and 70 mL of N,N-dimethylformamide for 1 h at 180°C. The resulting solution was cooled and the condenser rinsed with ethanol. The solution was then transferred to a separatory funnel and 100 mL of water was added to decompose the excess sodium methoxide. Dimethylalkylamines and other reaction products were collected by extracting with ether. Analysis of the dimethylalkylamines formed from dialkyldimethylammonium halides and alkyltrimethylammonium halide were conducted using a Pyrex glass column packed with 3% w/w JXR-Silicone on GasChrom Q (80 to 100 mesh). Analysis of the dimethylalkylamines formed from alkylbenzyldimethylammonium chloride required a Pyrex glass column packed with 3% OV-17 on Chromosorb-W (60 to 80 mesh) to separate the target amines from other products. The other GC conditions required were a linearly programmed column temperature from 100 to 300°C at 10°C/min, a helium flow rate of 50 mL/min, and FID and injection port temperatures of 280°C. Recoveries for each sample type were good with relative standard deviations of 3 to 6%. Additionally, the method can be applied to the analysis of several amphoteric quaternary ammonium salts.

In a subsequent paper [41], the same group describe a simplified degradation scheme. Samples are diluted in 3% w/v potassium hydroxide in methanol and introduced into an injection port fitted with a 9 cm × 3 mm ID Pyrex glass reaction chamber. The Hofmann degradation occurs in the injection port and the products are analyzed as described previously [40], except that the injection port and detector temperatures are reduced to 170°C. Recoveries and standard deviations were again satisfactory for alkyltrimethylammonium halides, alkylbenzyldimethylammonium chlorides, and dialkyldimethylammonium halides.

As an alternative approach to the analysis of quaternary ammonium compounds, the salts may be pyrolyzed without previous treatment [42–45]. Christofides and Criddle [42], for example, have noted that the direct injection of aqueous solutions of quaternary ammonium compounds led to the formation of tertiary amines. Amine formation was a function of both injection temperature and counter anion, but was reproducible enough to allow for the analysis of hexadecylpyridinium chloride in mouthwash with a recovery of 102%.

Cybulski [43] has determined alkyl chain distributions of various benzalkonium chloride raw materials using: a 0.9 m × 2 mm ID glass column packed with 3% OV 17 on Chromosorb W HP (80/100 mesh), a flow rate of 40 mL/min nitrogen, an injection port temperature of 250°C, a FID temperature of 350°C, and a temperature program running from 100 to 280°C at 8.0°C/min. Samples of benzalkonium chloride were introduced as 0.5% solutions in methanol and pyrolyzed to yield dimethylalkylamines and benzylmethylalkylamines. Both series of products formed were used to determine the alkyl chain distributions of benzalkonium chloride raw materials. Using the dimethylalkylamine peaks a sample was found to contain: 41.2 (± 0.21)% C_{12}, 50.6(± 0.23)% C_{14} and 8.2 (± 0.06)% C_{16} (for a total recovery of 101.6 $\pm$ 2.00%). Monitoring the benzylmethylalkylamine peaks of the same mixture, a composition of 42.0 (± 0.05)% C_{12}, 50.0 (± 0.15)% C_{14}, and 8.0 ($\pm$ 0.18)% C_{16} (for a total recovery of 102.8 $\pm$ 2.44%) was obtained. Other samples, of different alkyl chain distributions, were analyzed in a similar fashion and provided similar recoveries and precision. It was, however, noted that before reproducible results could be obtained, the column had to be silylated, and repeated injections of sample had to made to prevent amine adsorption.

Ng et al. [44] have used capillary GC to determine the alkyl chain distributions of commercial benzalkonium chlorides with good agreement to accepted values. The capillary GC method uses a 15 m × 0.25 mm ID DB-5 fused silica capillary column, a helium carrier gas flow rate of 1.0 mL/min, a makeup gas flow rate of 60 mL He/min, FID, and a temperature program consisting of a 2 min hold at 100°C, followed by a linear increase of 10°C/min to 280°C. The optimum injection temperature for the formation of alkyldimethylamines was determined experimentally to be 250°C. Calibration curves for the C_{12} to C_{18} homologs were all linear (correlation coefficient > 0.999) in the working concentration of 1 to 8 μg/μL, and the response factors were almost the same for each homolog.

David and Sandra [45] have applied pyrolysis GC to the simultaneous analysis of alkyltrimethyl-, dialkyldimethyl-, and trialkylmethylammonium salts. Samples are pyrolyzed in a vaporizing injector operated at temperatures above 360°C and analyzed using a 5 m × 0.1 mm ID OV-1 capillary column, hydrogen carrier gas at a pressure of 2 bar, and a temperature program from 100 to 350°C at 10°C/min. Selective detection was provided by a NP detector.

C. Supercritical Fluid Chromatography

David and Sandra [45] have also used SFC for the analysis of quaternary ammonium salts, converted into tertiary amines by off-line treatment with phosphoric acid at elevated temperatures. The tertiary amines resulting from the treatment of trimethylhexadecylammonium, dimethyldihexadecylammonium, and methyltrihexadecylammonium halides are separated using a 20 m × 0.1 mm ID OV-73 column, at a temperature of 150°C. Carbon dioxide pressure is held at 15 MPa for 15 min, then ramped linearly at 0.35 MPa/min to a final pressure of 25 MPa. The analyses additionally requires the use of a valve split injector and a NP detector.

D. Thin-Layer Chromatography

Cationic surfactants have been separated and detected with thin-layer chromatographic techniques in many applications. Volume 40 in this series [2] presents a general summary in table format. As is the case with HPLC, quaternary ammonium materials make up the majority of applications in TLC separations of cationic surfactants.

Mangold and Kammereck [46] reported the thin-layer separation and detection of mixed C_{12} to C_{18} alkyltrimethyl-, dialkyldimethyl-, and trialkylmethylammonium chlorides. A silica gel G adsorbent was used with an eluting solvent consisting of 90 volumes of acetone and 10 volumes of concentrated (14 N) aqueous ammonia. Detection was achieved by spraying with a 0.2% alcoholic solution of 2',7'-dichlorofluorescein and viewing yellow-green fluorescent spots under UV light, or by charring with a saturated solution of potassium dichromate in concentrated sulfuric acid. The resulting R_f values were 0.26 for the mono alkyl-, 0.43 for dialkyl-, and 0.65 for trialkylammonium salts.

Qualitative, systematic separations of surfactants in shampoos, bubble bath formulations, and soaps, as reported by Mattisek [47], employed silica gel 60 with a solvent mixture of 9:1 (v/v) ethanol/glacial acetic acid [48] to analyze for quaternary ammonium components. Pinacryptol yellow in water was used for visualization under long-wavelength UV light (366 nm). The quaternary ammonium spots had an R_f value of 0.08. An alternative system was also employed: silica gel G (impregnated with ammonium sulfate) [46] was developed with a mixture of 80:19:1 (v/v) chloroform/methanol/0.1 N H$_2$SO$_4$. R_f values of 0.24 were observed for quaternary ammonium components using pinacryptol yellow as the visualizing agent. Several other materials coelute with the quats under these conditions, so caution is advised when making unknown identifications based on R_f values and/or visualization color alone.

A method for the analysis of distearyldimethylammonium chloride (DSDMAC) in waters and waste, as described by Osburn [49], used a modification of the Waters and Kupfer [50] Disulfine Blue method. After extensive sample treatment/cleanup, including extractions and ion-exchange procedures, the materials were spotted along with known amounts of DSDMAC on a silica gel G plate. Upon completion of the elution with 75:23:3 (v/v/v) chloroform/methanol/water, the spots were visualized by spraying and charring with 25% sulfuric acid. The R_f value of DSDMA$^+$ was 0.48. Typical recoveries of added DSDMAC ranged from 75 to 122% in sample types including waste influent, final effluent, river water, river sediment, and primary sludge. Reproducibility data are also presented. Values obtained represent the semiquantitative estimate of the percentage of Disulfine Blue active substances attributable to the DSDMA$^+$ cation.

A similar body of work reported by Michelsen [51] employed cetylpyridinium bromide and myristyldimethylbenzylammonium chloride as the cations. Silica gel 60 was used with a developing solvent of 4:3:3 (v/v/v) ethyl acetate/acetic acid/water; visualization resulted from spraying with Dragendorff reagent. Wavelength setting for densitometric quantitation in transmitted light was at 525 nm, which included a yellow filter 303 (560 nm). The method was designed to determine concentrations of 0.002 to 1 mg/L, and its validity is supported by statistical data.

Distearyldimethylammonium chloride is one of several quaternary ammonium compounds separated by Hohm [52], who used primulin (CI 49000 Direct Yellow 59) to effect fluorescence detection (365- or 435-nm excitation and 390- or 460-nm emission, respectively). A mixture of 7:2:1 (v/v/v) 2-butanol/water/acetic acid was used for the development on silica gel 60 HPTLC plates. Applications were aimed at shampoos and shower/bath preparations. Additional quaternary ammonium salts analyzed include ricinoleicacidamidopropyltrimethyl-, cocoamidopropyldimethylacetamido-, stearylamidopropyldimethylmyristylacetate-, cetyltrimethyl-, and behenyltrimethyl-. Quantitative determinations via densitometric scanning are reported to be accurate to within 2 to 3%.

Iodoplatinate reagent was the visualizing agent of choice in the TLC work by de Zeeuw et al. [53] for the analysis of a number of quaternary ammonium compounds. Most of those studied are drug substances or herbicides, but the inclusion of benzalkonium chloride is noteworthy for our applications. Silica gel 60 plates were eluted with either 0.5 M NaBr in methanol or with 20:80 (v/v) chloroform/methanol containing 0.5 M NaI. R_f values for benzalkonium chloride were 0.84 with NaBr/methanol and 0.94 with NaI/chloroform/methanol. Migration may be influenced by substituting Cl^- for the Br^- and/or I^-, since chloride results in more retention than bromide, which, in turn, is retained more than iodide. Dipping the plate in methanol containing 0.5 M NaBr (blotting and drying for 30 min at 105°C) for impregnation prior to development, resulted in an R_f value of 0.74 for benzalkonium chloride after development with 70:30 (v/v) chloroform/methanol. Detection limits in the range 10 to 100 ng were obtained, especially when iodoplatinate spraying was followed by Dragendorff spraying.

Alumina adsorbants were used by McLean and Jewers [54] for qualitative separations of several quaternary ammonium salts, including C_{12}, C_{14}, C_{16}, and C_{18} alkyltrimethyl-. Using basic alumina, R_f values of 0.42 were obtained after elution with 85:15 (v/v) acetone/water, and 0.44 with 60:30:10 (v/v/v) chloroform/methanol/ammonia. Using acid alumina with 85:15 (v/v) chloroform/methanol, an R_f of 0.56 was obtained. Chain-length separation was minimal. The materials were visualized with either modified Dragendorff reagent or 2% iodine in methanol, with an overspray of saturated sodium nitrate solution in each case for intensification of spots.

A unique two-dimensional TLC separation of surfactants by Armstrong and Stine [55] uses both silica gel and C_{18} reversed-phase sorbents. Surfactants were spotted on the reversed-phase strip for class separation and developed with 75% ethanol. The anionics traveled to the solvent front, cationics remained near the origin and nonionics separated between the anionics and cationics. The plate was then cut into three sections and developed in the second (perpendicular) direction which incorporates the silica gel portion. Cetyltrimethylammonium and cetylpyridinium salts were separated in the second direction using 8:1:0.75 (v/v/v) methylene chloride/methanol/acetic acid with resultant R_f values of 0.20 and 0.27, respectively. Visualization was with I_2 vapor. Scanning densitometry was performed directly in the absorbance–reflectance mode at 215 nm, although detection limits were lower when using I_2 and scanning at 405 nm in the absorbance–transmittance mode. It was also reported that sensitivity and selectivity can be enhanced by using a variety of visualization or charring techniques [56,57]. A calibration plot of peak areas versus the amount of cetyltrimethylammonium chloride chromatographed indicated linearity from 3 to 12 μg scanned.

E. Capillary Electrophoresis

Since its introduction (arguably by Jorgenson and Lukacs in 1981 [58]), capillary electrophoresis (CE) has enjoyed rapid growth and commercialization to become a useful method of chemical separation, particularly for the analysis of ions. CE has been used by Weiss et al. [59], to separate C_{12}, C_{14}, C_{16}, and C_{18} alkylbenzyldimethylammonium halides. As shown in Fig. 5, the separation was achieved using a 24 cm × 50 μm ID capillary,

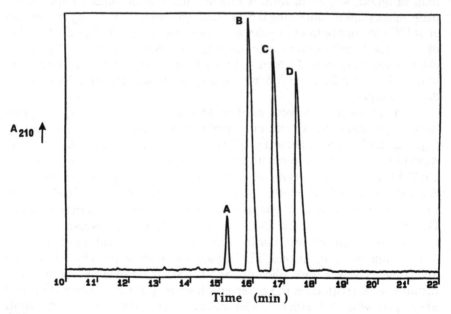

FIG. 5 Separation of (A) C_{12}, (B) C_{14}, (C) C_{16}, and (D) C_{18} alkylbenzyldimethylammonium salts by capillary electrophoresis. (From Ref 59.)

a buffer consisting of 57.5:42.5 (v/v) THF/water with 44 mM phosphate, an electric field strength of 25 kV/m, and a UV detection wavelength of 210 nm. Calibration of the system showed linearity from 10 to 80 ppm, a limit of detection of less than 1 ppm, and a precision of approximately 5%. Application of the method to the analysis of benzalkyl quaternary ammonium salts allowed for the separation of ethylbenzylalkyl and benzalkyl compounds and for the separation of ethylbenzylalkyl salts as ortho and para isomers.

Separation and indirect detection of alkyltrimethylammonium salts was achieved by changing the buffer to contain 3 mM C_{12} benzyl quaternary ammonium chloride, 3 mM sodium dodecyl sulfate, and 8 mM sodium dihydrogen phosphate in 57.5:42.5 (v/v) THF/water and by reducing the operating voltage to 18 kV/m. Precision, calibration, and sensitivity criteria were similar to those obtained by direct detection. The latter method was used to quantitate alkyltrimethylammonium halides in an experimental disinfectant.

III. LONG-CHAIN AMINES

A. High-Performance Liquid Chromatography

Long-chain amine determinations by HPLC are best accomplished with derivatization prior to separation to enhance detectability (sensitivity and selectivity) and to minimize peak tailing due to amino group adsorption. Although all references cited in this section do not specifically target applications with chain lengths greater than C_{10}, fatty amines (C_{12} to C_{20}) would easily be amenable to the conditions described with minor adjustments to mobile-phase composition.

Stable, intensely fluorescent amide derivatives were formed by the reaction of amines with 1,2-naphthoylenebenzimidazole-6-sulfochloride in a study by Jandera et al. [60]. The reaction is rapid and quantitative and can be performed even with aqueous solutions of amines. A 300 × 4.2 mm column packed with C_{18} on Lichrosorb Si-100 (10 μm) was used with linear gradient elution from 70 to 100% methanol in water (v/v) over 18 min. Fluorimetric detection of primary aliphatic amine homologs was measured with 365 nm excitation and emission at > 400 nm to produce detection limits corresponding to a concentration of approximately 3×10^{-9} mol/L of the derivative in a 10 μL sample volume (i.e., 1 to 3 pg of the amine).

Hohaus [61] used 2,2-diphenyl-1-*oxa*-3-oxonia-2-boratanaphthalene (DOOB) to form chelatelike, fluorescent, 3-alkyl-2,2-diphenyl-1-*oxa*-3-oxonia-2-boratanaphthalenes prior to HPLC separations of primary *n*-alkylamines. With a 10 μm Lichrosorb RP-8 column (250 × 4.6 mm) and 90:10

(v/v) methanol/water or an RP-2 column and 80:20 (v/v) methanol/water, excellent peak shapes were obtained with retention times between 7 and 9 min (at a flow rate of 1 mL/min) for n-decylamine. Excitation wavelengths of 366 or 405 nm were used with emission cutoff filters of 418 or 450 nm. Detection limits were in the range of 10 ng in 50 μL injected. Using a concentration of 5×10^{-5} mol/L that produced an average peak area ($n = 16$) of 402.1 mm^2, the relative standard deviation was 3.28%.

C_{10} monoamine and C_{12} diamine were included in the study by Gennaro et al. [62] using reversed-phase HPLC and dansylation to effect fluorimetric detection. Extensive details were reported to describe the influence of varying experimental conditions with derivatizing agents, column packings and mobile-phase composition. Derivatizing reagents tested included 4-nitrobenzoylchloride (NBC), 2,4,6-trinitrobenzenesulfonic acid (TNBS), 4-dimethylaminoazobenzene-4'-sulfochloride (Dabsyl), and 5-dimethylaminonaphtalene-1-sulfochloride (Dansyl). Dansyl was the precolumn derivatization agent of choice. The operating conditions for the reaction were optimized to be carried out as follows: 5.0 mL of 0.25 M NaHCO$_3$, 1.0 mL of 0.02 M Dansyl chloride, and 0.5 mL of amine solution were diluted to 10.5 mL with acetone and mixed in a vessel, which in turn was hermetically sealed and thermostatted at 60°C in a water bath for 20 min. Using a 250 $\times$ 4 mm Merck Hibar Lichrosorb C$_{18}$ (10 μm) column and isocratic elution, log k' for Dansyl-decyl amine varied from 1.75 with 70:30 (v/v) methanol/water to 0.46 with 90:10. Substituting the equivalent C$_8$ column resulted in log k' variation from 1.10 with 70% methanol to 0.03 with 90% methanol. Isocratic work with Dansyl-diamines on the C$_8$ column generated log k' values which ranged from 1.06 to 0.23 for 1,10-diaminodecane using 70% and 85% methanol in water, respectively. 1,12-Diaminododecane had log k' values of 1.37 and 0.38 under the same conditions. Gradient elution methods were also reported for separating complex mixtures of these compounds in less than 20 min. Fluorescence detection wavelength ranges of 340 to 380 nm (excitation) and 430 to 450 nm (emission) were chosen.

Mentasti et al. [63] followed the work cited above [62] with an investigation of *ortho*-phthalaldehyde (OPA) and 4-chloronitrobenzofurazane (CNB) derivatives. The reaction of OPA with primary amines to form an alkylthioisoindole was almost instantaneous in the presence of an alkylthiol (ethanethiol was much favored over mercaptoethanol in reactions studied in this work). CNB required about 20 min heating to form the derivative but produced derivatives with both primary and secondary amines, while OPA allowed for the determination of primary amines only (both reagents also reacted with aromatic amines). To form the OPA derivative: 1.0 mL of OPA solution (0.07 M in ethanol) was mixed with 0.3 mL of ethanethiol

(2.5 M in ethanol). The mixture was then diluted to 10.0 mL with 0.5 M pH 10.0 phosphate buffer. Next, 1.0 mL of the resulting solution was mixed with 0.6 mL of sample containing the amine and 2.4 mL of ethanol. The mixture was left to stand at room temperature for 3 min and then injected into the HPLC. The CNB derivatives were formed as follows: 1.0 mL of 0.05 M methanolic CNB was mixed with 0.5 mL of borate buffer (0.4 M boric acid brought to pH 8.5 with NaOH) and 1.0 mL of sample containing the amines to be derivatized was added. The mixture was then warmed in a water bath at 60°C for 30 min. At the end, a 0.5 mL portion of 1 M HCl was added to block any side reaction and the vial was kept in an ice-water bath. Fluorescence detection was performed with 340 nm excitation and 440 nm emission for OPA derivatives (UV detection is also used at 254 nm), and 470 nm excitation and 530 nm emission for CNB (UV-visible detection at 470 nm). Extensive data were reported for changes in mobile-phase compositions (isocratic and gradient elution) and column packing materials. Decylamine, derivatized with OPA, eluted in 14.6 min on a 5 μm Merck Lichrosphere RP-18 column when a water/acetonitrile gradient program was used (75 to 95% acetonitrile in 5 min at a flow rate of 1.0 mL/min). When derivatized with CNB, decylamine eluted in 12.2 min on the same column (55 to 90% acetonitrile in 6 min at a flow rate of 1.0 mL/min).

B. Gas Chromatography

Insofar as quaternary ammonium salts are frequently degraded to the corresponding amines for GC analysis, the conditions presented in Sec. IIB should also be applicable to the analysis of amines. Additional methods specific for the analysis of long-chain amines have, however, also appeared in the literature. Several of these papers address the primary obstacle of applying GC to the analysis of amines, which is minimization of adsorption effects between the amines and the column packing, which results in severe peak tailing [64]. For (fused) silica, the adsorption of amines is attributed to free silanol groups on the surface participating in hydrogen bonding with the free electron pair of the amine's nitrogen atom. The adsorption tendency is in the order: primary > secondary > tertiary amines [65].

To overcome adsorptive effects, Di Corcia and Samperi [66], for example, have used carbon black coated with KOH and polyethylene glycols as stationary phases. A separation of a standard mixture containing decylamine, dodecylamine, cyclododecylamine, tetradecylamine, and hexadecylamine was achieved using a 1.4 m × 2 glass column packed with 0.3% KOH and 1.3% PEG-20M on 60 to 80-mesh carbon black. The chromatography was performed isothermally at 220°C with a linear gas velocity of 8.3 cm/s helium.

Grossi and Vece [67] have separated primary, secondary, and tertiary fatty amines using E301 silicone grease coated on glass beads as the stationary phase. Separations of each homologous series and of commercial samples were achieved using a linear temperature program. Link et al. [68] have used deactivated Chromosorb W coated with either silicone grease or Apiezon L to separate fatty amines with carbon chain lengths ranging from C_8 to C_{22}. The latter method was shown to be applicable for the analysis of commercial mixtures of hexadecylamine. Excellent recoveries were obtained for known mixtures containing C_{12} to C_{18} amines.

Campeau et al. [69] have used GC with FID to analyze for decyl and dodecyl amines in commercial glycine-type amphoteric surfactants containing N-alkylaminopropylglycines, N-alkylamines, N-alkylaminopropyl amines, and di- and triacids. Separations were conducted using a 6-ft 3% SE-30 column, temperature programmed from 60 to 240°C at 15°C/min. The nitrogen carrier gas flow rate was 30 mL/min. A chromatogram of monoamines together with diamines and diamino acids is shown in Fig. 6 to illustrate the symmetrical alkylamine peak shapes obtained using these conditions, and the relative retention characteristics of surfactants with similar functionalities.

Abdel-Rehim et al. [70] have used ammonia as a carrier gas to dynamically deactivate the stationary phase. Drastically improved peak shapes were obtained using ammonia versus nitrogen as the carrier gas for aliphatic and aromatic primary and secondary amines on both polyethylene glycol- and methylphenylcyanopropylsilicone–coated capillary columns. As a result of the improvements in peak shapes, detection limits were improved from 20 ng in nitrogen to 2 to 4 ng in ammonia using FID. Furthermore, no detrimental effects on either the chromatography or the columns were observed over a period of 1 year.

Metcalfe and Martin [71] have separated long-chain amines not only by carbon chain length but also by the isomeric position of the amino group. A comparison of dimethylamine, trimethylsilyl, acetamide, and trifluoroacetamide derivatives revealed that the best separations are achieved for the dimethylamines. The conversion of primary amines to dimethylamines was achieved by (1) mixing 0.5 to 1 mL of primary amine with 2 mL of 37% formaldehyde and 2 mL of 90% formic acid, (2) heating the mixture in a steam bath until vigorous bubbling stopped (5 min), and (3) turning the solution basic with 10% sodium hydroxide (phenolphthalein indicator). The tertiary amine was isolated from the reaction mixture by adding sodium chloride solution to float it to the top, and was dissolved in hexane for chromatographic analysis. Chromatography was performed using a 200 ft × 0.02 in. stainless steel capillary column coated with SF-96 silicone oil (modified with trioctadecylmethylammonium bromide), a column temperature program from 100 to 200°C at 10°C/min (and then held isother-

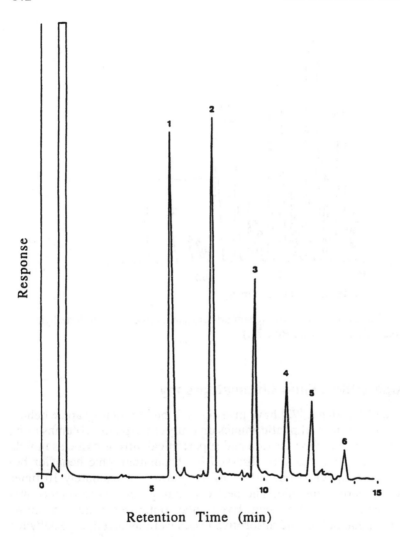

Retention Time (min)

FIG. 6 Gas chromatogram of a standard mixture of (1) decylamine, (2) dodecylamine, (3) decylaminopropylamine, (4) dodecylaminopropylamine, (5) decylam-
inopropylaminoacetic acid methyl ester, and (6) dodecylaminopropylaminoacetic
acid methyl ester. (From Ref. 69.)

mally), a carrier gas flow of 6 mL/min, and injection port and FID temperatures of 255 and 350°C, respectively. For C_{11} to C_{15} alkyl chain lengths,
(partial) separations were observed for all the isomers, although resolution
decreased as the position of substitution increased (Fig. 7). For larger
homologs, the number of isomers that can be separated is limited.

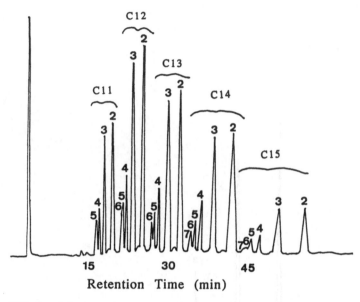

FIG. 7 GC separation of C_{11} to C_{15} isomeric primary amines as their dimethylamine derivatives. (Redrawn from Ref. 71.)

C. Supercritical Fluid Chromatography

Baastoe and Lundanes [72] have investigated the chromatographic behavior of several primary aliphatic amines on different capillary columns using both supercritical carbon dioxide and supercritical nitrous oxide as mobile phases. Little difference was observed in chromatographic behavior between the two systems, but FID sensitivity was approximately 10 times better using carbon dioxide. The best chromatographic performance was obtained using an OV-1 column. Even under optimized conditions, however, some peak tailing and adsorptive losses were incurred, especially for primary amines.

D. Thin-Layer Chromatography

Mangold and Kammereck's [46] thin-layer chromatographic work discussed in Sec. II.D includes several applications for long-chain amines as well. Trialkyl (C_{12} to C_{18}) tertiary amine was applied to a silica gel G plate, and an R_f value of 0.14 was obtained (dialkyl secondary amine remained at the origin in this case) after development with a solvent system prepared as follows: 100:10 (v/v) benzene/1 N aqueous ammonia were equilibrated at 20°C, and the organic layer was used as the eluant. The same adsorbent was used with an alternative solvent system to separate trialkyl tertiary (R_f = 0.83), dialkyl secondary (R_f = 0.46), and alkyldimethyl tertiary (R_f = 0.29) amines, as well as the primary amine (R_f = 0.08). The latter eluant was prepared as follows: 10:1 (v/v) chloroform/1 N aqueous ammonia was equilibrated at 20°C. A mixture of 97 volumes of this ammoniacal chloroform and 3 volumes of methanol was used as the developing system. Visualization was by charring with chromic sulfuric acid solution in both cases.

Barry and Saunders [35] have used TLC to check alkyltrimethylammonium bromides for tertiary amine content. Separations were performed on 20 cm × 20 cm glass plates coated with 0.25 mm of aluminum oxide G. The developing solvent was 100:5 acetone/water (v/v), and Dragendorff's reagent was used to color the analytes orange. Quaternary ammonium compounds move only a small distance from the point of application, allowing for the quantitation of the tertiary amines.

Another procedure for monitoring a cationic surfactant raw material for residual amine content was reported by Pelka and Metcalfe [73], who employed silica gel G–coated microscope slides to determine dimethylcocoamine in tertiary amine oxide. The procedure was designed for process control work and has advantages over other methods since its cost is nominal, it offers very good precision without a sacrifice in time, and it requires very little bench space for the equipment. An experienced operator can perform 25 to 30 determinations in an hour. The developing solvent system was similar to that recommended by Mangold and Kammereck [46] and was prepared as follows: 80:20 (v/v) chloroform/concentrated ammonium hydroxide was equilibrated for 1 h. The organic layer was separated and mixed with methanol in the ratio of 97:3 (v/v). After development, the plates were sprayed with a 0.05% solution of 2′,7′-dichlorofluorescein in methanol and observed under UV light. Quantitative data were presented to compare results obtained by TLC to those determined using a potentiometric determination.

Fatty amines are among 56 aliphatic amines separated by normal-phase and reversed-phase TLC in a comprehensive study by Prandi [74]. Normal-phase separation was accomplished on Merck silica gel G with a solvent system consisting of 82.5:15.5:2 (v/v/v) chloroform/methanol/17% aqueous ammonia. R_f values of 0.50 to 0.60 were obtained for the carbon chain homologs from tetradecylamine to stearylamine. Reversed-phase separation on Merck silanized silica gel H was achieved with a developing solvent mixture of 70:30 (v/v) acetone/17% aqueous ammonia. Under these conditions R_f values from 0.13 to 0.22 were reported for the homologs from stearylamine to tetradecylamine. Detection schemes included spraying the plates with 2,5-dimethoxytetrahydrofuran (1% buffered solution, pH 6.6), heating at 110°C for 5 min, and then spraying again with *p*-dimethylaminobenzaldehyde solution (1% in 3% HCl). Alternatively, the plates were sprayed with ninhydrin (1% in 95:5 ethanol/acetic acid). Both of the visualization reagents are particularly useful only for primary amines. Iodine (25% methanolic solution) or 1:1 1% potassium permanganate/1% potassium persulfate sprays were also employed, although permanganate is preferred because of its ease of application and better spot color stability. Only ninhydrin can be used with both normal- and reversed-phase adsorbents; the others are suitable only with silica gel G.

Armstrong and Stine's [55] two-dimensional separation also discussed in Sec. II.D, includes determination of dodecylamine and octadecylamine. The R_f values were 0.42 and 0.55, respectively, when detected on the silica gel portion of the plate after development with 8:1:0.5 (v/v/v) methylene chloride/methanol/acetic acid. Variations in visualization and scanning densitometry were performed with I_2 vapor, as noted previously.

The reaction of *n*-decylamine with 1,2-naphthoylenebenzimidazole-6-sulfochloride by Jandera et al. [60], described in Sec. III.A, produced a fluorescent derivative that was also amenable to separation by TLC. Silufol silica gel plates were used with eight different solvent systems. The two eluants that appeared most noteworthy were 2:1 (v/v) *n*-hexane/dioxane and 8:3 (v/v) benzene/ethyl acetate, which produced R_f values of 0.78 and 0.49, respectively.

Reversed- and normal-phase TLC [75] have been employed for separations of the chelatelike, fluorescent, 3-alkyl-2,2-diphenyl-1-*oxa*-3-oxonia-2-boratanaphthalene derivatives described by Hohaus [61] in Sec. III.A. Detection limits using the thin-layer technique were 30 ng and the RSD was 4.25%. In comparison, HPLC detection limits were 10 ng and the RSD 3.28%.

IV. FATTY ACID AMIDES

A. High-Performance Liquid Chromatography

Jasperse [76] has developed a simple isocratic HPLC method for the separation of fatty amides by chain length (Fig. 8). The separation is conducted with a 70:20:10 (v/v/v) hexane/chloroform/acetic acid eluant, using a 250 × 4.6 mm 5 μm Nucleosil 100 column and refractive index detection. The method was shown to be applicable to the determination of fatty amide in fatty nitrile at the 1000 ppm level. The quantitative accuracy of the method was verified by spiking fatty nitrile with stearamide and oleamide. Recoveries were 99 ± 8% for fatty amides added at the 0.05 to 1.2% level.

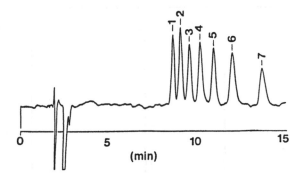

FIG. 8 Separation of the homologs of saturated fatty amides by HPLC. Peak identities: (1) arachidamide, (2) stearamide, (3) palmitamide, (4) myristamide, (5) lauramide, (6) capramide, and (7) caprylamide. (From Ref. 76.)

B. Gas Chromatography

Our current method for the analysis of fatty amides is a two-step process in which samples are acid hydrolyzed to form fatty acids and then methylated with boron trifluoride in methanol to form fatty acid methyl esters (Colgate-Palmolive Company internal method). Specifically, 0.4 g of fatty acid amide is refluxed in 20 mL of 1:3 (v/v) HCl/water for 1 h, and the resulting fatty acids are isolated by collecting the pentane layer from a 30 mL pentane/50 mL saturated NaCl extraction. Following evaporation of the pentane layer, the sample is converted to fatty acid methyl esters by refluxing with 25 mL of boron triflouride in methanol for 15 min. Fatty acid methyl esters are recovered by collecting the isooctane layer from a 25 mL isooctane/50 mL water extraction. Chromatographic analysis is performed on an aliquot of the isolated fatty acid methyl esters diluted 1:25 with isooctane. Columns for the analysis of fatty acid methyl esters are commercially available from many sources. Using the conditions specified in the figure legend, fatty acid methyl esters are separated by alkyl chain length and degree of unsaturation (Fig. 9).

Numerous authors have reported methods for separating fatty acid amides directly. Morrissette and Link [77] have separated native fatty amides with alkyl chain lengths of up to 24 carbons using packed column gas–liquid chromatography with thermal conductivity detection. Separations of fatty amide from the high-cut coconut fraction, tallow amide, rapeseed amide, and N,N-dimethyl alkyl amides were obtained on 4 to 6 ft × 0.25 in. OD aluminum columns packed with 5% (w/w) or 20% (w/w) Versamid on 60 to 80-mesh or 100 to 120-mesh Gas Chrom P. Quantitation of individual homologs using area percents was possible since the

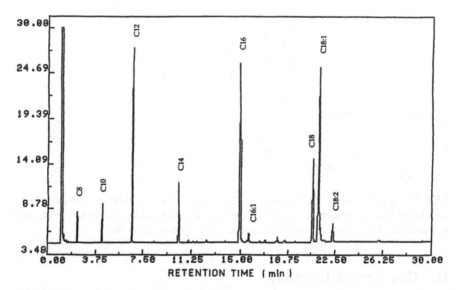

FIG. 9 Separation of fatty acid methyl esters derived from fatty amides. Conditions: column, 25 m × 0.32 mm ID CP Wax 57 CB (df = 0.2 μm); temperature program: (1) 105°C for 3 min, (2) ramped at 25°C/min to 140°C, (3) held for 3 min, (4) ramped at 4°C/min to 160°C, (5) held for 3 min, (6) ramped at 4°C/min to 200°C, (7) held as needed; linear gas velocity = 30 cm/s helium; FID; cool on-column injection. (Courtesy of Colgate-Palmolive Company.)

method showed no discrimination. However, all components of a given chain length, saturated or unsaturated, were eluted as single peaks.

Separation of stearamide and oleamide (saturated and unsaturated amides with the same carbon chain lengths) was achieved by Frisina et al. [78] using 1 or 3 m × 2 mm ID glass columns packed with 3% Dexil 300 GC on Gas-Chrom Q (80 to 100 mesh). The method was applied to the analysis of palmitamide, stearamide, oleamide, and erucylamide in polyolefin samples. Using external standard calibration, accuracy was acceptable and precision was ± 5% (95% probability) for amide concentrations of 500 ppm. Raw materials have also been separated by both alkyl chain length and degree of unsaturation by Wang and Metcalfe [79]. Using a 3 ft × ⅛ in. stainless steel column packed with a 5% mixed cyanopropyl silicone on Chromsorb W AW (DMCS treated, 100 to 120 mesh), separations of coco amide and erucic amide were obtained isothermally at 230 and 250°C, respectively. Analysis times were less than 15 min.

Brengartner [80] obtained separation not only by alkyl chain length and degree of unsaturation, but also managed to separate cis-trans isomers of

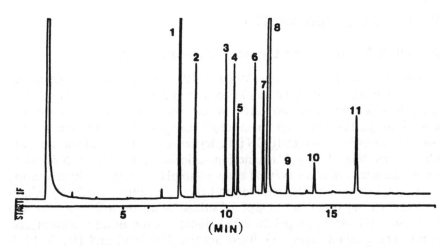

FIG. 10 GC separation of (1) benzamide, (2) myristamide, (3) palmitamide, (4) palmitelaidamide, (5) palmitoleamide, (6) stearamide, (7) elaidamide, (8) oleamide, (9) linoleamide, (10) linolenamide, and (11) erucamide. (From Ref. 80.)

9-hexadecenamide and 9-octadecenamide (Fig. 10). Analyses were conducted using a 30 m × 0.32 mm ID fused silica capillary coated with 0.2 μm of 90% biscyanopropyl/10% phenylcyanopropyl polysiloxane (SP-2330). The column was temperature programmed to initiate at 200°C for 4 min, the temperature was then ramped at 10°C/min to 260°C and held for 10 min. Injection was performed in the split mode and detection was by flame ionization. The method was used to determine compositional differences in commercial oleamide samples and to identify the composition of amides extracted from polymers. Polymer extractions were performed using 2-propanol spiked with benzamide as the internal standard. Accuracy was estimated to within 10% and submicrogram sensitivity was obtained for each species.

C. Thin-Layer Chromatography

Silica gel G–coated plates visualized by charring with chromic sulfuric acid solution as reported by Mangold and Kammereck [46] have been employed to separate and detect C_{12} to C_{18} amides. The eluant was prepared as follows: 10:1 (v/v) chloroform/1 N aqueous ammonia was equilibrated at 20°C and 97 volumes of the ammoniacal chloroform was mixed with 3 volumes of methanol. Using this solvent system, an R_f value of 0.38 was obtained.

V. ETHOXYLATED AMINES

A. High-Performance Liquid Chromatography

Schreuder et al. [81] have developed a HPLC method applicable to the characterization of ethoxylated alkylamines in pesticide formulations. The analysis is performed in two parts. In one assay the compounds are separated on the basis of alkyl chain length using a 250×4.6 mm column packed with Polygosil 60 D-10 CN (alkylcyano-modified silica) and a mobile phase consisting of 10 g/L magnesium chloride hexahydrate in 60:30:10 water/methanol/dioxane (v/v/v) adjusted to pH 2.5 with 1 M hydrochloric acid and 35 mg/L 9,10-dimethoxyanthracene-2-sulfonate (DAS). Components are separated by degree of ethoxylation on a 250×4.6 mm Hypersil APS (aminopropyl-modified silica) column using a solvent gradient. The gradient uses (A) 70:30 hexane/THF(v/v) and (B) 90:10 2-propanol/water (v/v) mixed linearly from 5 to 60% B in 60 min. Detection for both separations was accomplished using a postcolumn ion-pair extraction system, using DAS as the pairing ion and fluorometric detection (excitation 400 nm, emission 475 nm cutoff).

Using these methods, components with alkyl chain lengths of C_{10}, C_{12}, C_{14}, C_{16}, $C_{16:1}$, C_{18}, $C_{18:1}$, and $C_{18:2}$ were separated and ethoxylated amines with a mean EO content of up to 15 mol EO/mol were characterized. Furthermore, due to the specificity and sensitivity of the detection scheme, the ethoxylated alkylamines could be characterized in the presence of ethoxylated alkylphenols, alcohols, and esters at a detection limit of 25 ng.

B. Gas Chromatography

Slagt et al. [82] have converted ethoxylated amines to dibromoethane and N,N-di(2-bromoethyl)alkylamines and have analyzed the products by GC to determine the alkyl chain distribution and total ethylene oxide content. Quantitative conversions were obtained when the fission was conducted in 40% hydrobromic acid in glacial acetic acid in a sealed glass tube at 145°C for 2 h. Extraction of the desired products was obtained by collecting the carbon tetrachloride layer from a cold alkali/carbon tetrachloride extraction. Chromatographic analyses of N,N-di(2-bromoethyl)alkylamines were conducted using: an 80 cm $\times$ 2.5 mm ID glass column packed with 10% OV-1 on Chromosorb W, AW, DMCS (80 to 100 mesh); a linear temperature program ranging from 120 to 320°C at 6°C/min; a nitrogen flow rate of 15 mL/min; FID; and an injector temperature of 200°C. Analyses for dibromoethane were conducted using the same system, except that the column was operated isothermally at 50°C and the injector temperature was decreased to 150°C. The method was used to determine the alkyl chain distribution and ethylene oxide content of several commercial amine adducts. Mass recoveries ranged from 96.7 to 101.7%.

Szewczyk et al. [83] have formed trimethyl silyl derivatives of ethoxylated alkylamines by reacting 50 mg of the latter with 0.5 mL of di(trimethylsilyl)acetamide in a glass reaction vessel (70°C for 1 h). Analyses of the derivatives were conducted using 40 to 170 cm × 2.7 mm ID stainless steel columns packed with 1 to 3% OV-17 on Chromosorb G-AW-DMC5 (60 to 80 mesh). Nitrogen was used as a carrier gas at a flow rate of 20 mL/min, and injector and FID temperatures were set at 370°C. The initial column temperatures varied from 80 to 170°C, depending on product composition. After an initial hold time of 1 min, the temperatures were ramped at 6°C/min to 320°C and held constant. A sample chromatogram of a dodecylamine mixture containing up to 14 ethoxy units is shown in Fig. 11. Similar separations were obtained for other ethoxylated amines with single carbon chain lengths to allow for arithmetic retention indices to be computed. While calculated retention indices indicate that coelution

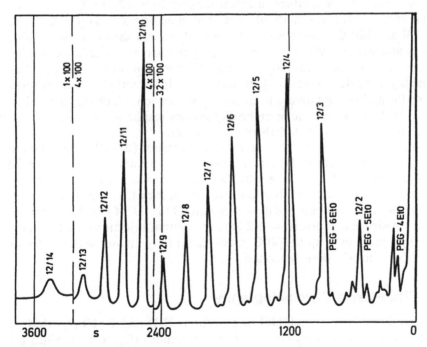

FIG. 11 Gas chromatogram of ethoxylated n-dodecylamines as their trimethyl silyl derivatives. The numbers on the peaks designate the number of moles of ethylene oxide. (From Ref. 83.)

of some ethoxylated alkylamines is inevitable, they did allow for the identification of individual species in commercial blends containing C_{14}, C_{16},
and C_{18} alkyl chain lengths. Separation by both chain length and degree
of ethoxylation was achieved for components containing up to 12 ethoxy
units. It was, however, not possible to distinguish between ethoxylated
alkylamines with saturated and unsaturated carbon chains.

Polyoxyethylene 4-alkylphenylamines have been converted to trimethylsilyl derivatives, separated, and identified on the basis of arithmetic retention indices, using a similar method [84]. The method was shown to be
applicable for compounds with 1 to 16 carbons in the alkyl groups and
average degrees of ethoxylation from 1 to 8. Positional isomers with the
same total number of oxyethylene groups were not resolved, but the compounds of interest were separated from polyoxyethylene glycol impurities.
N-Oligooxyethylene decyl and dibutylamine mixtures have also been separated [85], both as trimethyl silyl derivatives and directly (dissolved in
ethanol). Separations were conducted using a 1.8 m × 2.7 mm ID stainless
steel column packed with 3% OV-17 on Chromosorb G AW DMCS (60
to 80 mesh); a temperature program ranging from 120 to 320°C at 4°C/
min; argon carrier gas at 15 mL/min; and injector and FID temperatures
of 270 and 320°C, respectively. Direct analysis was shown to be possible
for di and monoalkylamines containing up to 6 oxyethylene units. As discussed, compounds with larger degrees of ethoxylation were analyzed as
trimethyl silyl derivatives [83]. However, while trimethyl silyl derivatives
were formed quantitatively from N-polyoxyethylene dialkylamines, silyl-
ation of N-polyoxyethylene monoalkylamines results in the formation of
both mono- and bis(trimethylsilyl) derivatives.

Venema [86] has used capillary GC to separate (directly) the reaction
mixture obtained from the ethoxylation of a fatty amine in less than 22
min (column = 20 m × 0.28 mm ID SE-30, df = 0.2 μm; carrier gas =
hydrogen; pressure = 0.2 bar; on-column injection). Using slightly modified conditions (column = 8 m × 0.28 mm ID SE-30, df = 0.1 μm; carrier
gas = hydrogen; pressure = 1.2 bar; on-column injection), and temperature programming up to 350°C, ethoxylated fatty amines with molecular
weights up to 923 were separated.

C. Thin-Layer Chromatography

Mangold and Kammereck [46] (see Sections II.D and IV.C) reported an
R_f value of 0.21 for long-chain N-2-ethoxy amides (C_{12} to C_{18}). Silica gel
G–coated plates were used and the eluant was prepared as follows: 10:1
(v/v) chloroform/1 N aqueous ammonia was equilibrated at 20°C, and 97
volumes of this ammoniacal chloroform and 3 volumes of methanol were

mixed and used as the developing system. Visualization was accomplished by charring with chromic sulfuric acid solution.

VI. AMINE OXIDES

A. Gas Chromatography

Gas chromatography (GC) methods for the determination of amine oxides are usually based on the conversion of the title compounds to 1-olefins. Lew [87], for example, effected the on-column pyrolysis of amine oxides to their respective 1-olefins using a 10 ft × ¼ in. stainless steel column packed with 20% Apiezon on Chromosorb W (HMDS); a helium flow rate of 50 mL/min; an injector temperature of 220°C; and a column temperature programmed from 180 to 280°C at 4°C/min. When the method was applied to the analysis of detergent mixtures, however, samples had to be treated with anion-exchange resins prior to chromatography since any alkyl sulfates present would have partially decomposed to olefins.

Devinsky and Gorrod [88] have used pyrolysis GC with flame ionization detection to separate N,N-dimethylalkylamine N-oxides with carbon chain lengths ranging from 6 to 16. The method provided a sensitivity of 10 to 15 nmol of compounds injected and was applicable to the analysis of amine oxides in the presence of parent tertiary amines. The separations were conducted on a 1 m × 3 mm ID glass column of which approximately 93 cm was packed with 8% Apiezon L on Chromosorb G AW DMCS (80 to 100 mesh) and 10% potassium hydroxide (conditioned for 48 h before use). The remaining 7 cm of the column (the portion inserted into the injection port) was packed with nontreated support. For analysis the column was held isothermally at 50°C for 10 min and then ramped at 4°C/min to 220°C. An injection port temperature of 450°C was used to assure quantitative pyrolysis to 1-olefins and N,N-dimethylhydroxylamine. Figure 12 shows the separation of the 1-olefins and parent tertiary amines.

As an alternative approach, Langley et al [89] have reduced amine oxides to tertiary amines by refluxing with triphenylphosphine in glacial acetic acid for 1 to 1.5 h. Tertiary amines are recovered by extraction into ether following neutralization with sodium hydroxide, and are separated on a 3-ft Carbowax/KOH column using a temperature program of 170 to 220°C at 4°C/min.

B. Thin-Layer Chromatography

Pelka and Metcalfe [73] employed silica gel G–coated microscope slides to analyze tertiary amine oxide (see Sec. III.D) for dimethyl coco amine.

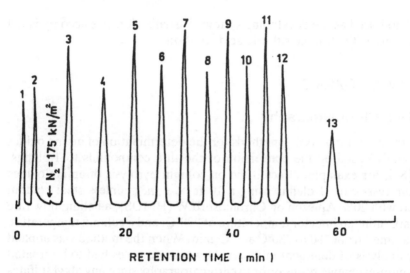

FIG. 12 GC separation of the pyrolysis products of N,N-dimethylalkylamine N-oxides (1-olefins) and N,N-dimethylalkylamine standards. Peak identities: (1) N,N-dimethylhydroxylamine, (2) 1-hexene, (3) hexyldimethylamine, (4) 1-octene, (5) octyldimethylamine, (6) 1-decene, (7) decyldimethylamine, (8) 1-dodecene, (9) dodecyldimethylamine, (10) 1-tetradecene, (11) tetradecyldimethylamine, (12) 1-hexadecene, and (13) hexadecyldimethylamine. (From Ref. 88.)

The developing solvent system was prepared as follows: 80:20 (v/v) chloroform/concentrated ammonium hydroxide was equilibrated for 1 h. The organic layer was separated and mixed with methanol in the ratio 97:3 (v/v). After development, the plates were sprayed with a 0.05% solution of 2',7'-dichlorofluorescein in methanol and observed under UV light. The amine oxide moved only a short distance from the origin.

ACKNOWLEDGMENTS

The authors with to thank Mark Demouth for running the chromatogram shown in Fig. 9, and Susan Friedman, Mike Knapp, and Nick Omelczenko for critical review of this manuscript.

REFERENCES

1. J. M. Richmond, ed., *Cationic Surfactants: Organic Chemistry,* Surfactant Science Series, Vol. 34, Marcel Dekker, New York, 1990.
2. T. M. Schmitt, ed., *Analysis of Surfactants,* Surfactant Science Series, Vol. 40, Marcel Dekker, New York, 1992.

3. L. D. Metcalfe, *J. Am. Oil Chem. Soc. 61*: 363 (1984).
4. R. Gloor and E. L. Johnson, *J. Chromatogr. Sci. 15*: 413 (1977).
5. S. L. Abidi, *J. Chromatogr. 324*: 209 (1985).
6. G. Muto and A. Nakae, *Chem. Lett. 6*: 549 (1974).
7. A. Nakae, K. Kunihiro, and G. Muto, *J. Chromatogr. 134*: 459 (1977).
8. K. Nakamura, Y. Morikawa, and I. Matsumoto, *J. Am. Oil Chem. Soc. 58*: 72 (1981).
9. K. Nakamura and Y. Morikawa, *J. Am. Oil Chem. Soc. 59*: 64 (1982).
10. M. Kanesato, K. Nakamura, O. Nakata, and Y. Morikawa, *J. Am. Oil Chem. Soc. 64*: 434 (1987).
11. R. C. Meyer, *J. Pharm. Sci. 69*: 1148 (1980).
12. D. F. Marsh and L. T. Takahashi, *J. Pharm. Sci. 72*: 521 (1983).
13. G. Ambrus, L. T. Takahashi, and P. A. Marty, *J. Pharm. Sci. 76*: 174 (1987).
14. R. C. Meyer and L. T. Takahashi, *J. Chromatogr. 280*: 159 (1983).
15. P. Linares, M. C. Gutierrez, F. Lazaro, M. D. Luque De Castro, and M. Valcarcel, *J. Chromatogr. 558*: 147 (1991).
16. J. Kawase, Y. Takao, and K. Tsuji, *J. Chromatogr. 262*: 408 (1983).
17. S. Takano and K. Tsuji, *J. Am. Oil Chem. Soc. 60*: 870 (1983).
18. S. L. Abidi, *J. Chromatogr. 213*: 463 (1981).
19. S. L. Abidi, *J. Chromatogr. 362*: 33 (1986).
20. P. Helboe, *J. Chromatogr. 261*: 117 (1983).
21. M. Dinkert, L. Hackzell, G. Schill, and E. Sjogren, *J. Chromatogr. 218*: 31 (1981).
22. N. Parris, *J. Liquid Chromatogr. 3*: 1743 (1980).
23. J. R. Larson and C. D. Pfeiffer, *Anal. Chem. 55*: 393 (1983).
24. H. Small and T. E. Miller, Jr., *Anal. Chem. 54*: 462 (1982).
25. F. Spagnolo, M. T. Hatcher, and B. K. Faulseit, *J. Chromatogr. Sci. 25*: 399 (1987).
26. A. Nakae and K. Tsuji, *Comun. J. Com. Esp. Deterg. 14*: 133 (1983).
27. J. Kawase, Y. Takano, and K. Tsuji, *J. Chromatogr. 262*: 293 (1983).
28. C. B. Huang, *J. Liquid Chromatogr. 10(6)*: 1103 (1987).
29. R. Caesar, H. Weightman, and G. R. Mintz, *J. Chromatogr. 478*: 191 (1989).
30. V. T. Wee and J. M. Kennedy, *Anal. Chem. 54*: 1631 (1982).
31. A. J. Wilkes, G. Walraven, and J.-M. Talbot, *J. Am. Oil Chem. Soc. 69*: 609 (1992).
32. F. Van Damme and M. Verzele, *J. Chromatogr. 351*: 506 (1986).
33. H. P. Warrington, Jr., *Anal. Chem. 33*: 1898 (1961).
34. L. D. Metcalfe, *J. Am. Oil Chem. Soc. 40*: 25 (1963).
35. B. W. Barry and G. M. Saunders, *J. Pharm. Sci. 60*: 645 (1971).
36. S. L. Abidi, *J. Chromatogr. 200*: 216 (1980).
37. S. Suzuki, M. Sakai, K. Ikeda, K. Mori, T. Amemiya, and Y. Watanabe, *J. Chromatogr. 362*: 227 (1986).
38. S. Suzuki, K. Mori, T. Amemiya, and Y. Watanabe, *J. Chromatogr. 387*: 379 (1987).
39. S. Suzuki, Y. Nakamura, M. Kaneko, K. Mori, and Y. Watanabe, *J. Chromatogr. 463*: 188 (1989).

40. S. Takano, C. Takasaki, K. Kunihiro, and M. Yamanaka, *J. Am. Oil. Chem. Soc. 54*: 139 (1977).
41. S. Takano, M. Kuzukawa, and M. Yamanaka, *J. Am. Oil Chem. Soc. 54*: 484 (1977).
42. A. Christofides and W. J. Criddle, *Anal. Proc. 19*: 314 (1982).
43. Z. R. Cybulski, *J. Pharm. Sci. 73*: 1700 (1984).
44. L.-K. Ng, M. Hupe, and A. G. Harris, *J. Chromatogr. 351*: 554 (1986).
45. F. David and P. Sandra, *J. High Resolut. Chromatogr. "Chromatogr." Commun. 11*: 897 (1988).
46. H. K. Mangold and R. Kammereck, *J. Am. Oil Chem. Soc. 39*: 201 (1962).
47. R. Mattisek, *Tenside Deterg. 19*: 57 (1982).
48. D. S. Frahne, S. Schmidt, and H.-G. Kuhn, *Fette Seifen Anstrichm. 79*: 32, 122 (1977).
49. Q. W. Osburn, *J. Am. Oil Chem. Soc. 59*: 453 (1982).
50. J. Waters and W. Kupfer, *Anal. Chim. Acta 85*: 241 (1976).
51. E. R. Michelsen, *Tenside Deterg. 15*: 169 (1978).
52. G. Hohm, *Seife Oele Fette Wachse 116*: 273 (1990).
53. R. A. de Zeeuw, P. E. W. van der Laan, J. E. Greving, and F. J. W. van Mansvelt, *Anal. Lett. 9*: 831 (1976).
54. W. F. H. McLean and K. Jewers, *J. Chromatogr. 74*: 297 (1972).
55. D. W. Armstrong and G. Y. Stine, *J. Liquid Chromatogr. 6*: 23 (1983).
56. C. Yonese, T. Shishido, T. Kaneko, and K. Maruyama, *J. Am. Oil. Chem. Soc. 59*: 112 (1982).
57. G. Zweig and J. Sherma, *Handbook of Chromatography*, Vol. II, CRC Press, Cleveland, Ohio, 1972.
58. J. W. Jorgenson and K. D. Lukacs, *Anal. Chem. 53*: 1298 (1981).
59. C. S. Weiss, J. S. Hazlett, M. H. Datta, and M. H. Danzer, *J. Chromatogr. 608*: 325 (1992).
60. P. Jandera, H. Pechova, D. Tocksteinova, and J. Churacek, *Chromatographia 16*: 275 (1982).
61. E. Hohaus, *Z. Anal. Chem. 319*: 533 (1984).
62. M. C. Gennaro, E. Mentasti, C. Sarzanini, and V. Porta, *Chromatographia 25*: 117 (1988).
63. E. Mentasti, C. Sarzanini, O. Abollino, and V. Porta, *Chromatographia 31*: 41 (1991).
64. L. D. Metcalfe, *J. Chromatogr. Sci. 13*: 516 (1975).
65. M. Abdel-Rehim, M. Hassan, and H. Ehrsson, *J. High Resolut. Chromatogr. 13*: 252 (1990).
66. A. Di Corcia and R. Samperi, *Anal. Chem. 46*: 977 (1974).
67. G. Grossi and R. Vece, *J. Gas Chromatogr. 3*: 170 (1965).
68. W. E. Link, R. A. Morrissette, A. D. Cooper, and C. F. Smullin, *J. Am. Oil. Chem. Soc. 37*: 364 (1960).
69. D. Campeau, I. Gruda, Y. Thibeault, and F. Legendre, *J. Chromatogr. 405*: 305 (1987).
70. M. Abdel-Rehim, M. Hassan, and H. Ehrsson, *J. High Resolut. Chromatogr. 13*: 252 (1990).

71. L. D. Metcalfe and R. J. Martin, *Anal. Chem. 44*: 403 (1972).
72. M. B. Baastoe and E. Lundanes, *J. Chromatogr. 558*: 458 (1991).
73. J. R. Pelka and L. D. Metcalfe, *Anal. Chem. 37*: 603 (1965).
74. C. Prandi, *J. Chromatogr. 155*: 149 (1978).
75. E. Hohaus, *Z. Anal. Chem. 310*: 70 (1982).
76. J. L. Jasperse, *J. Am. Oil Chem. Soc. 65*: 1804 (1988).
77. R. A. Morrissette and W. E. Link, *J. Am. Oil Chem. Soc. 41*: 415 (1964).
78. G. Frisina, P. Busi, and F. Sevini, *J. Chromatogr. 173*: 190 (1979).
79. C. N. Wang and L. D. Metcalfe, *J. Am. Oil Chem. Soc. 61*: 581 (1984).
80. D. A. Brengartner, *J. Am. Oil Chem. Soc. 63*: 1340 (1986).
81. R. H. Schreuder, A. Martijn, H. Poppe, and J. C. Kraak, *J. Chromatogr. 368*: (1986).
82. C. Slagt, J. M. H. Daemen, W. Dankelman, and W. A. Sipman, *Z. Anal. Chem. 264*: 401 (1973).
83. H. Szewczyk, J. Szymanowski, and W. Jerzykiewicz, *Tenside Deterg. 19*: 287 (1982).
84. M. Wisniewski, J. Szymanowski, and B. Atamanczuk, *J. Chromatogr. 462*: 39 (1989).
85. J. Szymanowski, H. Scewczyk, J. Hetper, and J. Beger, *J. Chromatogr. 351*: 183 (1986).
86. A. Venema, *J. Chromatogr. 279*: 103 (1983).
87. H. Y. Lew, *J. Am. Oil Chem. Soc. 41*: 297 (1964).
88. F. Devinsky and J. W. Gorrod, *J. Chromatogr. 466*: 347 (1989).
89. N. A. Langley, D. Suddaby, and K. Coupland, *Int. J. Cosmet. Sci. 10*: 257 (1988).

11

Molecular Spectroscopy of Cationic Surfactants

FOAD MOZAYENI Applied Analytical Support, Akzo Chemicals, Inc., Dobbs Ferry, New York

I. INTRODUCTION

In today's marketplace, there are variety of cationic surfactants and their numbers will undoubtedly increase as the market conditions change or additional applications are found. Molecular spectroscopy provides powerful tools for identification and quantitative analysis of these various cationic surfactants. In this chapter, two techniques, nuclear magnetic resonance (NMR) and infrared (IR), are discussed.

In recent years, NMR spectroscopy has experienced exceptional growth and has become a technique of choice for structural elucidation and, to a certain extent, quantitative analysis of organic compounds. The application of NMR to cationic surfactants is illustrated in a number of examples. Various pulse sequences for structural identification and determination of spin lattice relaxation time (T_1) and nuclear Overhauser effect for quantitative analysis are also discussed.

In the past two decades, infrared spectroscopy has also experienced considerable growth. Fourier transform–infrared (FT-IR) equipment is now commonly available in most industrial laboratories and used routinely. The application of FT-IR and FT-IR accessories to cationic surfactants is described. Although IR is a valuable tool for quantitative analysis and structural identification, it lacks the resolution required for identification of cationic surfactants such as quaternary amines or formulated products containing a variety of cationic surfactants. In such cases, NMR will be the technique of choice.

II. NUCLEAR MAGNETIC RESONANCE SPECTROSCOPY

Since its discovery by Bloch et al. [1] and Purcell [2] in 1946, nuclear magnetic resonance (NMR) has become an important and indispensable tool for identification of organic and inorganic compounds in liquid and solid states. In recognition of their pivotal work, these two pioneers were awarded the Nobel Prize in Physics in 1952. The importance of the technique was not, however, generally realized until 1951, following further development by Arnold et al. [3]. Until early the 1960s, proton continuous-wave (CW) NMR was the only commercially available NMR instrument, but by the mid-1960s, the first carbon-13 NMR was marketed.

Originally, Bloch [4] suggested an alternative method for observation of NMR signals by using a short radio-frequency (RF) pulse [5], and Han [6] proved that this procedure results in a time-domain spectrum. It was not until the early 1970s, however, that pulse NMR spectroscopy became commercially available. In the ensuing years, with the availability of sophisticated, yet inexpensive computers and superconducting high-field high-resolution magnets, the field of NMR spectroscopy has flourished. Richard Ernst was awarded the 1991 Nobel Prize in Chemistry for his pioneering work in developing FT-NMR and for his contributions to the modern field of NMR.

A. Basis of Nuclear Magnetic Resonance

The theory of NMR spectrography has been described in detail elsewhere [7–12]. Only a very brief description of the basis of the technique is provided here.

All nuclei carry a charge, and in some nuclei the charge spins on the nuclear axis. This circulation of nuclear charge generates a magnetic dipole along the axis. The angular momentum of the spinning charge is described in terms of spin numbers I. The intrinsic magnitude of the dipole generated is expressed in terms of nuclear magnetic moment (μ). Nuclei that possess an odd spin quantum number (I) are detectable by NMR. The most common nuclei observed in NMR spectroscopy are hydrogen (proton) (100% natural abundance) and carbon-13 (1.1% natural abundance), both with a spin number of $\frac{1}{2}$. The most abundant carbon nuclei, ^{12}C, however, have a spin number of zero and are therefore not detectable by NMR. In general, the magnitude of I depends on the atomic number and mass number. Table 1 lists a number of commonly used nuclei and their natural abundance, relative intensity, and spin number.

When a sample containing nuclei with a spin number (I) of $\frac{1}{2}$ or greater is placed in a static magnetic field (i.e., between two poles of a magnet), the nuclei precess (spin) in a plane at a frequency (Larmor frequency) determined by the intensity of the magnetic field. When an alternating radio frequency is applied perpendicular to the static field at a frequency that matches the Larmor frequency of the nucleus to be observed, an absorption signal is recorded in the receiver of the NMR spectrometer. The frequency of precession of each nucleus in a molecule is determined by its electronic and chemical environment. Since nuclei in a molecule are experiencing different chemical and electronic environments, the resulting precession frequencies provide a spectrum of the molecule under observation. As mentioned previously, Bloch described two techniques for obtaining an NMR spectrum, although both techniques produce similar results. One is the continuous-wave (CW) method, in which the strength of the magnetic field or the frequency is varied slowly so that each group of nuclei is observed (into resonance) in turn. The second technique is the

TABLE 1 Some Commonly Used Nuclei in NMR Spectroscopy

Nucleus	Natural (%) abundance	Relative sensitivity	Spin, I
^1H	99.9844	1.000	$\frac{1}{2}$
^{13}C	1.108	0.0159	$\frac{1}{2}$
^{14}N	99.635	0.0010	1
^{15}N	0.365	0.0010	$\frac{1}{2}$
^{19}F	100	0.834	$\frac{1}{2}$
^{31}P	100	0.0664	$\frac{1}{2}$

Source: Ref. 13.

FT-NMR method, in which a pulse brings all of the nuclei in the molecule into resonance at one time and a time-domain spectrum called free-induction decay (FID) is obtained. Using fast Fourier transform [14] calculations, the time-domain spectrum is converted to the frequency domain. The latter method (FT-NMR) is about two orders of magnitude faster and more sensitive.

B. Pulse Sequences

FT-NMR provides pulse-sequencing capabilities that provide a wealth of additional information simply not available by the CW technique. All modern high-resolution NMR spectrometers sold today are FT-NMR. To illustrate the importance of pulse sequencing, following are descriptions of a number of pulse sequences and techniques from a glossary compiled by R. Bible and L. Johnson (unpublished) based on references given in Varian Instruments Company software and a book by Derome [15].

APT: Attached proton test experiment [16]

ATPC: Attached proton test that uses a composite 180° pulse width [17]

BINOM: Binomial water suppression [18]

COSY: Homonuclear correlated two-dimensional NMR, N-type peak selection [19]

COSY, J-Resolved: An extension of a normal COSY experiment, to which a third time variable has been added for J modulation of the magnetization [20]

DEPT: Distortionless enhancement by polarization transfer [21]

DEPT+: Modified DEPT, which eliminates distortions in coupled spectra [22]

HETEROCORRELATION (HETCOR): Heteronuclear chemical shift correlation [23–25]

HOHAHA: Homonuclear Hartman–Hahn experiment, which can excite subspectra selectively through J coupling [26]

INADEQUATE: Selective double quantum coherence excitation of ^{13}C–^{13}C satellite by phase-countercycled multiple refocusing [27,28]

INEPT: Insensitive nuclei enhanced by polarization transfer [29]

NOESY: Pulse sequence for homonuclear magnetization transfer correlated two-dimensional NMR [30,31]

Bible and Johnson also suggest a general approach to problem solving by NMR which is summarized in Table 2.

There are a number of pulse sequences that are now being used routinely in industrial laboratories, some of which are described in more detail below. In all of these techniques, in addition to a proton RF source, the decoupler frequency synthesizer is used as the second source of frequency.

TABLE 2 Problems and Related Techniques

Problem	Techniques
Purity	1-D ^{1}H, 1-D ^{13}C, APT, INEPT, DEPT
^{13}C–^{13}C connectivities	1-D ^{1}H, 1-D ^{1}H decoupling, COSY, INADEQUATE
^{13}C multiplicities	APT, INEPT, DEPT (135°, 90°, and 45° decoupler pulse angle)
^{1}H–^{1}H coupling	1-D ^{1}H, 1-D ^{1}H decoupling, COSY

Source: R. Bible and L. Johnson (unpublished).

INEPT (insensitive nuclei enhanced polarization transfer) was originally designed [32] to enhance the signal of less abundant nuclei such as ^{15}N. INEPT therefore enhances the intensity of all signals, and the increased intensity is generally higher than nuclear Overhauser effect (a measure of dipole–dipole relaxation). It has, however, been shown [32] that INEPT can be used to determine the multiplicity of each signal in a ^{13}C NMR spectrum. The pulse sequence for INEPT is shown in Fig. 1. As indicated in Fig. 1, the delay τ is fixed and equal to $\frac{1}{4}J_{CH}$, and Δ, the waiting period prior to detection, is varied. J is the average carbon-proton coupling constant. In an INEPT spectrum, if $\tau = \Delta$, methyl, methylene, and methine carbons are observable. If $\Delta = \frac{1}{2}J_{CH}$, only methine carbons are observed,

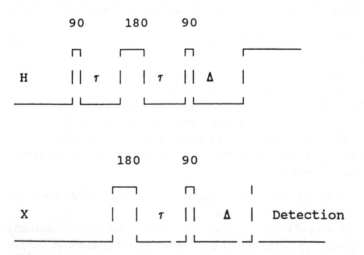

FIG. 1 Pulse sequence for INEPT experiment. The delay τ is fixed and equal to $\frac{1}{4}J_{CH}$ and Δ is varied. X can be ^{13}C or other nuclei, such as ^{15}N. (From Ref. 32.)

and if $\Delta = \frac{1}{4}J_{CH}$, methine and methyl carbons are positive and methylene carbons are negative in respect to each other.

Although this technique is valuable and has been used extensively for determination of carbon type, it suffers from its sensitivity to carbon (or X nuclei)-proton coupling constants, which causes skewed intensities for carbons with different coupling constants. The other disadvantage is that in coupled INEPT the relative intensity within a multiplet does not follow a binomial relationship [33]. The binomial multiplets are for —CH, − 1:1; for —CH$_2$ − 1:0:1; and for —CH$_3$, − 1: − 1:1:1. The negative sign indicates negative intensity. Some of these difficulties have been addressed by INEPT$^+$ or other pulse sequences, such as DEPT.

DEPT (distortionless enhancement by polarization transfer) is another commonly used method for identification of carbon type. Figure 2 shows the pulse sequence for DEPT. In a DEPT spectrum, if $\theta = 45°$, methyl, methylene, and methine carbons are observable and positive. If $\theta = 90°$, only methine carbons are observed, and if $\theta = 135°$, methyl and methine are positive and methylene carbons are negative in respect to each other. In both INEPT and DEPT, quaternary carbons are nulled. The advantage of DEPT is that it is insensitive to the coupling constant (J) and therefore a quantitative spectrum can be obtained. Using spectral editing (subtraction of spectra with different θ value), it is possible to produce a spectrum for methyl or methylene or methine carbons only. The disadvantage of DEPT is its sensitivity to the accuracy of the 90° and 180° pulse width.

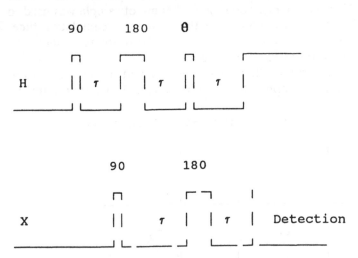

FIG. 2 Pulse sequence for DEPT experiment. The delay τ is fixed and equal to $\frac{1}{4}J_{CH}$ and θ (decoupler pulse angle) is varied. (From Ref. 33.)

Two-Dimensional NMR (2D NMR) is another powerful technique for structural elucidation of organic compounds. There are a number of pulse sequences for this purpose, some of which are discussed briefly and illustrated below by an example of a fatty amine precursor to a commercially available cationic surfactant. There are generally three types of 2D NMR spectroscopy [34], known as resolved 2D, correlated 2D, and exchange 2D. Resolved 2D is a homonuclear experiment and provides separation between overlapping signals. Correlated 2D allows correlation of spins to each other as a result of J coupling. Exchange 2D also allows correlation, but as a result of chemical exchange or the nuclear Overhauser effect.

In 2D NMR, a pulse sequence, consisting of preparation, evolution, mixing (when necessary), and detection periods [35,36], is applied and the response is detected as a function of two variables. The first of these, t_1, is changed systematically and a signal is acquired as a function of second variable, t_2, for each t_1 value. A general scheme of pulse sequence for a 2D NMR experiment is given in Fig. 3.

As an example to illustrate the utility of the 2D NMR technique, a series of 2D NMR spectra of hydrogenated tallow (HT) imidazoline alcohol (Fig. 4) are shown (Figs. 5 to 10). The methyl sulfate quaternary salt of this compound is a commercially available cationic fabric softener. By correlating information from 2D NMR, 1D proton, and NMR information, the proton and ^{13}C chemical shifts of the seven carbons indicated in Fig. 4 have been assigned. The samples were dissolved in deuterated benzene (20% wt/vol) and run on a JEOL GSX 270-MHz NMR spectrometer. In the case of 2D NMR experiments, about 50 to 300 mg of sample was used to increase signal-to-noise ratio and lower the required accumulation time. Figures 5 and 6 are 1D proton and ^{13}C spectra of imidazoline alcohol. The proton NMR shows considerable overlap in the 3.0 to 4.0 ppm region. The ^{13}C NMR spectrum, on the other hand, has resolved resonances, two of which, at 168.4 and 59.8 ppm, can easily be assigned to carbon numbers

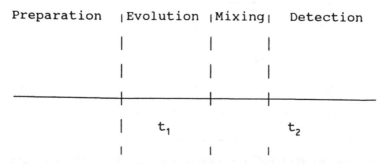

FIG. 3 General scheme for 2D NMR pulse sequence. (From Ref. 35.)

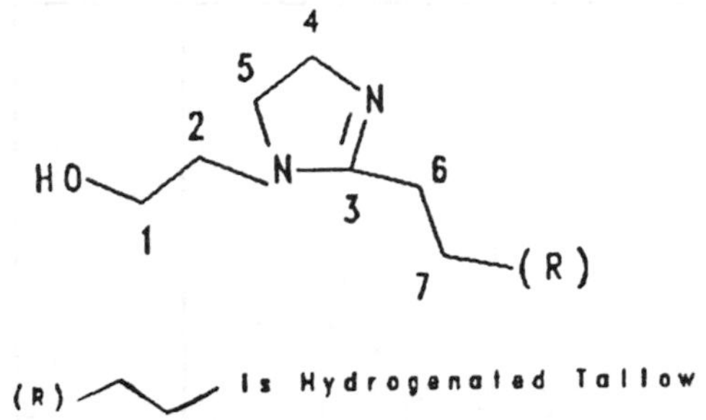

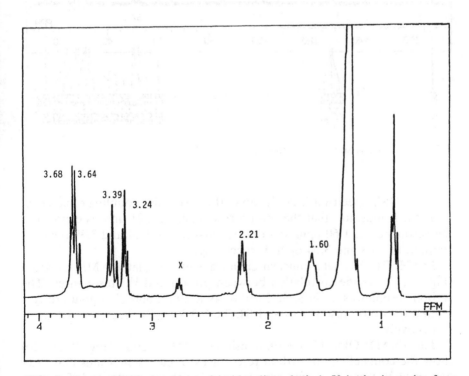

FIG. 4 Chemical structure of imidazoline alcohol and the carbons of interest (number 1 to 7).

3 and 1, respectively. There are, however, three resonances between 49.0 and 51.0 ppm and two other resonances in the alkyl chain region at 28.2 and 27.8 ppm which are more difficult to assign.

FIG. 5 Proton NMR spectrum of imidazoline alcohol. X is the impurity from amidoamine starting compound.

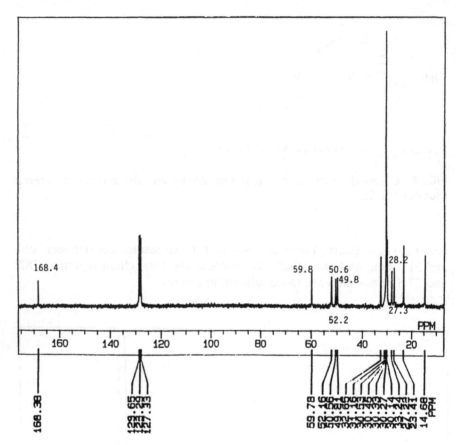

FIG. 6 ^{13}C NMR spectrum of imidazoline alcohol.

The COSY spectrum (Fig. 7) shows the first-order coupling of adjacent protons. It indicates that the proton resonance at 2.21 ppm is coupled to the resonance at 1.60 ppm and that resonances at 3.24 and 3.39 ppm are coupled to 3.68 and 3.64 ppm, respectively.

The HETCOR, heteronuclear chemical shift correlated NMR spectrum (Fig. 8), shows the correlation between proton and ^{13}C resonances. The carbon resonances at 59.8. 52.2, 50.6, 49.8, 28.2, and 27.3 ppm are correlated to proton resonances at 3.68, 3.64, 3.39, 3.24, 2.21, and 1.6 ppm, respectively.

The INADEQUATE two-dimensional NMR spectra (Fig. 9 and 10) show the connectivity of carbon resonances. Figure 9 indicates that resonances at 59.8 and 52.2 ppm are connected to those at 49.8 and 50.6 ppm. The resonances at 28.2 and 27.3 are also connected. Figure 10 shows that

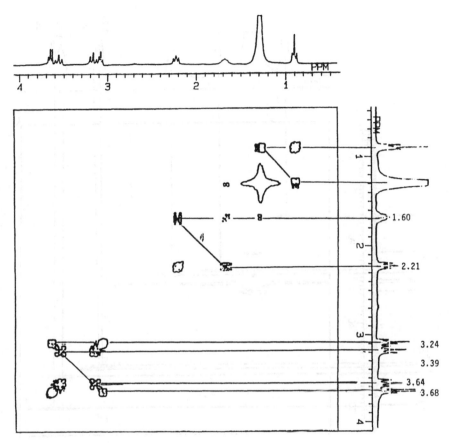

FIG. 7 Homonuclear COSY 2D NMR spectra of imidazoline alcohol. Correlated protons are shown.

the resonance at 168.4 ppm is connected to that at 28.2 ppm. The ^{13}C and proton chemical shifts for the seven carbons indicated in Fig. 4 are listed in Table 3.

G. A. Morris [37] suggests that a homonuclear (^{1}H,^{1}H) COSY experiment for 2 to 10 mg of sample, depending on the desired resolution, requires 4 to 60 min to produce an acceptable spectrum. A heteronuclear (proton, ^{13}C) COSY for 10 mg of sample requires 4 min and a 2D INADEQUATE requires at least 60 min to obtain an acceptable spectrum from 100 mg of sample. The time and the amount of sample required depend on the type of molecule. For example, for cationic surfactants, a homonuclear COSY requires about 20 min for 50 mg of sample, Heter-

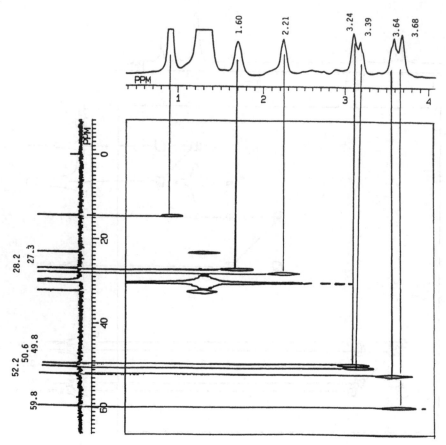

FIG. 8 Heteronuclear COSY spectra of HT imidazoline alcohol. The correlations of the proton to the carbon are indicated.

onuclear COSY about 1 h for 100 mg of sample and 2D INADEQUATE 24 to 48 h for 200 to 300 mg of sample.

There are many advantages in using NMR spectroscopy for identification and quantitative analysis. These include absence of the need for calibration, the nondestructive nature of the technique (sample can be recovered), and the ability to permit both identification and quantitative analysis from the same spectrum. It is also a relatively fast technique (for proton NMR) and components may not necessarily need to be separated prior to analysis. The apparent disadvantages are overlap of resonances and sensitivity. The power of higher magnetic field instruments (e.g., 750 MHz NMR) and their commercial availability [38] will effectively compensate for any shortcom-

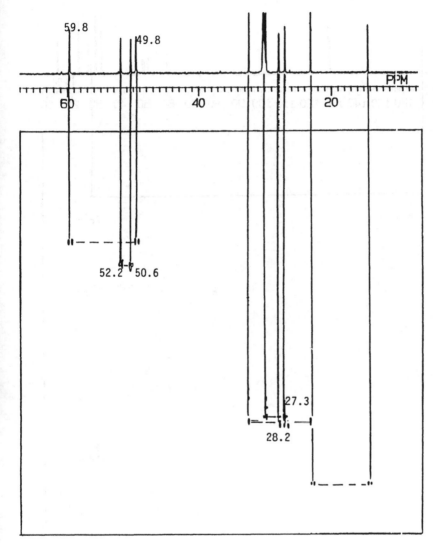

FIG. 9 INADEQUATE 2D NMR spectra of HT imidazoline alcohol (10 to 70 ppm). The ^{13}C connectivities are indicated.

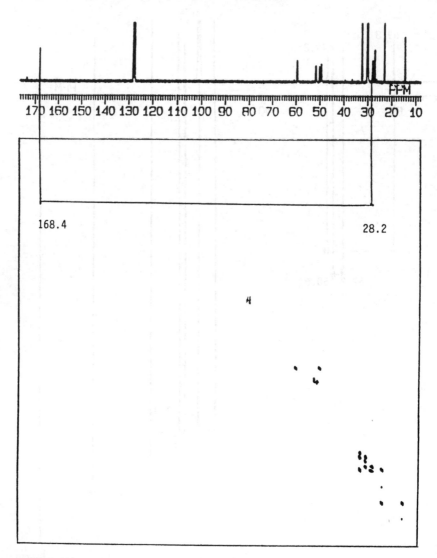

FIG. 10 INADEQUATE 2D NMR spectra of HT imidazoline alcohol (10 to 170 ppm). The ^{13}C connectivity is indicated.

TABLE 3 Proton and ^{13}C Chemical Shift Assignment of Imidazoline Alcohol

Carbon	^1H (ppm)	^{13}C (ppm)
1	3.68	59.8
2	3.24	49.8
3	None	168.4
4	3.64	52.2
5	3.39	50.6
6	2.21	28.2
7	1.60	27.3

ings. The price of the instrumentation is, however, still relatively high in comparison to that of an IR or GC instrument.

C. Proton and ^{13}C NMR of Cationic Surfactants

Spectra produced by the two most commonly used nuclei (^1H and ^{13}C) and their application to analysis of cationic surfactants and their fatty amine precursors are discussed in this section.

1. Proton (^1H) NMR of Cationic Surfactants

High-field NMR, 400 MHz and higher, provides adequate separation of frequencies due to protons attached to the α carbon to a nitrogen nucleus. At lower magnetic field strengths, however, there is a considerable amount of overlap and lack of resolution in separation of signals from various amines or their quaternary ammonium salts. In general, protons on the α carbons to the nitrogen nuclei of amines have chemical shifts between 2.3 and 2.8 ppm and of quaternary ammonium salts between 3.2 and 3.6 ppm. The α protons in amine salts (such as amine hydrochloride) show a shift to a lower field, similar to that exhibited by α protons of quaternary ammonium salts.

2. Quantitative Analysis by Proton NMR

The resonances of amines and their salts in proton NMR are quantitative and are routinely integrated for quantitative information. The quality of this information is limited by the degree of resolution of proton signals. Procedures have been reported to enhance the resolution of proton NMR in the absence of high-magnetic-field NMR (400 MHz and higher) and therefore allow quantitative measurements. Chelating agents such as Eu(DPM)$_3$, Pr(DPM)$_3$, and Yb(DPM)$_3$ (DPM stands for "dipivalomethanto") are not very useful with fatty amines due to their broadening effect. There are, however, other methods for identifying special features and groups [39].

Following is a method [40] for determination of primary, secondary, and tertiary amines by proton NMR: To about 25 mg of a sample containing primary, secondary, and tertiary amines, 0.5 mL of CF_3COOH is added. After the sample has dissolved completely, the excess acid is evaporated on a steam bath with air or nitrogen blowing through the sample. The residue is then dissolved in $CDCl_3$ (10 to 15% wt/vol) and the spectrum is obtained. The range of the chemical shifts for the NH_3^+ peak of a primary amine is 7 to 8 ppm, for NH_2^+ of a secondary amine is 8 to 9 ppm, and for NH^+ of a tertiary amine is 10 to 11 ppm. The mole percent value of each component is then calculated.

3. ^{13}C NMR Analysis of Cationic Surfactants

^{13}C NMR has been applied successfully to identification and quantification of cationic surfactants [41–43]. Due to the well-resolved resonances, ^{13}C NMR provides a simple method of analysis for amine mixtures, which are generally difficult to analyze with conventional separative methods. The spectra shown in Fig. 11 and 12 illustrates the separation power of ^{13}C NMR. In the synthesis of cationic surfactants, normal production steps are splitting tallow or other types of fat to fatty acid, and hence via primary amide, nitrile, and primary amines, to tertiary amines, followed by quaternization of the tertiary amine by methyl chloride or dimethyl sulfate. In the following tables, the chemical shifts of carbon α to functional groups such as nitrile, amines, and quaternary ammonium salts are provided. In most cases, the alkyl chain of the compounds reported is hydrogenated tallow (HT). The alkyl chain length commonly used for cationic surfactants (C_{12} to C_{22}) do not affect the reported chemical shifts. The chemical shifts for nitrile, various amides, and amines are provided in Tables 4 and 5, expressed in ppm from TMS and for the ^{13}C underlined.

The α carbons in amine salts show upfield shift, contrary to their proton chemical shifts. Anions, such as those formed from fatty acid or hydrochloric acid, have no effect on the chemical shift of the α carbons (Table 6). In quaternary ammonium salts, the α carbons exhibit a downfield shift, consistent with their proton chemical shift. Anions such as chloride or methyl sulfate have no effect on their chemical shift (Table 7). Ethoxylated and propoxylated amines and their quaternary ammonium salts are another class of cationic surfactants which are commonly available commercially. Table 8 shows the chemical shift for the α carbons in these compounds. The important point to note in these shifts is that the methylene in ethoxylated amines and methine in propoxylated amines α to —OH are shielded when the amines are converted to quaternary ammonium salts.

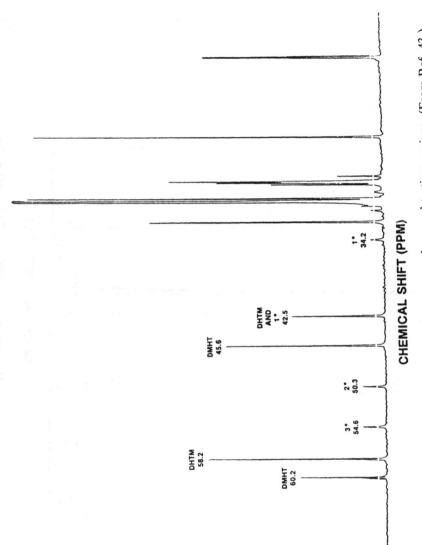

FIG. 11 ¹³C spectrum (0 to 2000 Hz) of primary, secondary, and tertiary amines. (From Ref. 43.)

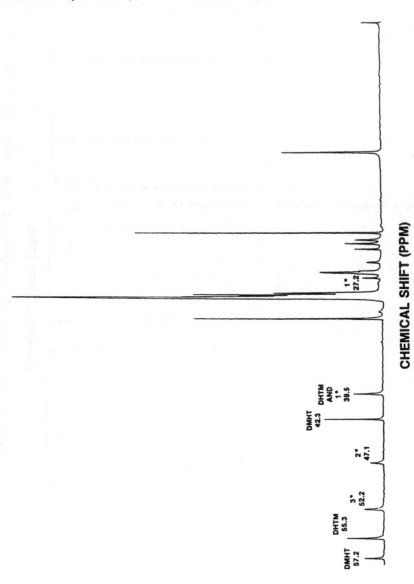

FIG. 12 ^{13}C spectrum (0 to 1333 Hz) of primary, secondary, and tertiary amine salts. (From Ref. 43.)

TABLE 4 ^{13}C Chemical Shifts for Nitrile and amides[a]

Compound	Chemical shift
(R)—$\underline{C}H_2$—CN	18.5
(R)—$\overline{C}H_2$—CN	118.5
(R)—$\underline{C}H_2$—$\overline{C}(O)$—NH_2	36.1
(R)—$\overline{C}H_2$—C(O)—NH_2	175.9
(R)—$\underline{C}H_2$—$\overline{C}(O)$—NH—CH_2—(R)	36.9
(R)—$\overline{C}H_2$—C(O)—NH—CH_2—(R)	39.6
(R)—CH_2—C(O)—NH—$\overline{C}H_2$—(R)	173.1
(R)—$\underline{C}H_2$—$\overline{C}(O)$—N[—CH_2—(R)]$_2$	33.4
(R)—$\overline{C}H_2$—C(O)—N[—$\underline{C}H_2$—(R)]$_2$	46.1, 48.2
(R)—CH_2—$\underline{C}(O)$—N[—$\overline{C}H_2$—(R)]$_2$	172.8

[a]In ppm from TMS. The chemical shift is for the ^{13}C underlined.

The moles of ethylene (EO) or propylene (PO) oxide content can easily be calculated from proton and ^{13}C spectra for the amines and from ^{13}C spectra for the quaternaries. To calculate the moles of EO in an ethoxylated amine from proton NMR, for example, the area per proton is calculated from methylenes α to the nitrogen, the total number of protons in the 3.1- to 3.8-ppm region is determined. After addition of four protons to the

TABLE 5 ^{13}C Chemical Shifts of Primary, Secondary, and Tertiary Amines[a]

Compound	Chemical shift
(R)—$\underline{C}H_2$—NH_2	42.4
(R)—$\overline{C}H_2$—CH_2—NH_2	34.2
(R)—$\overline{C}H_2$—NH—CH_3—(R)	50.3
(R)—$\overline{C}H_2$—N—($\overline{C}H_2$—(R))$_2$	54.6
[(R)—$\overline{C}H_2$]$_2$—N—CH_3	58.2
[(R)—$\overline{C}H_2$]$_2$—N—CH_3	42.5
(R)—$\underline{C}H_2$—N—($\overline{C}H_3$)$_2$	60.2
(R)—$\overline{C}H_2$—N—$(CH_3)_2$	45.6
(R)—$\underline{C}H$—(R)—$\overline{C}H_2$—N(CH$_3$)—CH_2—(R)	36.2
(R)—$\overline{C}H$—(R)—CH_2—N(CH$_3$)—CH_2—(R)	63.2
(R)—CH—(R)—$\overline{C}H_2$—N(CH$_3$)—CH_2—(R)	43.1
(R)—CH—(R)—CH_2—N($\overline{C}H_3$)—$\underline{C}H_2$—(R)	58.8

[a]In ppm from TMS. The chemical shift is for the ^{13}C underlined.

TABLE 6 ^{13}C Chemical Shifts of Primary, Secondary, and Tertiary Amine Salts[a]

Compound		Chemical shift
(R)—C$\underline{H}_2$—NH$_3^+$	X$^-$	39.4
(R)—$\overline{C}$H$_2$—CH$_2$—NH$_3^+$	X$^-$	27.2
(R)—$\overline{C}$H$_2$—NH$_2^+$—CH$_3$—(R)	X$^-$	47.1
(R)—$\overline{C}$H$_2$—NH$^+$—($\overline{C}$H$_2$—(R))$_2$	X$^-$	52.2
((R)—$\overline{C}$H$_2$)$_2$—NH$^+$—CH$_3$	X$^-$	55.3
((R)—$\overline{C}$H$_2$)$_2$—NH$^+$—$\underline{C}$H$_3$	X$^-$	39.5
(R)—CH$_2$—NH$^+$—(C$\overline{H}_3$)$_2$	X$^-$	57.2
(R)—$\overline{C}$H$_2$—NH$^+$—($\underline{C}$H$_3$)$_2$	X$^-$	42.3

[a]In ppm from TMS. The chemical shift is for the ^{13}C underlined. X$^-$ may be any anion.

total protons and dividing the total by 4, the number of moles of EO is obtained. It is suggested that a few drops of D$_2$O should be added to the sample solution, normally in deuterated chloroform, prior to accumulation of the proton spectrum. Other major classes of cationic surfactants are amidoamine, imidazoline, and ester quaternary ammonium salts. Table 9, 10, and 11 list the ^{13}C chemical shifts in these products. Diamines also form an important class of cationic surfactants. The chemical shifts of the α

TABLE 7 ^{13}C Chemical Shifts of Quaternary Ammonium Salts and Amine Oxide[a]

Compound		Chemical shift
(R)—C$\underline{H}_2$—N$^+$—(CH$_3$)$_3$	Cl$^-$	66.3
(R)—$\overline{C}$H$_2$—N$^+$—(CH$_3$)$_3$	Cl$^-$	52.8
(R)—(CH$_2$)$_2$—N$^+$—(CH$_3$)$_2$	Cl$^-$	63.6
(R)—($\overline{C}$H$_2$)$_2$—N$^+$—(CH$_3$)$_2$	Cl$^-$	51.0
(R)—(CH$_2$)$_3$—N$^+$—$\overline{C}$H$_3$	Cl$^-$	61.5
(R)—$\overline{C}$H$_2$)$_3$—N$^+$—CH$_3$	Cl$^-$	48.7
(R)—CH—(R)—C$\overline{H}_2$—N$^+$(CH$_3$)$_2$—CH$_2$—(R)	Cl$^-$	32.8
(R)—$\overline{C}$H—(R)—CH$_2$—N$^+$(CH$_3$)$_2$—CH$_2$—(R)	Cl$^-$	68.7
(R)—CH—(R)—$\overline{C}$H$_2$—N$^+$(CH$_3$)$_2$—CH$_2$—(R)	Cl$^-$	51.1
(R)—CH—(R)—CH$_2$—N$^+$($\overline{C}$H$_3$)$_2$—CH$_2$—(R)	Cl$^-$	64.2
(R)—C$\underline{H}_2$—N(O)(CH$_3$)2		71.7
(R)—$\overline{C}$H$_2$—N(O)($\underline{C}$H$_3$)2		58.1

[a]In ppm from TMS. The chemical shift is for the ^{13}C underlined.

TABLE 8 ^{13}C Chemical Shifts of Ethoxylated and Propoxylated (2 mol) Amines and Their Quaternary Ammonium Salts[a]

Compound		Chemical shift
(R)—CH$_2$—N—(CH$_2$—CH$_2$—OH)$_2$		54.9
(R)—C̄H$_2$—N—(CH$_2$—CH$_2$—OH)$_2$		56.2
(R)—CH$_2$—N—(C̄H$_2$—CH$_2$—OH)$_2$		59.6
(R)—CH$_2$—N$^+$(CH$_3$)—(C̄H$_2$—CH$_2$—OH)$_2$	Cl$^-$	64.0
(R)—C̄H$_2$—N$^+$(CH$_3$)—(CH$_2$—CH$_2$—OH)$_2$	Cl$^-$	55.5
(R)—CH$_2$—N$^+$(C̄H$_3$)—(CH$_2$—C̄H$_2$—OH)$_2$	Cl$^-$	50.3
(R)—CH$_2$—N—(CH$_2$—CH(CH$_3$)—OH)$_2$		55.1, 56.1
(R)—C̄H$_2$—N—(CH$_2$—CH(CH$_3$)—OH)$_2$		62.8, 63.8
(R)—CH$_2$—N—(C̄H$_2$—CH(CH$_3$)—OH):i2		63.9, 64.9
(R)—CH$_2$—N—(CH$_2$—C̄H(CH$_3$)—OH)$_2$		20.7, 20.5
(R)—CH$_2$—N$^+$(CH$_3$)—(C̄H$_2$—CH(CH$_3$)—OH)$_2$	Cl$^-$	64.4
(R)—C̄H$_2$—N$^+$(CH$_3$)—(CH$_2$—CH(CH$_3$)—OH)$_2$	Cl$^-$	68.4
(R)—CH$_2$—N$^+$(CH$_3$)—(C̄H$_2$—CH(CH$_3$)—OH)$_2$	Cl$^-$	62.5
(R)—CH$_2$—N$^+$(CH$_3$)—(CH$_2$—C̄H(CH$_3$)—OH)$_2$	Cl$^-$	22.7
(R)—CH$_2$—N$^+$(C̄H$_3$)—(CH$_2$—CH(C̄H$_3$)—OH)$_2$	Cl$^-$	50.0

[a]In ppm from TMS. The chemical shift is for the ^{13}C underlined.

TABLE 9   Chemical shifts of Hydrogenated Tallow Amidoamine and Amidoamine Quaternary Salts[a]

Compound		Chemical shift
((R)CH$_2$—C(O)—NH—CH$_2$—CH$_2$)$_2$—NH		36.2
((R)C̄H$_2$—C(O)—NH—CH$_2$—CH$_2$)$_2$—NH		173.6
((R)CH$_2$—C̄(O)—NH—CH$_2$—CH$_2$)$_2$—NH		39.6
((R)CH$_2$—C(O)—NH—C̄H$_2$—CH$_2$)$_2$—NH		48.9
((R)CH$_2$—C(O)—NH—CH$_2$—C̄H$_2$)$_2$—N$^+$(CH$_3$)(CH$_2$—CH$_2$—OH)	X$^-$	36.2
((R)CH$_2$—C(O)—NH—CH$_2$—C̄H$_2$)$_2$—N$^+$(CH$_3$)(CH$_2$—CH$_2$—OH)	X$^-$	173.6
((R)CH$_2$—C̄(O)—NH—CH$_2$—CH$_2$)$_2$—N$^+$(CH$_3$)(CH$_2$—CH$_2$—OH)	X$^-$	37.6
((R)CH$_2$—C(O)—N—C̄H$_2$—CH$_2$)$_2$—N$^+$(CH$_3$)(CH$_2$—CH$_2$—OH)	X$^-$	61.6
((R)CH$_2$—C(O)—N—CH$_2$—C̄H$_2$)$_2$—N$^+$(CH$_3$)(CH$_2$—CH$_2$—OH)	X$^-$	50.3
((R)CH$_2$—C(O)—N—CH$_2$—CH$_2$)$_2$—N$^+$(C̄H$_3$)(CH$_2$—CH$_2$—OH)	X$^-$	64.1
((R)CH$_2$—C(O)—N—CH$_2$—C̄H$_2$)$_2$—N$^+$(CH$_3$)(C̄H$_2$—CH$_2$—OH)	X$^-$	56.2

[a]In ppm from TMS. The chemical shift is for the ^{13}C underlined. X is Cl$^-$ or CH$_3$SO$_4$$^-$.

TABLE 10 Chemical Shifts of Hydrogenated Tallow Imidazoline and Imidazoline Salt[a]

Compound	Chemical Shift
(R) $\underline{C}H_2$—C(O)—NH—CH$_2$—CH$_2$—N—CH$_2$—CH$_2$ / C(R)=N	36.6
(R) CH$_2$—$\underline{C}$(O)—NH—CH$_2$—CH$_2$—N—CH$_2$—CH$_2$ / C(R)=N	173.2
(R) CH$_2$—C(O)—NH—$\underline{C}H_2$—CH$_2$—N—CH$_2$—CH$_2$ / C(R)=N	38.5
(R) CH$_2$—C(O)—NH—CH$_2$—$\underline{C}H_2$—N—CH$_2$—CH$_2$ / C(R)=N	46.9
(R) CH$_2$—C(O)—NH—CH$_2$—CH$_2$—N—$\underline{C}H_2$—CH$_2$ / C(R)=N	50.3
(R) CH$_2$—C(O)—NH—CH$_2$—CH$_2$—N—CH$_2$—$\underline{C}H_2$ / C(R)=N	52.3
(R) CH$_2$—C(O)—NH—CH$_2$—CH$_2$—N—CH$_2$—CH$_2$ / $\underline{C}$(R)=N	167.2
(R) $\underline{C}H_2$—C(O)—NH—CH$_2$—CH$_2$—N$^+$H—CH$_2$—CH$_2$ / R—COO$^-$ C(R)=N	36.1
(R) CH$_2$—$\underline{C}$(O)—NH—CH$_2$—CH$_2$—N$^+$H—CH$_2$—CH$_2$ / R—COO$^-$ C(R)=N	173.2
(R) CH$_2$—C(O)—NH—$\underline{C}H_2$—CH$_2$—N$^+$H—CH$_2$—CH$_2$ / R—COO$^-$ C(R)=N	36.1
(R) CH$_2$—C(O)—NH—CH$_2$—$\underline{C}H_2$—N$^+$H—CH$_2$—CH$_2$ / R—COO$^-$ C(R)=N	42.9
(R) CH$_2$—C(O)—NH—CH$_2$—CH$_2$—N$^+$H—CH$_2$—CH$_2$ / R—COO$^-$ C(R)=N	45.5
(R) CH$_2$—C(O)—NH—CH$_2$—CH$_2$—N$^+$H—CH$_2$—$\underline{C}H_2$ / R—COO$^-$ C(R)=N	49.0
(R) CH$_2$—C(O)—NH—CH$_2$—CH$_2$—N$^+$H—CH$_2$—CH$_2$ / R—COO$^-$ $\underline{C}$(R)=N	170.4

[a]In ppm from TMS. The chemical shift is for the ^{13}C underlined.

348   **Mozayeni**

TABLE 11 Chemical Shifts of Hydrogenated Tallow Ester Amine and Its Quaternary Salt[a]

Compound		Chemical shift
(R)—C̲H₂—N(CH₃)—CH₂—CH₂—O—C(O)—(R)		55.8
(R)—C̲H₂—N(CH₃)—CH₂—CH₂—O—C(O)—(R)		42.6
(R)—CH₂—N(C̲H₃)—CH₂—CH₂—O—C(O)—(R)		56.8
(R)—CH₂—N(CH₃)—C̲H₂—CH₂—O—C(O)—(R)		62.0
(R)—CH₂—N(CH₃)—CH₂—C̲H₂—O—C(O)—(R)		172.3
(R)—CH₂—N⁺(CH₃)₂—CH₂—CH₂—O—C(O)—(R)	Cl⁻	62.2
(R)—C̲H₂—N⁺(CH₃)₂—CH₂—CH₂—O—C(O)—(R)	Cl⁻	51.8
(R)—CH₂—N⁺(C̲H₃)₂—CH₂—CH₂—O—C(O)—(R)	Cl⁻	65.1
(R)—CH₂—N⁺(CH₃)₂—C̲H₂—CH₂—O—C(O)—(R)	Cl⁻	58.2
(R)—CH₂—N⁺(CH₃)₂—CH₂—C̲H₂—O—C(O)—(R)	Cl⁻	172.2

[a]In ppm from TMS. The chemical shift is for the ¹³C underlined.

carbons for the most common diamine and its diquaternary ammonium salt are listed in Table 12.

4. Quantitative Analysis by ¹³C NMR

In general, quantitative analysis by ¹³C NMR requires consideration of two variables.

TABLE 12 Chemical Shifts of Hydrogenated Tallow Diamines and the Diamine Oxide[a]

Compound	Chemical shift
(R)—CH₂—NH—CH₂—CH₂—CH₂—NH₂	40.0
(R)—CH₂—NH—CH₂—CH₂—C̲H₂—NH₂	47.3
(R)—CH₂—NH—C̲H₂—CH₂—CH₂—NH₂	49.5
(R)—C̲H₂—N(CH₃)—CH₂—CH₂—CH₂—N(CH₃)₂	45.4
(R)—CH₂—N(CH₃)—CH₂—CH₂—CH₂—N(C̲H₃)₂	55.9
(R)—CH₂—N(CH₃)—CH₂—CH₂—C̲H₂—N(CH₃)₂	58.0
(R)—CH₂—N(CH₃)—C̲H₂—CH₂—CH₂—N(CH₃)₂	42.3
(R)—CH₂—N(C̲H₃)—CH₂—CH₂—CH₂—N(CH₃)₂	58.0
(R)—C̲H₂—N(O)(CH₃)₂—CH₂—CH₂—CH₂—N(O)(C̲H₃)₂	58.7
(R)—CH₂—N(O)(CH₃)₂—CH₂—CH₂—C̲H₂—N(O)(C̲H₃)₂	70.1
(R)—CH₂—N(O)(CH₃)₂—CH₂—CH₂—C̲H₂—N(O)(CH₃)₂	68.0
(R)—CH₂—N(O)(CH₃)₂—C̲H₂—CH₂—CH₂—N(O)(CH₃)₂	54.9
(R)—C̲H₂—N(O)(C̲H₃)₂—CH₂—CH₂—CH₂—N(O)(CH₃)₂	66.2

[a]In ppm from TMS. The chemical shift is for the ¹³C underlined.

(*a*)   *Spin-Lattice Relaxation Time* (T$_1$).   The spin-lattice relaxation arises from return of the excited nuclear spin to the ground state by dissipating energy to its environment lattice [44]. The finite time required for the full dissipation of this energy varies with different types of carbons (e.g., as in methyl, methylene, or methine). There are three methods—inversion recovery (IR), progressive saturation (PS), and saturation recovery (SR)—for measuring the relaxation time (T_1) (Fig. 13). The pulse sequence for the IR method is (180° − t − 90° − T), where t is the pulse interval and is varied for the measurement and T is the waiting period between each pulse (in general, 5 to 10 times the actual T_1). T_1 is calculated by the general equation $\ln[(A_\infty - A_t)/2A_\infty] = -t/T_1$ where A_∞ and A_t are the signal intensities at ∞ and pulse interval t, respectively. A plot of log $[A_\infty - A_t]$ versus t is made (Fig. 14) and the spin-lattice relaxation time is determined from the slope, $1/T_1$. In practice, A_∞ of the ^{13}C signal is obtained after a waiting period 5 to 10 times that of the actual T_1 (an estimation of this is necessary initially).

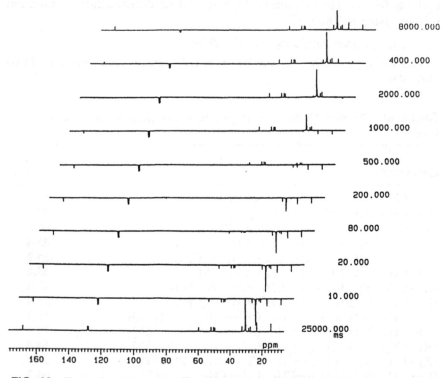

FIG. 13   Example of a plot of ^{13}C NMR spectra [chemical shifts versus time (t) in millisecond] for measurement of relaxation time of ^{13}C.

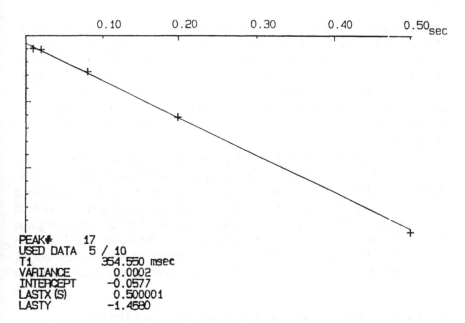

PEAK#   17
USED DATA 5 / 10
T1   354.550 msec
VARIANCE   0.0002
INTERCEPT   -0.0577
LASTX (S)   0.500001
LASTY   -1.4580

FIG. 14 Example of a plot of log of $[A_x - A_t]$ versus t and calculations of T_1. T_1 = 354 ms.

The magnitude of T_1 depends on the molecular size and on the chemical and electronic environment. In addition, field strength, temperature, solvent, and concentration are important parameters for these measurements. Using the three methods cited, the T_1 reported [45] for carbon 4 of phenol are 1.39, 1.51, and 1.46 by IR, PS, and SR, respectively. These reported measurements were made at 25.2 MHz, 30°C, with an 80% solution of phenol in hexadeuterated acetone. In addition to IR, PS, and SR there is the null method [46], which is more rapid but less accurate than the first three procedures. In this procedure, T_1 is equal to $1.443t_0$, where t_0 is the pulse interval corresponding to a null signal in the receiver.

The longest T_1 observed for cationic surfactants was that for the α carbon of primary amine. As alkyl chain substitution (primary, secondary, and tertiary) on nitrogen nuclei increased, T_1 dropped sharply. The values for methyl-substituted amines, however, show a lesser decrease. Increasing the chain length also lowers the T_1 value. For example, the T_1 of α carbons in a series of primary amines was shown to drop sharply from n-butyl amine (13.0 s to 12.6 s) to n-octyl amine (5.9 s to 5.3 s) and then to form a plateau from dodecyl amine (4.9 s to 4.0 s) to octadecyl amine (4.9 s to 3.9 s).

There are two ways to compensate for the T_1 relaxation time. One is to have a pulse delay 5 to 10 times higher than the T_1 value of the nuclei under observation, and the other is the use of a relaxation agent such as chromium acetylacetonate (CrAcAc). The latter method generally causes some peak broadening that may have to be taken into account. The measurements were made in a 20% (wt/vol) solution of $CDCl_3$ at 22.5 MHz and 28°C. Table 13 is a list of T_1 values for various type of amines. Three measurements were made, of which the lowest and the highest values are reported. The T_1 values rapidly drop off when amines are converted to their salts.

(b) *Nuclear Overhauser Effect.* The nuclear Overhauser effect (NOE) is the second major factor to consider in quantitative analysis by ^{13}C NMR. Since ^{13}C NMR spectra are accumulated under full decoupling conditions for protons, the dipolar energy created by this decoupling is transmitted to the ^{13}C nuclei under observation. The effect of NOE is, therefore, highest for —CH_3 carbon and zero for quaternary (^{13}C with no attached proton) carbon. The maximum value of NOE for ^{13}C nuclei is 1.99. In other words, some resonances in ^{13}C spectra may have artificially higher intensities due

TABLE 13 T_1 Measurements of Primary, Secondary, and Tertiary Amines, Amine Salts, and Quaternary Ammonium Salts[a]

Compound		T_1 (s)
(R)—$\underline{C}H_2$—NH_2		4.8–3.9
(R)—$\underline{C}H_2$—NH—CH_2—(R)		0.88–0.67
—$\underline{C}H_2$—N—$(CH_2)_2$—(R)		0.42–0.26
(R)—$(CH_2)_2$—N—$\underline{C}H_3$		1.0–0.9
(R)—$(\underline{C}H_2)_2$—N—CH_3		1.6–1.4
$(CH_3)_2$—N—$\underline{C}H_2$—(R)		3.2–3.6
$(\underline{C}H_3)_2$—N— CH_2—(R)		3.1–2.7
(R)—$\underline{C}H_2$—N^+H_3	Cl^-	0.18–0.17
(R)—$\underline{C}H_2$—N^+H_2—CH_2—(R)	Cl^-	0.19–0.16
(R)—$\underline{C}H_2$—N^+H—$(CH_2)_2$—(R)	Cl^-	0.13–0.12
(R)—$(\underline{C}H_2)$:i2—$N^+\underline{H}$—CH_3	Cl^-	0.25–0.21
(R)—$(\underline{C}H_2)_2$—N^+H—CH_3	Cl^-	0.87–0.85
(R)—$(CH_3)_2$—N^+H—$\underline{C}H_2$—(R)	Cl^-	0.60–0.49
$(CH_3)_2$—N^+H—CH_2—(R)	Cl^-	1.0–0.89
(R)—$\underline{C}H_2$—N^+—$(CH_3)_3$	Cl^-	0.37–0.32
(R)—$\underline{C}H_2$—N^+—$(CH_3)_3$	Cl^-	0.56–0.49
(R)—$(\underline{C}H_2)_2$—N^+—$(CH_3)_2$—(R)	Cl^-	0.18–0.14
(R)—$(\underline{C}H_2)_2$—N^+—$(CH_3)_2$—(R)	Cl^-	0.35–0.26
(R)—$(\underline{C}H_2)_3$—N^+—$\underline{C}H_3$	Cl^-	0.10–0.08
(R)—$(\underline{C}H_2)_3$—N^+—$\underline{C}H_3$	Cl^-	0.27–0.25

[a]The T_1 is for the ^{13}C underlined.

to the NOE rather than the nuclei quantitative population. NOE, a measure of dipole–dipole relaxation, could be avoided by accumulating the signals under gated decoupling conditions. Nearly all modern NMR instruments have such a capability, which turns the decoupler on for a fraction of a second during FID acquisition in order to suppress the proton and ^{13}C couplings. If the decoupler is not turned on at all, the spectra are fully proton coupled, and due to the large coupling constant (ca. 120 to 140 Hz), the signals generally overlap because of the multiplicity of ^3C signals and are therefore weak. For the practical reasons, chemists should consider the effect of NOE and whether or not it has a significant contribution to the quantitative results thus obtained. Mozayeni et al. [43] reported that if similar type of carbons (e.g., —CH_2^- α to nitrogen) are used for quantitative analysis of fatty amines, no significant effects on the quantitative data are observed. Table 14 shows the effect of NOE on quantitative calculations for various types of amines. A similar effect was observed with quaternary ammonium salts.

D. ^{14}N and ^{15}N NMR Spectroscopy

The two nuclei that seem to be the most suitable for NMR analysis of cationic surfactants are ^{15}N and ^{14}N. ^{15}N has a natural abundance of 0.36% and a spin number of $\frac{1}{2}$, and ^{14}N has natural abundance of 99.6% and a spin number of 1. ^{14}N seems to be the most appropriate nucleus. In addition to high natural abundance, it has a very short relaxation time (T_1), which should make it ideal for NMR spectroscopy. The major drawback, how-

TABLE 14 NOE Effect on Quantitative Analysis of Primary, Secondary, and Tertiary Amines and Quaternary Ammonium Salts

Type of amine[a]	Standard (wt %)	With NOE (wt %)	No NOE (wt%)
Primary amine (HT)	11.8	11.5	11.9
Secondary amine (2HT)	15.2	14.8	15.9
Tertiary amine (3HT)	22.9	23.6	23.1
Tertiary amine (DHTM)	29.7	29.0	29.5
Tertiary amine (DMHT)	20.2	20.9	20.5
Quaternary ammonium salts			
Monoalkyl trimethyl	29.3	29.0	29.4
Dialkyl dimethyl	31.3	30.8	31.1
Trialky monomethyl	39.5	40.1	39.6

[a]HT, hydrogenated tallow; DHTM, dihydrogenated tallow methyl; DMHT, dimethyl hydrogenated tallow.
Source: Ref. 43.

ever, is that like most nuclei with a spin number greater than $\frac{1}{2}$, it has an electric quadrupole moment that arises from a nonspherical electric charge distribution in the nucleus [47]. The characteristic properties of these types of nuclei are that they have a very short spin lattice relaxation time (T_1) but a very broad linewidth, from 10 to 1000 Hz. It is the latter property that limits the application of this nuclei to the analysis of cationic surfactants.

The observations of this author suggest that for amines, as cationic precursors, the linewidth is too broad to be useful in a commercial laboratory. The quaternaries, however, have relatively narrower linewidths and are better suited for identification and quantitative analyses. Figure 15 is an example of a ^{14}N spectrum of a mixture of dihydrogenated tallow, monohydrogenated tallow, and tetramethylammonium chloride quaternaries. The sample (20% solution in deuterated chloroform) was run in a JEOL GSX 270-MHz NMR spectrometer.

Spectra derived from ^{15}N show narrow linewidths, but this advantage is countered by its very low natural abundance (0.36%), negative NOE, and

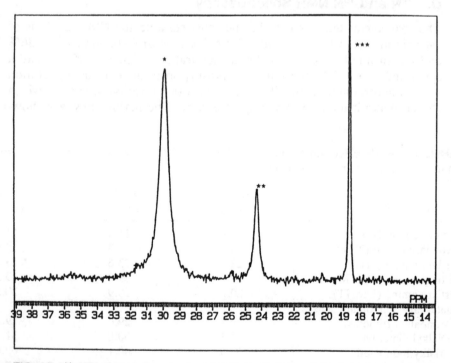

FIG. 15 ^{14}N spectrum of a mixture of dialkyl dimethyl quaternary (*), monoalkyl trimethyl quaternary (**), and tetramethyl quaternary (***) ammonium salts.

very long relaxation time (T_1). The negative NOE can sometimes be overcome by gated decoupling, but this process compounds the low-sensitivity problem of this nuclei. The high values of T_1 can be avoided by using paramagnetic relaxation reagents. More recently, INEPT polarization transfer sequences have been used to increase the sensitivity of these nuclei. Unfortunately, the use of these nuclei in analysis of cationic surfactants, with their invariably long alkyl chains (C_{12} to C_{22}), is difficult and time consuming. Even the application of the INEPT polarization transfer sequence is limited. The sensitivity of nitrogen nuclei such as -—NH_2 and —NH are enhanced, but this enhancement drops off very quickly once the alkyl chain length reaches eight carbon atoms.

III. INFRARED SPECTROSCOPY

Infrared spectroscopy is one of the oldest of the modern techniques used in industrial laboratories. The usefulness of infrared was recognized early in the twentieth century when Coblentz [48] described the C—H stretching modes of organic compounds. The first IR instrument, introduced in 1913 [49], was a pivotal step in providing a powerful tool for future chemists to use in determining the structural features of both organic and inorganic chemicals. Infrared is usually defined as electromagnetic radiation whose frequency covers the range between 14,000 and 20 cm^{-1}[50]. The importance of infrared in the development of organic chemistry cannot be overestimated. It was, and still is, one of the simplest techniques for identification of organic compounds. Its introduction provided chemists with a versatile and simple tool that led to the foundation of the science of spectroscopy. The IR absorption bands upon which analyses are based arise from dipole moment changes due to various types of molecular vibrations, principally stretching and bending.

There are volumes of literature dealing with this subject [51–53]. In recent articles in *Analytical Chemistry,* the history of the development of commercial IR spectrometers [54] and that of the supporting systems and infrastructure has been discussed [55]. Although the first IR instrument was introduced in 1913, the commercialization of IR spectrometer did not take place until the mid-1940s. Two instrument companies, Perkin-Elmer and Beckman, and two chemical companies, Dow and American Cyanamid, were the major contributors to this development.

A. FT-IR Instrument

Today, almost all commercial IR spectrometers are of the FT-IR type, the historical development of which has been described by Griffiths [56]. They utilize a double-beam interferometer, first described by A. A. Michelson

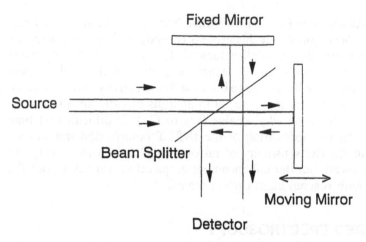

FIG. 16   Schematic diagram of Michelson interferometer.

100 years ago [57–59] but not incorporated into a commercially available instrument until the mid-1970s. Advances in the manufacture of stable, trouble-free interferometers and in the development of fast, inexpensive computers, coupled with the development of fast Fourier transform (FFT) by Tukey and Cooley and the demonstration of its application to interferometric spectroscopy by Forman [60], were major factors in the commercialization of FT-IR. In this section the advantages of this technique and some of its important applications are described briefly. FT-IR is based on the modulation of infrared waves, rather than the use of a prism or grating, to obtain absorption bands in the IR spectra. This modulation is accomplished by the use of an interferometer, which consists basically of a fixed mirror, a moving mirror, and a beam splitter (Fig. 16). The moving mirror allows modulation of the beam, and the beam splitter permits selection of the IR frequency of desired wavelength. Table 15 is a summary of various beam splitters currently in use.

TABLE 15   Common Beam Splitters Used in FT-IR Interferometer

	Region (cm^{-1})	Material	Coating
Near-IR	20,000–3000	Quartz	MgO
	10,000–2000	CaF_2	Si
Mid-IR	4000–400	KBr	Ge
	800–200	CSi	
Far-IR	650–100	Mylar	
	300–30	Wire mesh	

TABLE 16 Common Detectors for FT-IR

Detector[a]	Operating temperature	Detection range (cm^{-1})
TGS	Room temperature	5000–30
DTGS	Room temperature	5000–30
MCT	Liquid nitrogen	3000–700
PbS	Liquid nitrogen	20,000–3000
PbSe	Liquid nitrogen	20,000–3000

[a]TGS, triglycine sulfate; DTGS, deuterated triglycine sulfate; MCT, Hg–Cd–Te.

Another important feature of the FT-IR instrument is its adaptability to use with a variety of detectors which operate in different parts of IR region and with different sensitivities. Table 16 provides a list of some of these. The only moving part of the FT-IR instrument is the moving mirror. The total distance that the mirror travels (retardation) determines the instrument's resolution. Table 17 illustrates the relationship between the distance (one way) traveled by the mirror, the retardation (round trip, twice the mirror travel), and the resolution (in cm^{-1}) of the IR spectra.

Although the first application of the Michelson interferometer dates back to the World War II era and its original military application in the near-IR (NIR) region, it is still the most widely used design, with some modifications, to be found in modern instruments. Fifty percent of the infrared light beam is transmitted through the beam splitter to the moving mirror, and 50% is reflected by the beam splitter to the fixed mirror. The difference in path length traveled by the two parts provides modulation of the beam. When the two mirrors are at exactly the same distance from the beam splitter, no modulation of the light waves occurs and therefore no absorption will take place.

TABLE 17 Spectral Resolution in FT-IR

Mirror travel (cm)	Retardation (cm)	Resolution (cm^{-1})
0.06	0.125	8.000
0.25	0.500	2.000
1.00	2.000	0.500
4.00	8.000	0.125
8.00	16.000	0.060

There are five advantages associated with the FT-IR technique:

1. Fellgett or multiplex advantage
2. Jacquinot advantage
3. Connes advantage
4. Stray light advantage
5. No-heat advantage

The Fellgett advantage, also known as the multiplex advantage [61], is the speed at which data acquisition takes place in a FT-IR instrument. The Jacquinot advantage [62] is due to the high-energy throughput, which is calculated to be 40 to 50 times more than for a dispersive instrument; in practice, however, this factor may not be as high [63]. The Connes advantage refers to the high degree of wavelength accuracy obtained in the IR spectra, due to the use of monochromatic laser light. Finally, the reduction of problems associated with stray light, the absence of any heat effect (the sample is far removed from the IR radiation source), and the mechanical simplicity of FT-IR instruments offer additional advantages to chemists and spectroscopists.

B. FT-IR Accessories

Accessories that were extremely difficult to use with dispersive IR instruments are now being used routinely in FT-IR instruments. Perhaps the most critical advantage of FT-IR is that due to the Fellget advantage, which results in higher sensitivity and signal-to-noise ratio spectra. Recent advances in hyphenated FT-IR techniques such as GC/FT-IR, LC/FT-IR [64], microscope FT-IR, and the thermogravimetric (TGS) FT-IR are just a few examples of the manner in which use of this technique is expanding.

Other accessories, such as photoacoustic spectroscopy (PAS), diffuse reflectance of infrared Fourier transform spectroscopy (DRIFTS), attenuated total reflectance (ATR), and cylindrical internal reflection (circle cell), are additional dimensions of modern FT-IR in use in industrial laboratories.

1. *PAS.* In PAS the sample is placed in a chamber and the sound waves generated from the sample due to heat waves from IR light are detected by a powerful, focused microphone. The advantage of this technique is that the shape of the solid sample does not pose a problem for IR analysis. PAS has been used for surface analysis of different shapes and types of solid materials.

2. *DRIFTS.* DRIFTS allows IR analysis of small quantities of solid sample (microsampling) and in most cases, eliminates the use of KBr disk.

3. *ATR*. Although ATR has been used with dispersive IR instruments, its use with FT-IR results in better resolution and spectra of higher signal-to-noise ratio. Horizontal ATR is commonly used for IR analysis of powder samples and in at-line process analyses. The latter application is becoming more common in industrial laboratories since it permits following the formation of products and by-products during synthesis and kinetic studies of these processes.

4. *Circle cell*. Circle cell allows IR spectra of diluted samples (1) to 5% in water or other solvent to be obtained. It is particularly useful for examination of biological samples.

5. *FT-IR microscopy*. FT-IR microscopy was introduced in 1986 [65] with only limited application, but within the span of a few years has experienced enormous growth. IR microscopy performs three functions: it illuminates the sample with the IR source, collects the transmitted or reflected light, and allows the specific area under investigation to be observed. This technique is well suited to many types of difficult-to-handle samples. Quantitative analysis using IR microscopy is difficult and has to be done with great care. Some sources of error for quantitative analysis are associated with locating the right spot, diffraction and refraction, optical energy, and polarization and alignment.

6. *GC/FT-IR*. The potential importance of GC/FT-IR was demonstrated by Low and Freeman [66] in 1966. Although the initial instrumentation lacked the sensitivity and sophistication of that available today, it was considered to be a major advancement in the field of IR technology and opened the way to other important hyphenated techniques. The first commercial GC/FT-IR was introduced by the Digilab Company (now Bio-Rad Digilab Division) in 1969 and has advanced due primarily to the introduction of the MCT (mercury–cadmium–tellurium) detector, which has lowered the detection limit to the nanogram level, and to the construction of the gold-coated borosilicate glass light pipe. In a light pipe system, the analysis is done as different fractions are eluted from the GC column into the light pipe "on the fly."

In addition, other techniques, such as cryogenic freezing of the GC column eluant on a gold plate in solid-argon matrices and, more recently, Tracer (manufactured by Bio-Rad Digilab Division), which allows the freezing of column eluant on a cesium iodide plate at liquid nitrogen temperature, have lowered the level of detection. The latter technique also provides the spectrum of the solid rather than that of the vapor (light pipe) or crystalline (solid-argon matrices).

Other methods of collection of column eluant permit on-the-fly detection as well as detection of the frozen eluant at a later time with an increased number of scans. The chromatogram generated during on-the-fly analysis is known as a Gram–Schmidt (G–S) chromatogram [67]. It is computer reconstructed and qualitatively is similar to a flame ionization detector (FID) chromatogram, but differs in the quantitative sense due to variations in the molar absorbance of infrared light by different classes of organic compounds. A Gram–Schmidt chromatogram provides a means for computer assignment of each fraction to a specific spectrum.

C. Applications of FT-IR to Cationic Surfactants

1. GC/FT-IR

The utility of this technique and its application to cationic surfactants have been reported [68]. As examples, two different applications of GC/FT-IR are described here. The samples were run on a Digilab Model (FTS-60) GC/FT-IR with the light-pipe system and equipped with a Hewlett-Packard Model 5890 gas chromatograph. The column used was a fused silica capillary (DB1, 30 m, 1 μm film thickness, from J&W Scientific Inc.). The carrier gas was helium at 3.5 mL/min flow. GC is most commonly used for chain-length distribution of fatty amines. In the case of quaternary ammonium salts, this type of analysis becomes difficult, due to breakdown of the quaternary in the injection port. The use of cool, on-column injection is also unsatisfactory, due to the loss of resolution.

The G-S chromatogram (Fig. 17) of dicocodimethyl quaternary ammonium chloride indicates the number of peaks due to this breakdown of the quaternary. There are three types of decomposition by-products, illustrated in Figs. 18 to 20. Figure 18 is the vapor-phase IR spectrum (at 20.6 min retention time) of dimethylcocoamine with two characteristic bands at 2822 and 2776 cm^{-1} due to C—H stretching vibration of a methyl group α to nitrogen. Figure 19 is the vapor-phase IR spectrum (at 22.8 min retention time) of dicocomonomethyl amine with the characteristic shift of C—H stretching vibration at 2793 cm^{-1}. The third by-product (Fig. 20) is the vapor-phase IR spectrum (at 22.3 min retention time) of a hydrocarbon type of compound. A library (EPA vapor-phase library) search of this spectrum, however, indicates that the best match is with an alkyl chloride (Fig. 21). Another example of the utility of GC/FT-IR is analysis of a distillate from an impurity. Figure 22 is a Gram–Schmidt chromatogram of this distillate with complete identification of each peak.

2. IR Frequencies of Selected Cationic Surfactants

In general, IR spectra of cationic surfactants such as mono-, di-, or trialkyl quaternary ammonium chloride salts are not particularly informative. Due to the hydrophilic nature of these quaternaries, moisture bands will appear on the IR spectra if the sample is not dried properly prior to analysis. The

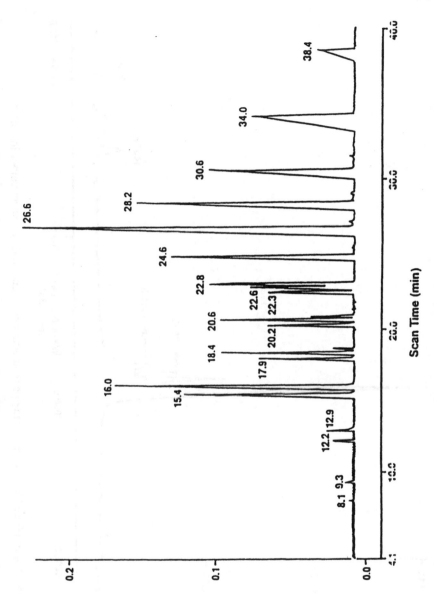

FIG. 17 Gram–Schmidt chromatogram of dicocodimethyl quaternary ammonium chloride salt.

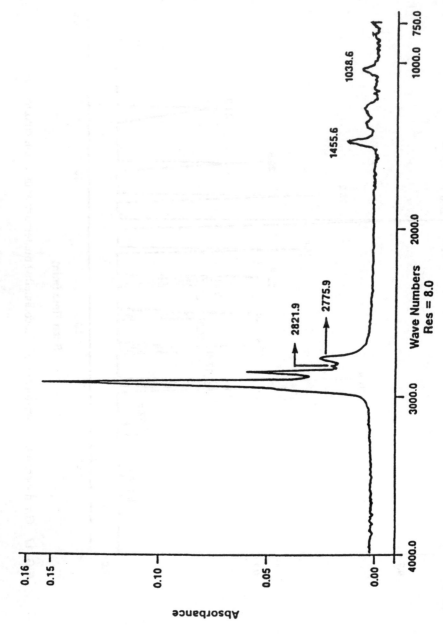

FIG. 18 Vapor-phase IR spectrum of the peak at 20.6 min retention time (wavenumber in cm^{-1}).

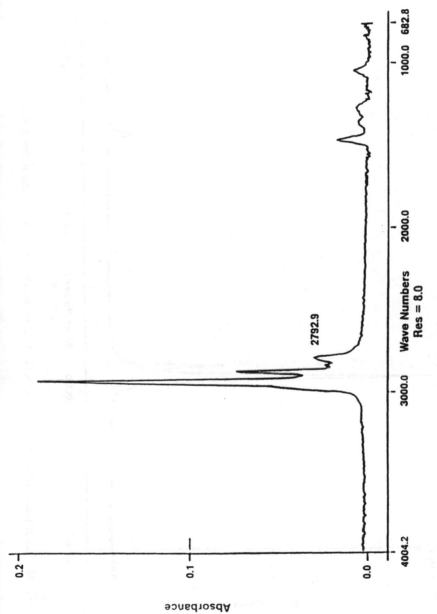

FIG. 19 Vapor-phase IR spectrum of the peak at 22.8 min retention time (wavenumber in cm^{-1}).

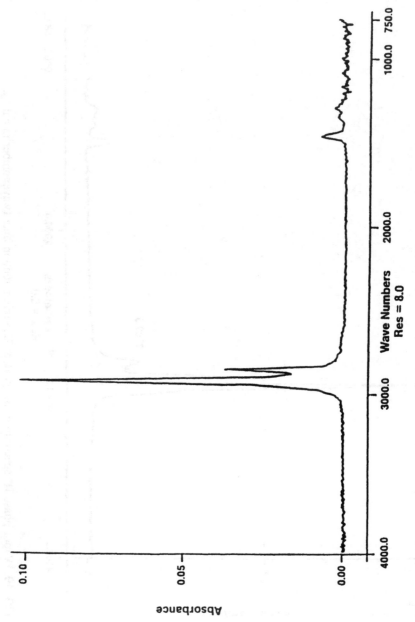

FIG. 20 Vapor-phase IR spectrum of the peak at 22.3 min retention time (wavenumber in cm^{-1}).

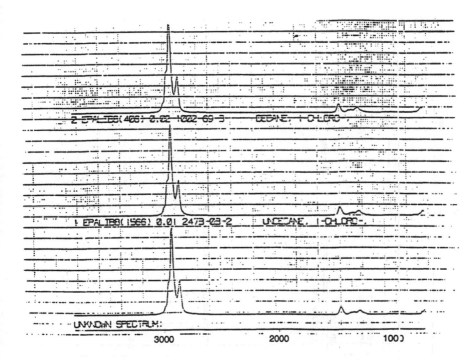

FIG. 21 EPA library search of the vapor-phase spectrum at 22.3 min retention time.

precursors to these quaternaries (e.g., fatty acids, amides, nitriles, and amines) have specific IR stretching bands associated with them which can be used for identification and quantitative analysis, particularly for process control and reaction monitoring. Group frequency bands in the finger-printing region of IR spectra are useful only when the materials are in a pure state. The following is a summary of specific frequencies for each of these compounds, run as a thin film on sodium chloride disks.

Fatty acids, in addition to the carbonyl (C=O) stretching band at 1725 to 1705 cm^{-1}, exhibit a broad hydrogen-bonded hydroxyl (—OH) band between 3200 and 2600 cm^{-1}. Primary fatty amides may be identified easily by the doublet carbonyl stretching bands at 1656 and 1630 cm^{-1} and the doublet —NH_2 stretching bands at 3358 and 3190 cm^{-1}. Monosubstituted amides have carbonyl stretching bands at 1655 to 1650 and 1550 to 1545 cm^{-1} and the —HN band at 3250 cm^{-1}. Fatty-based nitriles show a sharp characteristic band at 2260 to 2240 cm^{-1}. Primary fatty amines have their —NH stretching bands at 3400 to 3380 and 3340 to 3320 cm^{-1} and also a broad deformation band at 1650 to 1610 cm^{-1}. Secondary fatty amines

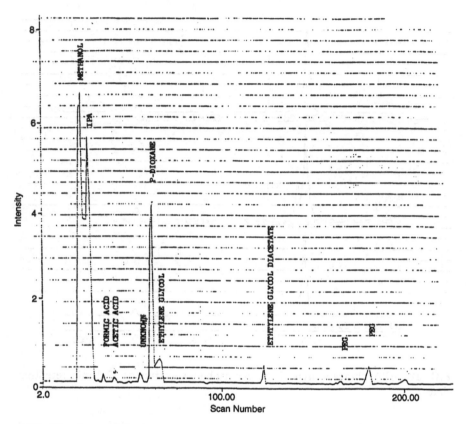

FIG. 22 Gram–Schmidt chromatogram of an impurity distillate.

have their —NH stretching band at 3350 to 3320 cm^{-1} and a C—N stretching band at 1150 cm^{-1}. Trialkyl fatty amines do not exhibit specific bands; however, dimethylmonoalkyl and monomethyldialkyl tertiary fatty amines have specific stretching bands due to C—H of a methyl group α to nitrogen at 2810 to 2760 cm^{-1}. The quaternization of these amines causes these bands to shift to a higher wavenumber and into a region where they usually overlap with the C—H stretching band of alkyl chain at 2920 to 2850 cm^{-1}. Cationic surfactants containing the —C≡N— group are identified by the C≡N stretching band at 1670 to 1665 cm^{-1}. Cationic surfactants containing a sulfate anion have characteristic S=O doublet bands at 1248 and 1217 cm^{-1}. Ethoxylated and propoxylated amines have their —OH and C—O stretching bands at 3372 and 1046 cm^{-1} respectively. Fatty acid salts of amines exhibit a characteristic band at 1550 cm^{-1} due to the carboxylate ion.

REFERENCES

1. F. Bloch, W. W. Hansen, and M. E. Packard, *Phys. Rev. 69*: 127 (1946).
2. E. M. Purcell, H. C. Torrey, and R. V. Pound, *Phys. Rev. 69*: 37 (1946).
3. J. T. Arnold, S. S. Dharmatti, and M. E. Packard, *J. Chem. Phys. 19*: 507 (1951).
4. F. Bloch, *Phys. Rev. 70*: 460 (1946).
5. E. D. Becker, *Anal. Chem. 65*: 295A (1993).
6. E. L. Han, *Phys. Rev. 8*: 580 (1950).
7. J. R. Dyer, *Application of Absorption Spectroscopy of Organic Compounds*, Prentice Hall, Englewood Cliffs, N.J., 1956, p. 96.
8. J. W. Emsley, J. Feeney, and L. H. Sutcliffe, *High Resolution Nuclear Magnetic Resonance Spectroscopy*, Pergamon Press, Elmsford, N.Y. 1967.
9. P. Lancelin, J. Champy, and G. Albert, *Methods Phys. Anal. 7* (2): 170 (1971).
10. D. E. Leyden and H. R. Cox, *Analytical Applications of NMR*, Wiley-Interscience, New York, 1977, p. 258.
11. E. Breitmaier and W. Voelter, *Carbon-13 NMR Spectroscopy*, Verlag Chemie, New York, 1978.
12. A. J. Montana, in *Nonionic Surfactants: Chemical Analysis* (J. Cross, ed.), Marcel Dekker, New York, 1986, pp. 295–331.
13. F. A. Bovey, *Nuclear Magnetic Resonance Spectroscopy*, Academic Press, New York, 1969, p. 3.
14. J. W. Colley and J. W. Tukey, *Math. Comput. 19*: 297 (1965).
15. A. E. Derome, *Modern NMR Techniques for Chemistry Research*, Pergamon Press, Elmsford, N.Y. 1987.
16. S. L. Patt and J. N. Shoolery, *J. Magn. Reson. 46*: 435 (1982).
17. G. Gray and S. L. Patt, *Magn. Moment 6*: 11 (1986).
18. P. J. Hore, *J. Magn. Reson. 55*: 183 (1983).
19. A. Bax, R. Freeman, and G. A. Morris, *J. Magn. Reson. 42*: 169 (1981).
20. G. W. Vuister and R. Boelens, *J. Magn. Reson. 73*: 328 (1987).
21. D. M. Doddrell, D. T. Pegg, and M. R. Bendall, *J. Magn. Reson. 48*: 323 (1982).
22. O. W. Sorensen and R. R. Ernst, *J. Magn. Reson. 51*: 477 (1983).
23. A. D. Bax and G. A. Morris, *J. Magn. Reson. 42*: 51 (1981).
24. A. D. Bax, *J. Magn. Reson. 53*: 517 (1983).
25. J. A. Wilde and P. H. Bolton, *J. Magn. Reson. 59*: 343 (1984).
26. S. Subramanian and A. Bax, *J. Magn. Reson. 71*: 325 (1987).
27. A. Bax, R. Freeman, and S. P. Kempsell, *J. Am. Chem. Soc. 102*: 4849 (1980).
28. A. Bax, R. Freeman, and S. P. Kempsell, *J. Magn. Reson. 41*: 349 (1980).
29. G. A. Morris and R. Freeman, *J. Am. Chem. Soc. 101*: 760 (1979).
30. A. Kumar, R. Ernst, and K. Wuthrich, *Brown Boveri Rev. Chem. 95*: 1 (1980).
31. A. Kumar, G. Wagner, R. Ernst, and K. Wuthrich, *Brown Boveri Rev. Chem. 96*: 1156 (1980).

32. C. Dybowski and R. L. Lichter, *NMR Spectroscopy Techniques,* Marcel Dekker, New York, 1987, p. 182.
33. Ref. 32, p. 184.
34. Ref. 32, p. 95.
35. Ad Bax, *Two-Dimensional Nuclear Magnetic Resonance in Liquids,* Delft University Press, D. Reidel, Dordrecht, The Netherlands, 1982.
36. J. Schraml and J. M. Bellama, *Two Dimensional NMR Spectroscopy,* Wiley, New York, 1988.
37. G. A. Morris, *Magn. Reson. Chem. 24*: 371 (1986).
38. *Chem. Eng. News 53* (Mar. 22, 1993).
39. N. F. Chamberlain, *The Practice of NMR Spectroscopy,* Plenum Press, New York, 1974.
40. F. Mozayeni, *Appl. Spectrosc. 33*: 520 (1979).
41. H. Eggert and C. Djerassi, *J. Am. Chem. Soc. 9*: 3710 (1973).
42. E. H. Fairchild, *J. Am. Oil Chem. Soc. 59*: 305 (1982).
43. F. Mozayeni, C. Plank, and L. Gray, *Appl. Spectrosc. 38*: 518 (1984).
44. F. W. Wehrli and T. Wirthlin, *Interpretation of Carbon-13 NMR Spectroscopy,* Heyden, Philadelphia, 1978, p. 130.
45. Ref. 44, p. 146.
46. M. L. Martin, J.-J. Delpuech, and G. J. Martin, *Practical NMR Spectroscopy,* Heyden, London, 1980, p. 258.
47. J. B. Lambert and F. G. Riddell, *The Multinuclear Approach to NMR Spectroscopy,* D. Reidel, Boston, 1982, p. 208.
48. W. W. Coblentz, *Investigation of Infra-red Spectra,* Part I, Carnegie Institute, Washington, D.C., 1905.
49. N. Sheppard, *Anal. Chem. 64*: 877A (1992).
50. C. L. Putzig, M. A. Leugers, M. L. McKelvy, G. E. Mitchell, R. A. Nyquist, R. R. Papenfuss, and L. Yurgo, *Anal. Chem. 64*: 270R (1992).
51. R. M. Silverstein and G. C. Bassler, *Spectrometric Identification of Organic Compounds,* Wiley, New York, 1968.
52. L. J. Bellamy, *The Infra-red Spectra of Complex Molecules,* Wiley, New York, 1975.
53. G. Rauscher, in *Nonionic Surfactants: Chemical Analysis* (J. Cross, ed.), Marcel Dekker, New York, 1986.
54. P. L. Wilkes, *Anal. Chem. 64*: 833A, (1992).
55. F. Miller, *Anal. Chem. 64*: 824A, (1992).
56. P. R. Griffiths, *Anal. Chem. 64*: 869A, (1992).
57. A. A. Michelson, *Philos. Mag. Ser. (5) 31*: 256 (1891).
58. A. A. Michelson, *Philos. Mag. Ser. (5) 34*: 280 (1892).
59. A. A. Michelson, *Light Waves and Their Uses,* University of Chicago Press, Chicago, 1902.
60. M. L. Forman, *J. Opt. Soc. Am. 56*: 978 (1966).
61. P. B. Fellgett, *Aspen Institute Conference on Fourier Spectrometry,* AFCRL-17-0019, 1970, p. 139.

62. P. Jacquinot, *17th Congress de GAMS*, Paris, 1954.
63. P. R. Griffiths, *Chemical Infrared Fourier Transform Spectroscopy*, Vol. 45, Wiley, New York, 1975.
64. K. Jinno and K. Kikan, *Sosetu 9*: 160 (1990).
65. J. E. Katon and A. J. Sommers, *Anal. Chem. 64*: 931A (1992).
66. M. J. D. Low and S. K. Freeman, *Anal. Chem. 39*: 194 (1967).
67. J. A. deHaseth and T. L. Isenhour, *Anal. Chem. 49*: 1977 (1977).
68. F. Mozayeni, *J. Am. Oil Chem. Soc. 65*: 1420 (1988).

Index

Skin barrier, 81, 83
Sludge, activated, 98, 103, 107,
    114–115, 125
Slurry walls, sorption by, 98
Snail, toxicity to, 117
Sodium dodecyl sulfate, 141, 146, 173
Sodium lauryl sulfate, 141, 146, 173
Sodium tetrakis (4-fluorophenyl)
    borate, 160
Sodium tetrophenylborate, 140, 154,
    174, 191
Solvent extraction of metals, 13
Spermicidal agent, 65
Spin-lattice relaxation, 349
Staphylococcus aureus, 39, 47
Staphylococcus epidermidis, 52
Stratum corneum:
    penetration of, 81–84
    vulvar, 89
Streptococcus faecalis, 47
Streptococcus lactis, 53
Sublation, 100, 240
Surfactants:
    definition, 4
    determination in seawater, 231
    determination in surface water, 232
    reaction with indicators, 141, 241

Tandem mass spectrometry, 280
Tensammetric peaks, 213

Tensammetric titrations, 227
Tensammetry, 201
Tetrabromophenolphthalein, 153, 246
Test liquors, quats in, 235
Thermospray technique, 276
Thin layer chromatography (TLC):
    amides, 318
    amine oxides, 322
    amines, 313
    ethoxylated amines, 321
    quats, 304
Titration:
    nonaqueous, 171
    photometric, 162
    potentiometric, 164–171, 191
    self-contained system, 161
    tensammetric, 227
    two-phase, 145, 156
Trace analysis, 235
Transepidermal water loss (TEWL),
    84
Trichophyton mentagrophytes, 41
Trichophyton rubrum, 41
Trickling filters, 103, 106, 125
Turkey, toxicity to, 64

Victoria blue B, 158
Virus, lipophilic, 41

Wound healing, effect on, 85